Spring Boot 3 核心技术与最佳实践

周红亮 | 著

電子工業出版社
Publishing House of Electronics Industry
北京•BEIJING

内容简介

本书是一本针对Java开发人员的图书，旨在帮助Java开发人员掌握Spring Boot的基本使用，以及深入了解Spring Boot的应用及原理。

本书内容由浅入深、循序渐进，第1～5章介绍Spring Boot的基础知识（基础入门、配置管理、Starter、自动配置、启动过程与扩展应用、日志管理），第6～9章介绍Spring Boot的综合应用（Web、数据访问、计划任务、缓存、消息队列），第10～12章介绍Spring Boot应用的附加能力（调试、单元测试、打包、部署、监控、报警），全面覆盖了Spring Boot的核心知识要点。

本书涵盖了笔者多年的研究和实践经验，从中提炼出了核心知识要点，从Spring Boot的基本概念和基础实践入手，再通过大量的知识点分析及代码实践，详细介绍如何利用Spring Boot简化开发过程，提高开发效率。

图书在版编目（CIP）数据

Spring Boot 3核心技术与最佳实践 / 周红亮著. —北京：电子工业出版社，2023.4
ISBN 978-7-121-45290-1

Ⅰ. ①S… Ⅱ. ①周… Ⅲ. ①JAVA语言—程序设计 Ⅳ. ①TP312.8

中国国家版本馆CIP数据核字（2023）第049830号

责任编辑：陈晓猛
印　　刷：三河市良远印务有限公司
装　　订：三河市良远印务有限公司
出版发行：电子工业出版社
　　　　　北京市海淀区万寿路173信箱　　邮编：100036
开　　本：787×980　1/16　　印张：31.75　　字数：711.2千字
版　　次：2023年4月第1版
印　　次：2023年5月第2次印刷
定　　价：158.00元

凡所购买电子工业出版社图书有缺损问题，请向购买书店调换。若书店售缺，请与本社发行部联系，联系及邮购电话：（010）88254888，88258888。

质量投诉请发邮件至zlts@phei.com.cn，盗版侵权举报请发邮件至dbqq@phei.com.cn。

本书咨询联系方式：（010）51260888-819，faq@phei.com.cn。

前　　言

写作背景

Spring 作为 Java 开发界的万能框架，曾经和 Struts2、Hibernate 组成 SSH 框架，成为 Java Web 开发领域的“三驾马车”。在 2013 年左右，Spring 又和 Spring MVC、MyBatis 组成 SSM，成为新一代 Web 开发框架全家桶，一直流行至今。

为了简化 Spring 的上手难度，Spring Boot 于 2014 年诞生，可以帮助开发者更加轻松、快捷地使用 Spring 的组件，它是 Spring、Spring MVC 等框架更上一层的框架，它需要依赖 Spring、Spring MVC 等原生框架，而不能独立存在。

笔者最初接触和学习 Spring Boot 是在 2016 年，工作之余经常和同事们聊聊最新的 Java 技术，Spring Boot 就是其中之一。那时候，Spring Boot 虽然应用还不是很广泛，但很火热，逐渐成为炙手可热的 Java 框架。

自 2016 年起，笔者有幸负责和参与公司的多个系统重构，其中就包括从 SSM 框架到以 Spring Boot 为基础的技术转型，也包括以 Spring Cloud 为基础的微服务系统的设计和研发，Spring Cloud 就是基于 Spring Boot 构建的。在多年的实际开发和架构设计工作中，笔者积累了大量 Spring Boot 的使用经验，也见证了 Spring Boot 1.x ～ 3.x 的发展历史。

虽然笔者熟练掌握了 Spring Boot，但在其不断发展的情况下，某些知识点更新了也可能浑然不知，直到在使用出错时才后知后觉。Spring Boot 至今已经发展了近十年，最新的版本已经发布到了 3.x 版本，其底层实现逻辑、使用方式发生了翻天覆地的变化，同时在 Spring Boot 各个版本接二连三地停止维护的情况下，大部分版本已经不能满足技术更新的需要了。

市面上的很多图书、博客都是基于 Spring Boot 低版本的，即使介绍了最新的 Spring Boot 技术，也不是很系统，并没有对最新的技术要点进行系统的梳理、解读及应用举例，如果还继续学习低版本的内容，就会给很多初学者带来不必要的知识更新成本，也会走很多弯路。鉴于市面上关于

Spring Boot 3.x 的系统性学习资料比较匮乏，笔者撰写了本书，以帮助国内 Java 开发人员正确地学习、理解和使用最新的 Spring Boot 技术。

笔者从 2017 年开始，陆续写了一些关于 Spring Boot 的文章，本以为出版此书会相当顺利，结果远没有想象中的简单。写博客没有什么太重的思想负担，所以写得比较随意、零乱，也不成体系，与成体系的图书相差甚远，而且博客中的部分内容已经过时，所以仅能作为一个基石，笔者又花费了近一年的时间，在此基础上深度打磨、加强和完善，并新增了 90% 以上的内容，最后形成了本书的全部内容。

本书章节安排

本书共 12 章，这是一个由浅入深、循序渐进的学习过程。

第 1 章为 Spring Boot 基础入门，主要介绍 Spring Boot 的基础知识，包括背景介绍、核心特性、核心模块、核心思想、版本选择，以及 Maven、Spring Boot CLI 等相关工具的使用，还会分享简单的应用案例。

第 2 章为 Spring Boot 配置管理，主要介绍 Spring Boot 应用中的各种配置技巧，包括配置类、配置文件、外部化配置、配置绑定、导入配置、随机值配置、多文档配置、Profile、加载机制、配置加密、配置迁移，有助于后续章节的融会贯通。

第 3 章为 Spring Boot Starter 与自动配置，主要介绍 Spring Boot Starter 及其应用，包括命名规范、分类、自动配置原理及扩展机制，还介绍了邮件 Starter 及其应用，以及如何自定义一个 Spring Boot Starter。

第 4 章为 Spring Boot 启动过程与扩展应用，主要介绍 Spring Boot 的启动过程，包括引导方式、启动入口方法、启动流程源码分析，以及启动过程中丰富的扩展应用，包括启动日志、启动图案、启动失败分析、启动事件和监听器、全局懒加载、启动运行器等。

第 5 章为 Spring Boot 日志管理，主要介绍 Spring Boot 日志的使用，包括日志格式、日志文件、日志级别、日志分组、日志归档、自定义日志配置文件、切换 Log4j2 日志框架、输出彩色日志等。

第 6 章为 Spring Boot Web 核心应用，这是 Spring Boot 项目最基本、最核心的部分，包括嵌入式容器、Web 组件注册、静态资源处理、模板引擎、异常处理、参数校验、国际化、跨域、分布式会话、安全性、REST 服务调用等常用的 Web 技术的原理介绍、集成与应用。

第 7 章为 Spring Boot 数据访问，主要介绍 Spring Boot 与关系型数据库及非关系型数据库的集成应用，包括数据源、连接池、事务管理、Spring Data JPA、MyBatis、MyBatis-Plus、Redis、MongoDB、Elasticsearch 的原理、集成与应用。

第 8 章为 Spring Boot 计划任务，主要介绍计划任务在 Spring Boot 中的应用，包括 Spring 中的计划任务和 Quartz 计划任务的原理、集成与应用。

第 9 章为 Spring Boot 缓存与消息队列，主要介绍缓存和消息队列在 Spring Boot 中的应用，缓存机制的原理及 Redis 缓存的实现与集成，消息队列的原理及 ActiveMQ、RabbitMQ、Kafka 的原理、集成与应用。

第 10 章为 Spring Boot 调试与单元测试，主要介绍 Spring Boot 应用的调试方法、开发者工具的详细使用，以及如何在 Spring Boot 应用中做单元测试。

第 11 章为 Spring Boot 打包与部署，主要介绍 Spring Boot 应用的打包方式，以及如何将 Spring Boot 应用包以不同的方式运行、以不同的方式部署到 Linux 和 Docker 容器中，包括 Spring Boot 3.0 支持的构建 GraalVM 原生镜像的应用方式。

第 12 章为 Spring Boot 监控与报警，主要介绍 Spring Boot Actuator 监控模块及其应用，包括端点、指标，以及 Spring Boot 3.0+ 支持的可观测性技术的应用，还包括第三方监控平台 Spring Boot Admin、Prometheus+Grafana 的集成与应用。

本书特色

本书以最新的 Spring Boot 3.0 为基础，解读 Spring Boot 核心技术，包括最新的核心知识点、技术原理、应用方式与第三方主流技术集成的方法等，再到服务的测试、调试、部署和监控等，帮助读者一站式学习和掌握最新的 Spring Boot 核心技术。本书具有以下特色。

1. 全面

本书全面介绍了 Spring Boot，覆盖 Spring Boot 所有主流知识点，包括基础入门、配置管理、Starter 与自动配置、启动过程与扩展应用、日志管理、Web 核心应用、数据访问、计划任务、缓存与消息队列、调试与单元测试、打包与部署、监控与报警等。

2. 全新

本书的讲解和实战均基于 Spring Boot 最新主版本 3.0.0，书中带有“Spring Boot 3.0 新变化”“Spring Boot 3.0+”等标识，方便读者对 Spring Boot 3.0 的新特性和变更项有一个更清晰的认识，读者从 Spring Boot 2.x 迁移到 Spring Boot 3.x 也可以有一个对比和参考。

3. 实用

本书不仅介绍了 Spring Boot 的理论知识，还提供了大量的底层原理分析，并为核心知识点、第三方主流技术的集成与应用提供了大量实战案例，理论和实践相结合，清晰易懂，使读者可以更好地吸收和理解。

4. 权威

本书以 Spring Boot 的官方文档和框架源码作为主要参考依据，包括里程碑版本和正式版本的文档、框架源码，再辅以知识点实践和验证，以确保知识点的权威性和正确性。

实战源代码

本书提供了所有实战源代码，虽然本书介绍的内容全部基于 Spring Boot 3.0.0，但是 Spring Boot 后续发布的新版本、新特性，也会进行更新和适配，读者可以"Star"、克隆、下载，学习最新的 Spring Boot 技术。

扫码关注微信公众号"Java 技术栈"，在后台回复关键字"boot"，可获取本书实战源代码。

勘误和交流

因能力、时间和精力有限，本书难免存在纰漏之处，恳请广大读者朋友批评指正，笔者会第一时间在源代码仓库 Issues 一栏中发布最新勘误。读者可以通过以下方式向笔者反馈勘误或进行技术交流。

- 博客：https://www.javastack.cn。
- 邮箱：admin@javastack.cn。
- 微信公众号：Java 技术栈（ID：javastack）。

致谢

本书能顺利出版，离不开电子工业出版社的编辑人员，特别是陈晓猛编辑，他在笔者写作过程中提供了很多建议和帮助，修正了不少错误。在此，特别感谢他们的默默付出。还要感谢笔者的家人，因为在这一年时间中，撰写此书占用了大量本该和他们共处的美好时光，导致没能好好陪伴他们，感谢家人对笔者写书的大力支持、包容和理解。

周红亮

目　　录

第 1 章 Spring Boot 基础入门

本章是 Spring Boot 基础入门章节，笔者会介绍 Spring Boot 的一些基础知识和应用，包括 Spring Boot 的诞生背景、基本介绍、安装和集成的方式，以及使用 Maven 的方式安装和集成 Spring Boot，并快速开发一个 Spring Boot 接口。另外，本章也会介绍如何在 Spring Boot 中免安装、快速使用 Maven 和 Gradle，以及 Spring Boot CLI 的使用，读者可以对 Spring Boot 框架有一个基本的认识，并且初步使用 Spring Boot。

1.1 Spring Boot 概述

1.1.1 诞生背景

在 Java 后端框架繁荣发展和遍地开花的今天，Spring 框架无疑是最火热，也是必不可少的开源框架，更是稳坐 Java 后端框架的“龙头”位置，它几乎能与任何 Java 主流框架进行结合。

Spring 能流行的原因是它的两把利器：IoC 和 AOP。IoC 可以帮助我们管理对象的依赖关系，极大减少对象的耦合性，而 AOP 的切面编程功能可以使我们更方便地使用动态代理来实现各种动态方法功能，如数据库事务、缓存、日志等。

传统的 Spring 框架，不管是 Spring 旗下的组件，还是第三方框架，如果要基于 Spring 集成使用，就必须用到 XML 配置文件，或者注解式的 Java 代码配置。但无论是使用 XML 配置文件还是代码配置方式，都需要对相关组件的配置有足够的了解，再编写大量冗长的配置代码，这显然加大了开发者对 Spring 的使用难度。

所以，为了降低 Spring 框架的使用难度，让 Spring 组件能实现开箱即用，让开发者能快速上手，Spring Boot 框架便诞生了！

1.1.2 基本介绍

Spring Boot 是 Spring 全家桶项目中的一个子项目，也是 Spring 组件应用一站式解决方案，主要是为了降低 Spring 框架的使用难度，简省繁重的配置，所以，Spring Boot 现在也成了后端标准开发框架。

Spring Boot 提供了各种技术组件的一站式启动器（Starter），开发者只要定义好对应技术组件的配置参数，Spring Boot 就会自动配置，让开发者能快速搭建基于 Spring 生态的 Java 应用。

Spring Boot 不但能创建传统的 war 包应用，还能创建独立的、不依赖于任何外部 Servlet 容器的应用（内置 Servlet 容器），使用 java -jar 命令就能启动一个应用，Spring Boot 3.0 更是添加了对 GraalVM 原生镜像的支持，大大提升了 Spring 应用的使用体验。

1.1.3 核心思想

Spring Boot 框架的核心思想是：约定大（优）于配置（Convention over Configuration），即按约定进行编程，这是一种软件设计范式，旨在减少软件开发人员自主配置的数量，主要体现在以下两个方面：

- 约定并提供一些推荐的默认配置参数。
- 开发者只需要定义约定以外的配置参数。

这样做的好处是，如果约定的默认配置符合开发要求，则不用配置，反之，再进行额外的自定义参数配置，这样就能大大减少开发者的配置量。

关于约定大于配置，在后续的章节中都会有体现，比如 Spring Boot 中提供了默认的配置文件、配置格式，再到默认值自动配置，或者仅提供少量的个性化配置，这些都可以看出约定大于配置在 Spring Boot 框架中带来的便利性。

1.1.4 基本特性

1. 独立运行

Spring Boot 框架的诞生，改变了 Java 应用部署的格局。

Spring Boot 框架最大的亮点就是内嵌了各种 Servlet 容器，包括 Tomcat、Jetty、Undertow，应

用不再需要被打成 war 包再部署到 Tomcat 等 Servlet 容器中了，Spring Boot 应用可以被打包成一个可执行的 jar 包，所有的依赖都在一个 jar 包内，使用 java -jar 命令就能独立运行。

> 更多的 Spring Boot 部署和运行的方式可以参考第 11 章 Spring Boot 打包和部署。

2. 简化配置

传统的 Spring 应用中充斥着各种配置，比如各种技术组件的依赖配置、参数配置，以及 Spring 的各种配置文件，配置琳琅满目，不是专业的技术人员、架构师，去搭建一个 Spring 应用还是有一定难度的。

Spring Boot 提供了各种开箱即用的"Starter"一站式启动器，只要引入对应技术组件的 Starter 就能简化很多依赖配置，另外，Spring Boot 自动配置包中提供了各种技术的默认底层实现组件、默认的组件配置实例、默认的配置参数等，这些都有默认的实现。

以 Web 开发为例，只要添加一个 spring-boot-starter-web 启动器依赖，不需要其他任何配置，应用就能拥有 Spring Web 的能力，极大减少了开发人员对技术组件的配置管理。

> 关于 Spring Boot Starter 及自动配置原理可以参考第 3 章 Spring Boot Starter 与自动配置。

3. 自动配置

对于 Java 相关技术框架的集成，传统的 Spring 应用需要开发人员手动配置，包括应用所依赖的各种技术组件的核心配置类、Spring Bean 等组件，如果不是非常了解这些技术的配置组件，那么配置起来相当有难度。

Spring Boot 提供了自动配置能力，只要引入对应技术组件的 Starter，然后通过 Spring Boot 提供的默认配置参数，或者应用自定义的配置参数，就能自动配置，自动配置后这些技术就能快速集成并开箱使用。

自动配置功能可以推断应用可能需要加载哪些 Spring Bean，比如，如果类路径下有一个 HikariCP 连接池依赖包，并且此时应用没有提供任何有效的连接池配置类，那么 Spring Boot 就能推断应用可能需要一个连接池，并根据 HikariCP 依赖提供相应的连接池进行配置。如果应用已经配置了其他连接池，那么 Spring Boot 会放弃自动配置。

对于 Java 常用的主流技术，Spring Boot 官方都提供了自动配置能力及相关的 Starter，如果是官方不支持的第三方技术，它们一般也会提供带有自动配置的 Spring Boot Starter 启动器依赖，比如 MyBatis 提供的 mybatis-spring-boot-starter 启动器等。

> 如果没有特别复杂的定制需求，Spring Boot 默认自动配置后的组件就能轻松上手使用，如果不符合需求，Spring Boot 完全支持自定义并覆盖默认的自动配置。更多内容可以参考第 3 章 Spring Boot Starter 与自动配置。

4. 无代码生成和无须 XML 配置文件

Spring Boot 配置过程中无代码生成，意思就是 Spring Boot 不是通过生成代码的方式来完成自动配置的，也无须 XML 配置文件，而是通过 Spring & Spring Boot 中的各种条件注解完成的，Spring Boot 应用在运行时根据各种指定的条件、上下文环境参数等完成自动配置。

5. 各种生产级特性

Spring Boot 提供一系列生产级别的特性，比如端点、指标、健康检查、应用监控等，这对部署在生产环境中的 Spring 应用特别有帮助。

> 详细介绍可以参考第 12 章 Spring Boot 监控与报警。

1.1.5 核心模块

和 Spring 框架一样，Spring Boot 框架也是由许多核心模块组成的，每个模块负责不同的功能点，笔者整理了以下 10 大 Spring Boot 核心模块。

1. spring-boot

这是 Spring Boot 框架的主模块，也是支持其他模块的核心模块，主要功能如下：

- 提供了一个启动 Spring 应用的主类，并提供了一个相当方便的静态方法，它的主要作用是负责创建和刷新 Spring 容器的上下文。
- 提供了内嵌式的并可自由选择搭配的 Servlet 应用容器，如 Tomcat、Jetty、Undertow 等。
- 提供了一流的配置外部化支持。
- 提供了一个很方便的 Spring 容器上下文初始化器，包括对合理记录日志默认参数的支持。

2. spring-boot-autoconfigure

这个模块提供了常用的 Java 主流技术的自动配置组件，其提供的 @EnableAutoConfiguration 注解就能启用 Spring Boot 的自动配置功能，它能根据类路径下的内容决定是否自动配置。

3. spring-boot-starters

这个模块是所有 Starter 启动器的基础依赖，Starter 主要包括一系列技术组件依赖，它可以一站式开发 Spring 及相关技术应用，而不需要你到处找依赖和示例配置代码，它都帮你做好了。

例如 spring-boot-starter-web 这个 Starter 启动器，它引入了 spring-boot-starters 基础依赖及其相关的其他技术依赖。

4. spring-boot-cli

这是 Spring Boot 提供的命令行工具，也是另外一种创建 Spring 应用的方式，支持编译和运行 Groovy 应用，所以，它可以十分简单地编写并运行一个应用。

5. spring-boot-actuator

这是 Spring Boot 提供的监控模块，比如，它提供了健康端点、环境端点、Spring Bean 端点等端点，可以更好地帮助开发者监控应用并和应用进行交互。

6. spring-boot-actuator-autoconfigure

这个模块是为 spring-boot-actuator 监控模块提供自动配置的模块。

7. spring-boot-test

这是模块是 Spring Boot 的测试模块，为应用提供了许多非常有用的单元测试功能，包含了单元测试所需要的核心组件及注解等。

8. spring-boot-test-autoconfigure

这个模块是为 spring-boot-test 测试模块提供自动配置的模块。

9. spring-boot-loader

这个模块用于将 Spring Boot 应用构建为一个单独可执行的 jar 包，使用 java -jar 就能直接运行，一般不会直接使用这个来打包，而是使用 Spring Boot 提供的 Maven 或者 Gradle 插件。

10. spring-boot-devtools

这是 Spring Boot 开发者工具模块，主要用于 Spring Boot 应用的开发阶段，它提供了一些显著提升开发效率的特性，如修改了代码就自动重启应用等，以帮助开发者获得更流畅的应用开发体验。

这个模块的功能是可选的，只限于本地开发环境，当打成整包运行时，这些功能会自动被禁用。

1.1.6　Spring Boot 与 Spring MVC、Spring 之间的关系

Spring Boot、Spring MVC、Spring 三者之间是互相依存的关系，如下图所示。

```
∨ spring-boot-starter-web : 3.0.0 [compile]
  > spring-boot-starter : 3.0.0 [compile]
  > spring-boot-starter-json : 3.0.0 [compile]
  > spring-boot-starter-tomcat : 3.0.0 [compile]
  > spring-web : 6.0.2 [compile]
  ∨ spring-webmvc : 6.0.2 [compile]
    > spring-aop : 6.0.2 [compile]
      spring-beans : 6.0.2 [compile]
    > spring-context : 6.0.2 [compile]
      spring-core : 6.0.2 [compile]
    > spring-expression : 6.0.2 [compile]
      spring-web : 6.0.2 [compile]
```

spring-boot-starter-web 依赖了 spring-webmvc，spring-webmvc 又依赖了 spring-beans、spring-core 等 Spring 底层组件，所以 Spring 还是底层的框架，Spring Boot、Spring MVC 只是其上层的封装。

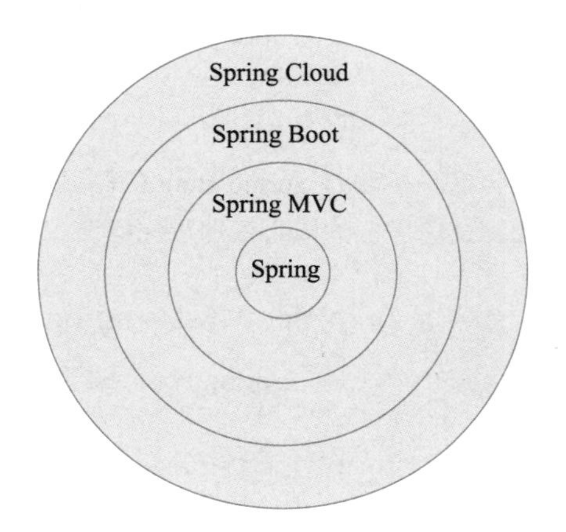

Spring Boot 既不是完全摒弃，也不是用来代替 Spring MVC、Spring 框架，Spring Boot 只是简化了它们的使用而已。

另外，再说说微服务框架 Spring Cloud，其本身也是基于 Spring Boot 框架进行构建的，Spring Cloud 也不能脱离以上任何一个 Spring 组件而独立存在。

这几个框架的依赖关系如右图所示。

1.2 Spring Boot 安装集成

1.2.1 版本周期

Spring Boot 项目同时维护了几条版本线，其中又包括不同意义的版本，比如 CURRENT、GA、PRE、SNAPSHOT 等代号，它们都代表着不同生命周期的版本，如下图所示。

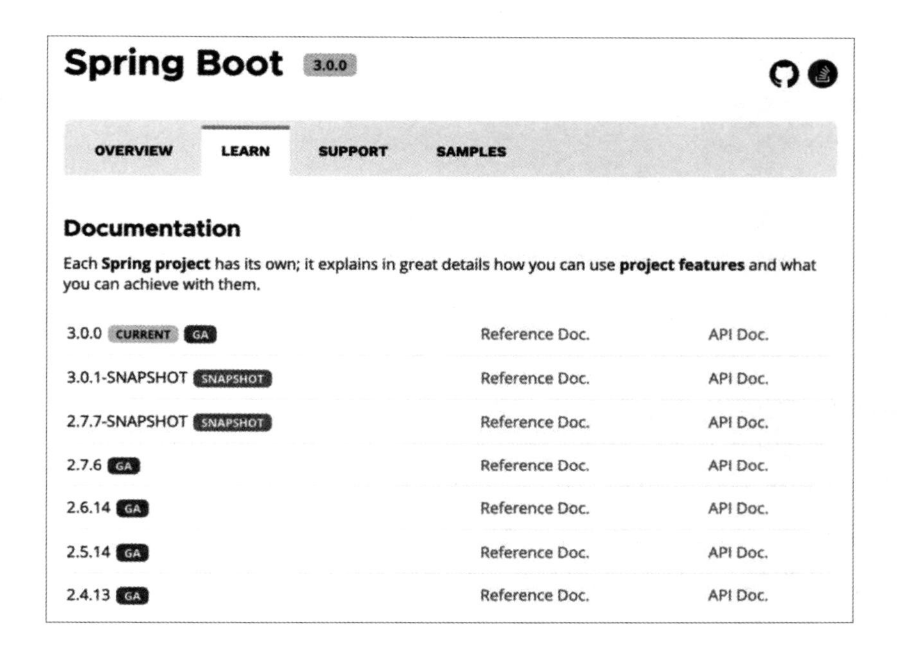

1. GA（正式版本）

GA 全称为 General Availability，表示一般可用的版本，即已经向大众公开发行的正式版本，一旦发布，日后就不会修改该发行版本的任何内容，如果发现 Bug，则会在下一个版本中修复。GA 版本号一般不会带任何标识后缀，比如 3.0.0、2.7.2，但是 2.4.x 之前的版本是带 .RELEASE 标识的，比如 2.3.12.RELEASE。

在生产项目中必须要使用 GA 版本，我们从 Maven 中央仓库中获取的依赖也都是 GA 版本。

2. CURRENT（最新正式版本）

CURRENT 表示最新的 GA 版本，如上图所示，Spring Boot 3.0.0 正式发布的时候，其右侧显示了 CURRENT 标识，说明它就是目前最新的可用发行版本。

如果是学习研究使用，则可以使用 CURRENT 版本，但如果是生产项目，那么还是要谨慎使用，因为最新的发行版本没有多少用户量，可能存在潜在的问题及漏洞，所以生产项目最好和最新发行版本保持几个版本差。

3. SNAPSHOT（快照版本）

SNAPSHOT 即快照版本，一般每晚进行构建以包含最新的变更。一般每个版本只有一个 SNAPSHOT 快照版本，比如 3.0.0-SNAPSHOT、2.7.3-SNAPSHOT 等，如果发现 Bug，那么也会在当前 SNAPSHOT 快照版本中修复。

SNAPSHOT 快照版本是最新测试版本，可能包含大量的 Bug，开发者可以拿来学习研究使用，强烈不建议用于生产项目。

4. PRE（预览版本）

PRE表示 Pre Release 版本，即预览版本，预览版本又有两个不同的阶段：

- **里程碑版本：**比如 3.0.0-M3，这里的 M 指的是 milestone，即里程碑版本，里程碑版本的版本号从 M1 开始递增。
- **候选发行版本：**比如 3.0.0-RC2，这里的 RC 指的是 Release Candidate，即候选发行版本，正式发行前的候选版本，和里程碑版本一样，从 1 开始递增，比如 Spring Boot 3.0.0 经历了 3.0.0-RC1 和 3.0.0-RC2。

预览版本发布后也是不会改变的，比如在 M1 中发现 Bug 时会在 SNAPSHOT 快照版本中修复并发布 M2，以此类推，当里程碑版本逐渐稳定之后就会发布候选发行版本，候选发行版本逐渐稳定之后就会发布正式版本。

PRE预览版本也是测试版本，同样不建议在生产项目中使用，它可能会包含少量 Bug，开发者可以提前学习研究即将发布的 Spring Boot 最新正式版本的特性，比如笔者撰写本书的初衷是要基于最新的 Spring Boot 3.0，但不可能等它发布后再去写，那时就太迟了，所以笔者就基于 3.0.0-M* 里程碑系列版本开始撰写，然后又更新到了 3.0.0-RC2，最后等 3.0.0 正式版本发布后再进行复核。

SNAPSHOT 快照版本和 PRE 预览版本没有发布到 Maven 中央仓库，仅发布在 Spring 仓库中，如果要使用这两个版本，则需要在 Maven 配置文件中配置 Spring 仓库：

```
<?xml version="1.0" encoding="UTF-8"?>
<project xmlns="http://maven.apache.org/POM/4.0.0" xmlns:xsi="http://www.w3.org/
2001/XMLSchema-instance"
    xsi:schemaLocation="http://maven.apache.org/POM/4.0.0 https://maven.apache.org/xsd/
maven-4.0.0.xsd">
    <modelVersion>4.0.0</modelVersion>
    ...
    <repositories>
        <repository>
            <id>spring-snapshots</id>
            <url>https://repo.spring.io/snapshot</url>
            <snapshots><enabled>true</enabled></snapshots>
        </repository>
        <repository>
            <id>spring-milestones</id>
```

```
            <url>https://repo.spring.io/milestone</url>
        </repository>
    </repositories>

    <pluginRepositories>
        <pluginRepository>
            <id>spring-snapshots</id>
            <url>https://repo.spring.io/snapshot</url>
        </pluginRepository>
        <pluginRepository>
            <id>spring-milestones</id>
            <url>https://repo.spring.io/milestone</url>
        </pluginRepository>
    </pluginRepositories>

</project>
```

这些版本的发布流程如下图所示。

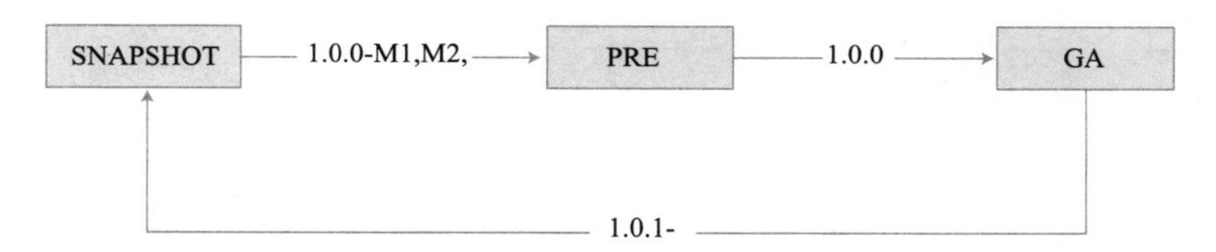

从 SNAPSHOT 快照版本开始，到多个 PRE 预览版本，再到正式版本。一旦正式版本发布，就不再维护相同版本的 SNAPSHOT 快照版本和 PRE 预览版本了，然后下一个版本就会继续从 SNAPSHOT 快照版本开始，如此往复，从 Spring Boot 版本图中可以看到，所有的 GA 版本都是没有 SNAPSHOT 快照版本和 PRE 预览版本的，只有还未正式发布的版本才会有这两个版本。

Spring Boot 官方源代码仓库的发布记录如右图所示。

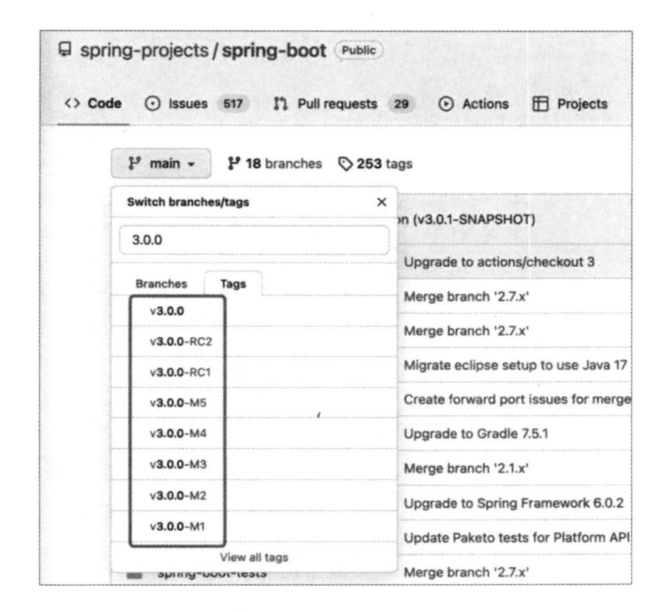

发布时间从近到远由上到下排列。

这也见证了 Spring Boot 从预览初始版本 M1 到 v3.0.0 正式发行版本的发布过程。

1.2.2 支持版本

Spring Boot 版本线如下表和下图所示。

版本线	发布日期	停止维护时间	商业支持停止时间
3.0.x	2022-11-24	2023-11-24	2025-02-24
2.7.x	2022-05-19	2023-11-18	2025-02-18
2.6.x	2021-11-17	2022-11-24	2024-02-24
2.5.x	2021-05-20	2022-05-19	2023-08-24
2.4.x	2020-11-12	2021-11-18	2023-02-23
2.3.x	2020-05-15	2021-05-20	2022-08-20
2.2.x	2019-10-16	2020-10-16	2022-01-16
2.1.x	2018-10-30	2019-10-30	2021-01-30
2.0.x	2018-03-01	2019-03-01	2020-06-01
1.5.x	2017-01-30	2019-08-06	2020-11-06

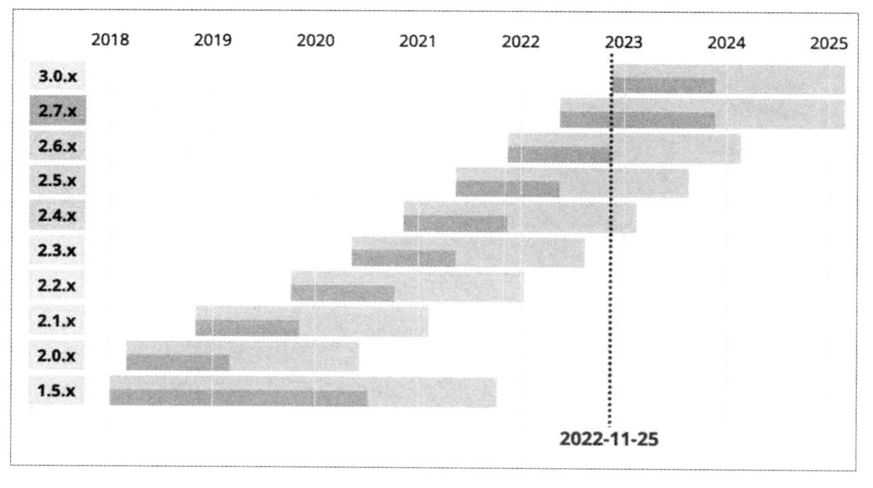

以上数据和图表来自 Spring Boot 官方文档（链接 1）。

随着 Spring Boot 3.0.0 的发布，免费受支持的版本就只有 Spring Boot 2.7+ 了，受商业支持的版本为 Spring Boot 2.4+，其他的低版本不建议使用了。

如果只是学习或者测试使用，使用哪个版本都没有问题，如果是公司实际生产环境项目，建议不要使用最新的版本，避免“踩坑”，也不要使用已经停止维护的版本，避免带来潜在的漏洞和重大 Bug。如果只是公司内部系统，不会暴露到外网，在不存在重大 Bug 的情况下，选用哪个版本理论上都是没问题的。

1.2.3　环境要求

不同的 Spring Boot 版本对 Java 开发环境的要求是不一样的，以及对支持的 Servlet 版本和内置容器版本也是不一样的，以下以 Spring Boot 不同三条版本线的最新版本为例列出环境要求的对比。

> 笔者编写本书时，Spring Boot 2.5 已经不再继续维护，所以这里不再列出。另外，随着 Spring Boot 3.0.0 的正式发布，Spring Boot 2.6.x 也到了停止维护的时间。

对 Java 开发环境的要求对比如下表所示。

Spring Boot	JDK	Spring	Maven	Gradle
3.0.0	17 ～ 19	6.0.2+	3.5+	7.5+
2.7.6	8 ～ 19	5.3.24+	3.5+	6.8.x、6.9.x、7.x
2.6.14	8 ～ 19	5.3.24+	3.5+	6.8.x、6.9.x、7.x

Spring Boot 3.0 最低要求 JDK 17，并向上兼容支持 JDK 19，虽然 Oracle 宣布现在 Java 17+ 是可以免费使用及商用的，但不确定在免费时间截止后是否会继续收费，未来使用 Spring Boot 3.0+ 可以转战开源的 OpenJDK 或者其他开源的 JDK 版本。

Spring Boot 3.0 最低支持的 Spring 框架也变成了 Spring 6.0.2+，虽然是框架自动集成依赖的，但需要注意这一点，因为在 Spring Boot 3.0 发布之前的 Spring 6.0 也有不少的底层升级。

对 Servlet 容器的版本要求对比如下表所示。

Spring Boot	Servlet	Tomcat	Jetty	Undertow
3.0.0	5.0+	10+	11+	2.2+
2.7.6	3.1+	9.0	9.4/ 10.0	2.0
2.6.14	3.1+	9.0	9.4/ 10.0	2.0

Spring Boot 3.0 最低支持 Servlet 5.0，并已将所有底层依赖项从 Java EE 迁移到了 Jakarta EE API，基于 Jakarta EE 9 并尽可能地兼容 Jakarta EE 10。因为早在几年前，Java EE 已经正式更名为 Jakarta，所有相关的名称都变了，包括包名，所以升级 Spring Boot 时要考虑兼容性。

1.2.4　集成方式

Spring Boot 有两种集成方式：

- 使用 Maven/Gradle 插件，具体见 1.2.5 节。
- 使用 Spring Boot CLI 命令行工具，具体见 1.5 节。

大部分 Java 开发者日常的主要开发工作还是基于 Maven 进行的，Maven 也是 Java 开发最主流的项目构建工具，这从各种开源项目的演示教程和使用文档中也可以看出来，所以本书也是以 Maven 作为标准进行实战的。

> 笔者并不会介绍 Maven 的具体使用，读者可以关注微信公众号“Java 技术栈”，在后台回复关键字“mvn”获取最新完整版高清教程。

1.2.5 使用 Maven 集成 Spring Boot

使用 Maven 方式集成又有两种方式：

1. 继承 spring-boot-starter-parent 父项目

Spring Boot 应用继承 spring-boot-starter-parent 父项目依赖：

```
<!-- 从Spring Boot 父项目中继承 -->
<parent>
  <groupId>org.springframework.boot</groupId>
  <artifactId>spring-boot-starter-parent</artifactId>
  <version>${spring-boot.version}</version>
</parent>
```

2. 导入 spring-boot-dependencies 依赖

Spring Boot 应用导入 spring-boot-dependencies 依赖：

```
<dependencyManagement>
  <dependencies>
    <dependency>
      <!-- Import dependency management from Spring Boot -->
      <groupId>org.springframework.boot</groupId>
      <artifactId>spring-boot-dependencies</artifactId>
      <version>${spring-boot.version}</version>
      <type>pom</type>
      <scope>import</scope>
    </dependency>
  </dependencies>
</dependencyManagement>
```

两种集成方式的区别：

前者需要继承 spring-boot-starter-parent 依赖，其实也继承了后者 spring-boot- dependencies，spring-boot-starter-parent 的 POM 配置文件如下：

```
<project xmlns="http://maven.apache.org/POM/4.0.0" xsi:schemaLocation="
http://maven.apache.org/POM/4.0.0 https://maven.apache.org/xsd/maven-4.0.0.xsd"
xmlns:xsi="http://www.w3.org/2001/XMLSchema-instance">
    <modelVersion>4.0.0</modelVersion>
    <parent>
        <groupId>org.springframework.boot</groupId>
        <artifactId>spring-boot-dependencies</artifactId>
        <version>3.0.0-M3</version>
    </parent>
    <artifactId>spring-boot-starter-parent</artifactId>
    <packaging>pom</packaging>
    <name>spring-boot-starter-parent</name>
    <description>Parent pom providing dependency and plugin management for applications
built with Maven</description>
    <properties>
        <java.version>17</java.version>
        <resource.delimiter>@</resource.delimiter>
        <maven.compiler.source>${java.version}</maven.compiler.source>
        <maven.compiler.target>${java.version}</maven.compiler.target>
        <project.build.sourceEncoding>UTF-8</project.build.sourceEncoding>
        <project.reporting.outputEncoding>UTF-8</project.reporting.outputEncoding>
    </properties>

  ...
```

前者 spring-boot-starter-parent 继承了 spring-boot-dependencies，定义了一些常用的项目相关的配置，比如 properties（参数）、plugins（插件）：

```
<properties>
    <java.version>17</java.version>
    <resource.delimiter>@</resource.delimiter>
    <maven.compiler.source>${java.version}</maven.compiler.source>
    <maven.compiler.target>${java.version}</maven.compiler.target>
    <project.build.sourceEncoding>UTF-8</project.build.sourceEncoding>
    <project.reporting.outputEncoding>UTF-8</project.reporting.outputEncoding>
</properties>

<build>
    <pluginManagement>
        <plugins>
```

```
            ...
            <plugin>
                <groupId>org.springframework.boot</groupId>
                <artifactId>spring-boot-maven-plugin</artifactId>
                <executions>
                    <execution>
                        <id>repackage</id>
                        <goals>
                            <goal>repackage</goal>
                        </goals>
                    </execution>
                </executions>
                <configuration>
                    <mainClass>${start-class}</mainClass>
                </configuration>
            </plugin>
        </plugins>
    </pluginManagement>
</build>
```

使用继承集成的方式有以下几个优点。

继承方式的优点一：

插件只需要定义即可，不需要进行额外配置，如下所示。

```
<build>
  <plugins>
    <plugin>
      <groupId>org.springframework.boot</groupId>
      <artifactId>spring-boot-maven-plugin</artifactId>
    </plugin>
  </plugins>
</build>
```

因为其他插件配置已经在父项目中定义好了，子项目直接定义插件坐标即可。如果使用导入 spring-boot-dependencies 依赖的集成方式，则需要额外配置该插件。

继承方式的优点二：

可以直接定义 properties 参数覆盖依赖的版本，比如，不想使用 Spring Boot 继承的版本，而想使用自定义的版本，那么直接定义对应的 properties 参数覆盖版本号即可，如下所示。

```
<slf4j.version>1.7.36</slf4j.version>
```

如果使用导入 spring-boot-dependencies 依赖的集成方式，则需要显式导入依赖并指定版本号：

```
<dependencies>
    <dependency>
        <groupId>org.slf4j</groupId>
        <artifactId>slf4j-api</artifactId>
        <version>${slf4j.version}</version>
    </dependency>
</dependencies>
```

继承集成的缺点就是不利于扩展，比如公司某个子项目需要继承某个父项目，因为只能单继承，所以一旦继承了 Spring Boot 的父依赖，就无法再继承公司的父项目了，扩展性不是很好。

两种集成方式各有优劣，笔者建议可以继承就继承，毕竟方便快捷。如果不能继承就使用导入依赖的方式。不过站在设计的原则上讲，一般推荐使用导入依赖的方式，避免后面出现多项目继承问题。笔者觉得，使用哪种方式还是要看具体的项目，集成方式本身并不会影响 Spring Boot 的正常使用。

1.3　快速开发一个 Spring Boot 接口

在软件编程领域，Hello World 经常用于某种编程语言、技术的快速入门体验，下面笔者就用 Spring Boot 框架开发一个最简洁的 Hello World 接口应用。

1.3.1　生成项目

Spring 提供了一站式生成 Spring 应用的网站（链接 2）。

首先在该网站上指定要生成的项目环境信息，也可以添加相关的依赖或者后续再导入，笔者生成的应用环境如下：

- Maven。
- Java。
- Spring Boot 3.0.0。
- JDK 17。

单击 GENERATE 按钮就能自动生成并下载该项目，如下图所示。然后打开 IDE 开发工具，通过 Maven 的方式导入该项目即可。

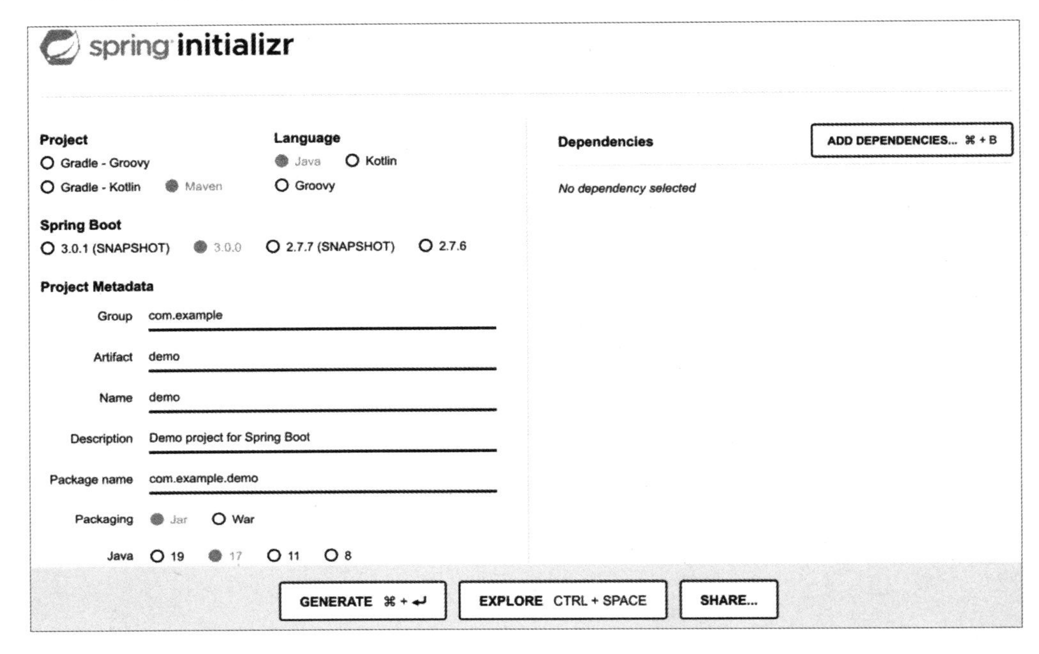

生成的 demo 项目结构如下图所示。

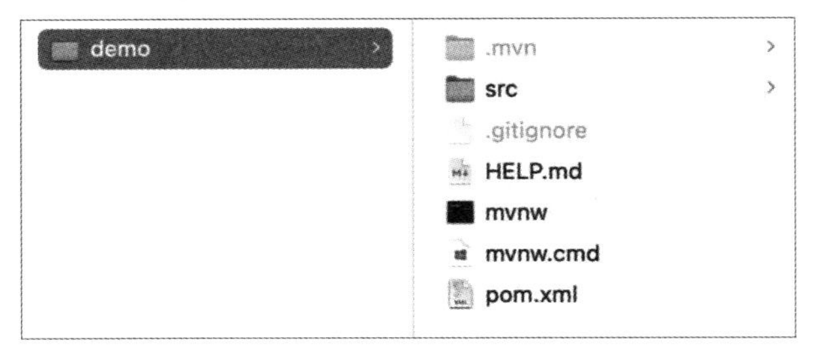

这是一个典型的 Spring Boot 应用结构：

```
.
├── HELP.md
├── mvnw
├── mvnw.cmd
├── pom.xml
└── src
    ├── main
    │   ├── java
    │   │   └── com
```

```
│   │           └── example
│   │               └── demo
│   │                   └── DemoApplication.java
│   └── resources
│       └── application.properties
└── test
    └── java
        └── com
            └── example
                └── demo
                    └── DemoApplicationTests.java
```

打开 pom.xml 配置文件：

```
<?xml version="1.0" encoding="UTF-8"?>
<project xmlns="http://maven.apache.org/POM/4.0.0" xmlns:xsi="http://www.w3.org/2001/
XMLSchema-instance"
        xsi:schemaLocation="http://maven.apache.org/POM/4.0.0 https://maven.apache.
org/xsd/maven-4.0.0.xsd">
    <modelVersion>4.0.0</modelVersion>
    <parent>
        <groupId>org.springframework.boot</groupId>
        <artifactId>spring-boot-starter-parent</artifactId>
        <version>3.0.0-M3</version>
        <relativePath/> <!-- lookup parent from repository -->
    </parent>
    <groupId>com.example</groupId>
    <artifactId>demo</artifactId>
    <version>0.0.1-SNAPSHOT</version>
    <name>demo</name>
    <description>Demo project for Spring Boot</description>
    <properties>
        <java.version>17</java.version>
    </properties>
    <dependencies>
        <dependency>
            <groupId>org.springframework.boot</groupId>
            <artifactId>spring-boot-starter</artifactId>
        </dependency>

        <dependency>
            <groupId>org.springframework.boot</groupId>
            <artifactId>spring-boot-starter-test</artifactId>
```

```
            <scope>test</scope>
        </dependency>
    </dependencies>

    <build>
        <plugins>
            <plugin>
                <groupId>org.springframework.boot</groupId>
                <artifactId>spring-boot-maven-plugin</artifactId>
            </plugin>
        </plugins>
    </build>
    <repositories>
        <repository>
            <id>spring-milestones</id>
            <name>Spring Milestones</name>
            <url>https://repo.spring.io/milestone</url>
            <snapshots>
                <enabled>false</enabled>
            </snapshots>
        </repository>
    </repositories>
    <pluginRepositories>
        <pluginRepository>
            <id>spring-milestones</id>
            <name>Spring Milestones</name>
            <url>https://repo.spring.io/milestone</url>
            <snapshots>
                <enabled>false</enabled>
            </snapshots>
        </pluginRepository>
    </pluginRepositories>

</project>
```

这种方式默认使用的是继承集成的方式，会添加默认的依赖：

- spring-boot-starter。
- spring-boot-starter-test。

还会添加一个插件：

- spring-boot-maven-plugin。

另外会生成一个启动类：

```
import org.springframework.boot.SpringApplication;
import org.springframework.boot.autoconfigure.SpringBootApplication;

@SpringBootApplication
public class DemoApplication {

    public static void main(String[] args) {
        SpringApplication.run(DemoApplication.class, args);
    }

}
```

这是 Spring Boot 应用的入口类，而 @SpringBootApplication 注解则用来标识 Spring Boot 应用的入口类，更多深入分析请参考第 4 章 Spring Boot 启动过程与扩展应用。

1.3.2 导入依赖

编写Spring Boot接口，如果在生成项目时没有选择spring-boot-starter-web依赖，则需要手动导入：

```
<dependencies>

    <dependency>
        <groupId>org.springframework.boot</groupId>
        <artifactId>spring-boot-starter-web</artifactId>
    </dependency>

</dependencies>
```

前面的章节也介绍了，导入一个这样的启动器依赖就能拥有 Spring Web 开发的能力。

1.3.3 编写接口

编写接口就和 Spring Boot 框架本身没有关系了，使用 Spring MVC 框架的相关注解即可，笔者这里编写了一个 /hello 的测试接口：

```
@SpringBootApplication
@RestController
public class Application {
```

```
    public static void main(String[] args) {
        SpringApplication.run(Application.class);
    }

    @RequestMapping("/hello")
    public String helloWorld() {
        return "hello world.";
    }

}
```

1.3.4 启动应用

在 IDE 中启动 Spring Boot 应用有两种方法：

1. 使用 Java 中的 main 方法启动

Spring Boot 应用可以直接运行应用启动类的 main 方法，Spring Boot 启动类一般位于根包下面，比如 demo 项目中的 DemoApplication 类。

2. 使用应用构建插件运行

如果应用集成了 Maven 插件 spring-boot-maven-plugin，就可以使用 Maven 命令启动：

```
mvn spring-boot:run
```

一般在 IDE 中运行 Maven 命令，可以省略 mvn 命令前缀，如下图所示。

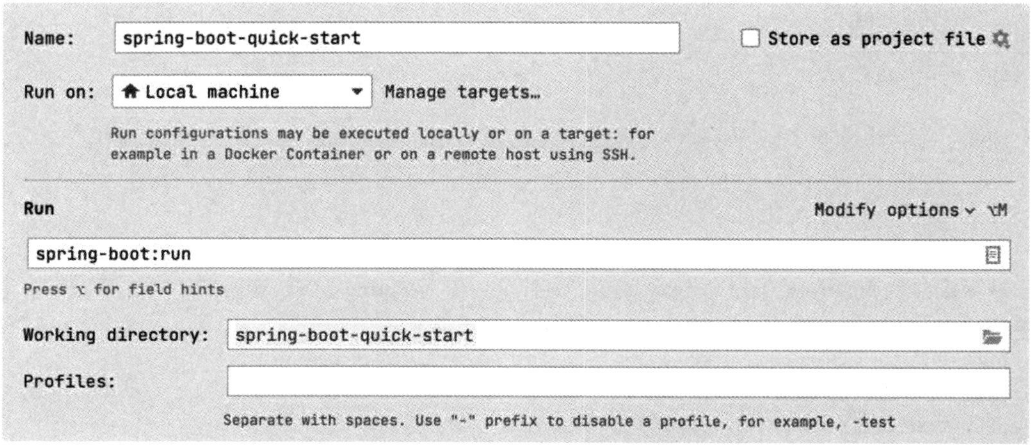

Name: spring-boot-quick-start
Store as project file
Run on: Local machine
Manage targets…
Run configurations may be executed locally or on a target: for example in a Docker Container or on a remote host using SSH.
Run
Modify options ⌥M
spring-boot:run
Press ⌥ for field hints
Working directory: spring-boot-quick-start
Profiles:
Separate with spaces. Use "-" prefix to disable a profile, for example, -test

Maven 应用建议使用这种启动方式，因为这样可以使用 Maven 的各种插件，如在使用资源插件时的配置文件中的占位符替换处理，如果使用第一种方法直接运行 main 方法，就会因为无法替换占位符而启动报错。

> Gradle 也有类似的插件，这里不再赘述。

应用启动日志，如下所示。

```
  .   ____          _            __ _ _
 /\\ / ___'_ __ _ _(_)_ __  __ _ \ \ \ \
( ( )\___ | '_ | '_| | '_ \/ _` | \ \ \ \
 \\/  ___)| |_)| | | | | || (_| |  ) ) ) )
  '  |____| .__|_| |_|_| |_\__, | / / / /
 =========|_|==============|___/=/_/_/_/
 :: Spring Boot ::                (v3.0.0)

。... : Starting Application using Java 17.0.4 with PID 85087 ...
。... : No active profile set, falling back to 1 default profile: "default"
。... : Devtools property defaults active! Set 'spring.devtools.add-properties' to
'false' to disable
。... : For additional web related logging consider setting the 'logging.level.web'
property to 'DEBUG'
。... : Tomcat initialized with port(s): 8080 (http)
。... : Starting service [Tomcat]
。... : Starting Servlet engine: [Apache Tomcat/10.1.1]
。... : Initializing Spring embedded WebApplicationContext
。... : Root WebApplicationContext: initialization completed in 1070 ms
。... : LiveReload server is running on port 35729
。... : Tomcat started on port(s): 8080 (http) with context path '/javastack'
。... : Started Application in 2.336 seconds (process running for 2.963)
```

这样，一个 Spring Boot 应用就运行起来了，默认的绑定端口是 8080。

1.3.5　测试接口

在浏览器中直接访问刚才编写的 /hello 接口，效果如下图所示。

```
http://localhost:8080/hello
```

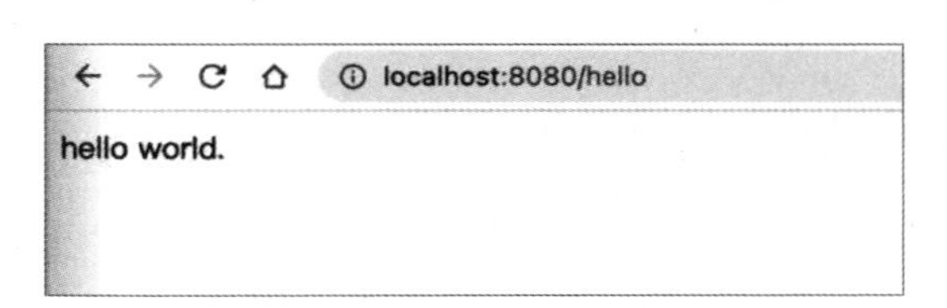

接口正常响应了，网页正常显示了 hello world.。

1.4 快速使用 Maven

Spring Boot 还提供了相关快捷脚本，可以免安装和快速使用 Maven、Gradle。

1.4.1 基本介绍

如下图所示，在生成的 Demo 项目中，除了常规的项目文件，还有两个特殊的脚本文件：

- mvnw（Linux 版本）。
- mvnw.cmd（Windows 版本）。

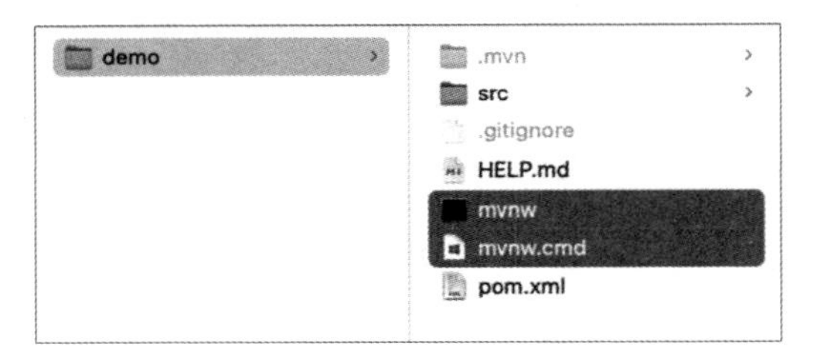

mvnw 的全称为 Maven Wrapper，使用 mvnw 可以快速将 Maven 集成到项目中，它的适用场景有以下两个：

- 想省心，不想自己安装配置 Maven 环境。
- 已有的 Maven 版本不合适，需要使用特定的 Maven 版本。

除了上面所说的 mvnw 脚本，在当前目录下会初始化一个 .mvn/wrapper 目录，如下图所示。

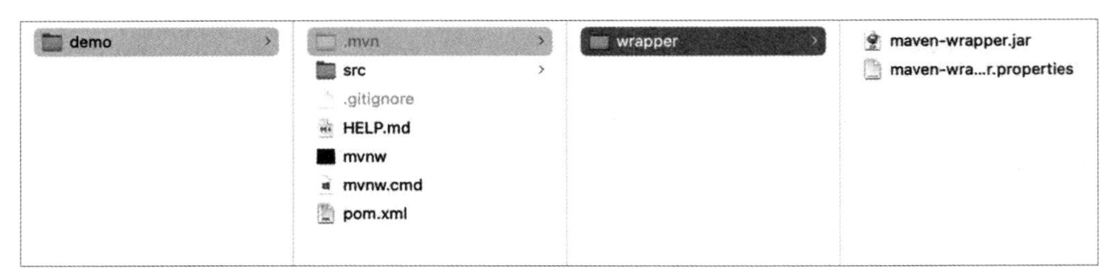

如果 Maven Wrapper 使用的并不是最新版本，但想使用最新的 Maven 3.8.5，则可以在 maven-wrapper.properties 配置文件中指定版本，如下图所示。

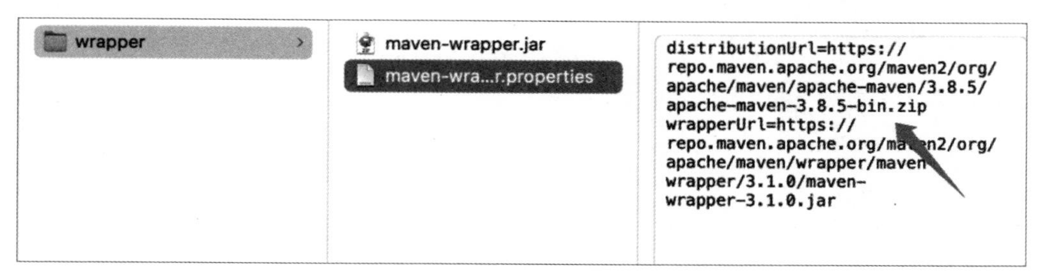

这意味着如果想使用 mvnw，只要复制这些生成的文件到对应的项目目录中，想使用哪个版本，改一改参数即可。

然后在 $USER_HOME/.m2/wrapper 目录中可以看到安装信息，如下图所示。

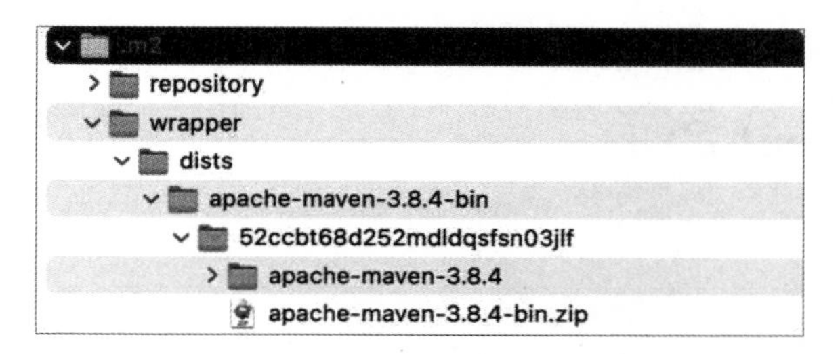

虽然不用自己另行安装，但本地仓库还是会自动下载对应版本的 Maven，只是省去了自己动手安装的流程。

1.4.2　使用方式

使用 Spring Initializr 网站一键生成的 Spring Boot 项目不需要单独安装 mvnw，生成后默认就带有 mvnw 系列文件，直接使用即可。

比如，我们切换到 demo 项目：

```
cd demo
```

运行项目清理安装命令：

```
./mvnw clean install
```

就像直接使用 mvn 命令一样，mvnw 只是在其基础上封装了一层而已，底层还是 mvn。所以，如果有多个 Maven 版本管理的需求，或者是不想自己动手安装 Maven 的场景，那么这个 mvnw 脚本还是可以用来试试的，但实际工作中很少会使用。

1.4.3 Gradle Wrapper

Gradle 也有类似的包装，如下图所示。

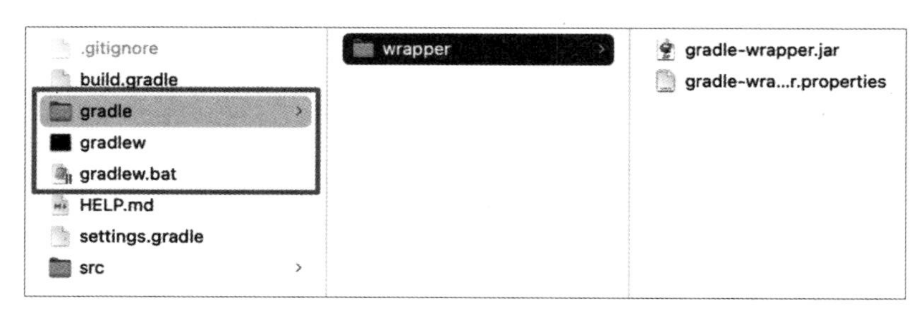

和 Maven 类似，这里就不再赘述了。

1.5 Spring Boot CLI

1.5.1 概述

Spring Boot CLI 的全称为 Spring Boot Command-Line Interface，是一个构造 Spring Boot 项目的命令行工具，可用于从 start.spring.io 网站构建新项目或用来加密。它可以运行 Groovy 脚本，它也是 JVM 系语言，拥有和 Java 类似的语法，但比 Java 要简洁得多，无须太多冗余的项目结构配置代码，从而可以快速构建项目。

Spring Boot 并不一定需要 Spring Boot CLI，但有了 Spring Boot CLI，可以在不需要 IDE 的情况下快速启动基于 Spring 的应用。

1.5.2 安装方式

Spring Boot CLI 要求 JDK 17 环境，Groovy 不需要安装，已经被打包在 Spring Boot CLI 包中了，已经安装的也会被忽略。

1. 手动安装

在 Spring 官网下载相应的版本 ZIP 包（链接 3）。

下载后，解压并复制到某个目录中，如下图所示。

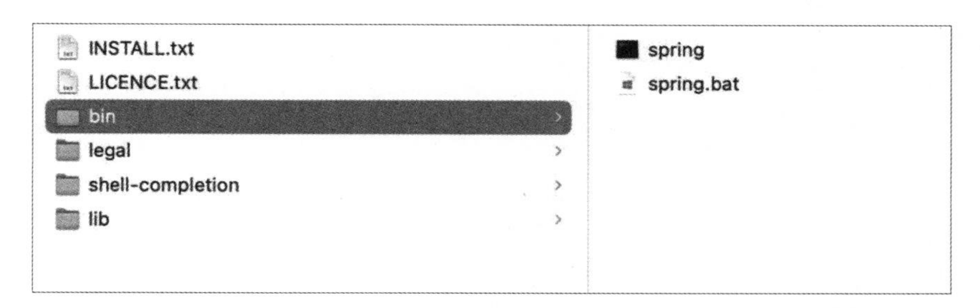

如上图所示，目录中有区别 Linux/ Windows 的运行程序，根目录中还有一个 INSTALL.txt 安装说明文件，主要介绍环境要求及如何设置 SPRING_HOME/PATH 环境变量：

- SPRING_HOME：设置环境变量指向 Spring Boot CLI 安装包解压后的根目录。
- PATH：设置环境变量指向 Spring Boot CLI 安装包解压后的 /bin 目录。

配置好之后，通过以下命令验证是否安装成功：

```
$ spring --version
Error: A JNI error has occurred, please check your installation and try again
Exception in thread "main" java.lang.UnsupportedClassVersionError: org/springframework/boot/loader/
JarLauncher has been compiled by a more recent version of the Java Runtime (class file version 61.0),
this version of the Java Runtime only recognizes class file versions up to 52.0
 at java.lang.ClassLoader.defineClass1(Native Method)
 at java.lang.ClassLoader.defineClass(ClassLoader.java:756)
 at java.security.SecureClassLoader.defineClass(SecureClassLoader.java:142)
 at java.net.URLClassLoader.defineClass(URLClassLoader.java:468)
 at java.net.URLClassLoader.access$100(URLClassLoader.java:74)
 at java.net.URLClassLoader$1.run(URLClassLoader.java:369)
 at java.net.URLClassLoader$1.run(URLClassLoader.java:363)
 at java.security.AccessController.doPrivileged(Native Method)
 at java.net.URLClassLoader.findClass(URLClassLoader.java:362)
 at java.lang.ClassLoader.loadClass(ClassLoader.java:418)
 at sun.misc.Launcher$AppClassLoader.loadClass(Launcher.java:355)
 at java.lang.ClassLoader.loadClass(ClassLoader.java:351)
 at sun.launcher.LauncherHelper.checkAndLoadMain(LauncherHelper.java:601)
```

上面显示抛出了异常，说明本机的 JDK 版本和 Spring Boot CLI 中运行的 JDK 版本不兼容，查看 /lib/spring-boot-cli-3.0.0-*.jar 包中的配置文件，实际用的是 JDK 17，和安装说明文件描述的 JDK 1.8 有出入，所以需要调整环境变量到 JDK 17，如下图所示。

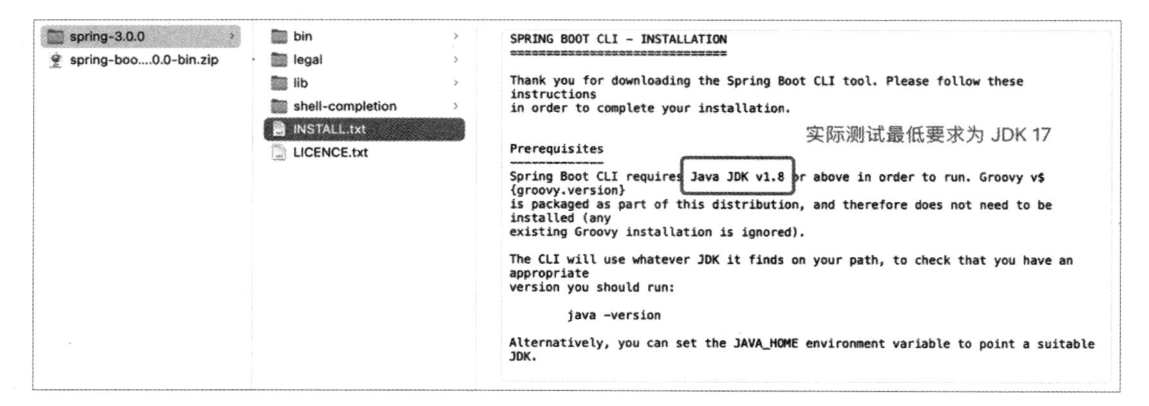

调整 JDK 环境变量后再来验证一下：

```
$ spring --version
Spring CLI v3.0.0.RELEASE
```

显示版本号就说明安装成功了，Spring Boot 3.0.0 最低要求 JDK 17，这一点也体现在 Spring Boot CLI 工具上了。

2. 通过 SDKMAN安装

SDKMAN 的全称为 Software Development Kit Manager，是一个用于在大多数基于 UNIX 的系统上管理多个软件开发工具包的并行版本的工具，它提供了一个方便的命令行界面（CLI）和 API，可用于应用安装、切换、删除、列出候选版本等。SDKMAN 就支持包括 Groovy 和 Spring Boot CLI 的安装，本节就通过 SDKMAN 的方式来安装。

首先需要前往 SDKMAN 网站下载 SDKMAN（链接 4）。

SDKMAN 官网提供了一键安装脚本：

```
curl -s "https://get.sdkman.io" | bash
```

SDKMAN 安装完成之后，使用以下命令安装 Spring Boot CLI：

```
$ sdk install springboot
$ spring --version
Spring CLI v3.0.0.RELEASE
```

3. OSX Homebrew/ MacPorts

Spring Boot CLI 还支持 macOS 操作系统下 OSX Homebrew、MacPorts 等安装方式。

OSX Homebrew：

```
$ brew tap spring-io/tap
$ brew install spring-boot
```

MacPorts：

```
$ sudo port install spring-boot-cli
```

1.5.3　命令行自动补全

Spring Boot CLI 提供了命令行自动补全 Shell 脚本，支持 BASH 和 ZSH 集成，Spring Boot CLI 是通过手动下载压缩包安装的，需要自行建立符号链接，参考命令如下：

```
$ ln -s ./shell-completion/bash/spring /etc/bash_completion.d/spring
$ ln -s ./shell-completion/zsh/_spring /usr/local/share/zsh/site-functions/_spring
```

如果是通过 SDKMAN安装的，则使用以下参考命令：

```
$ . ~/.sdkman/candidates/springboot/current/shell-completion/bash/spring
```

如果是通过 Homebrew 或者 MacPorts 方式安装的，则会注册自动补全 Shell 脚本。

设置好之后，输入 spring 命令、空格，按 tab 键，会自动出现命令补全提示：

```
$ spring 按 tab 键
encodepassword  -- Encode a password for use with Spring Security
grab            -- Download a spring groovy script's dependencies to ./repository
help            -- Get help on commands
init            -- Initialize a new project using Spring Initializr (start.spring.io)
install         -- Install dependencies to the lib/ext directory
jar             -- Create a self-contained executable jar file from a Spring Groovy script
run             -- Run a spring groovy script
shell           -- Start a nested shell
uninstall       -- Uninstall dependencies from the lib/ext directory
version         -- Show the version
war             -- Create a self-contained executable war file from a Spring Groovy script
```

比如这里的 encodepassword 指令是使用 Spring Security 框架对密码进行加密的，这样在某些使用 Spring Security 框架需要手动生成密码的场合，就不需要在代码中写 main 方法生成密码了。

1.5.4 快速开始

前面介绍了Spring Boot CLI可以运行Groovy脚本以快速构建Spring项目，本节来快速实战一下，以下是笔者在桌面创建的一个简单的 app.groovy 脚本文件：

```
@RestController
class ThisWillActuallyRun {

    @RequestMapping("/hi")
    String home() {
       "Hello Spring Boot!"
    }

}
```

然后使用 Spring Boot CLI 运行命令就能直接运行该脚本文件了，第一次运行会比较慢一点，因为要下载必要的依赖项，后续运行会很快。

运行 app.groovy 脚本参考命令：

```
$ spring run app.groovy
Resolving dependencies....................

.   ____          _            __ _ _
/\\ / ___'_ __ _ _(_)_ __  __ _ \ \ \ \
( ( )\___ | '_ | '_| | '_ \/ _` | \ \ \ \
\\/  ___)| |_)| | | | | || (_| |  ) ) ) )
'  |____| .__|_| |_|_| |_\__, | / / / /
=========|_|==============|___/=/_/_/_/
:: Spring Boot ::                (v3.0.0)

...

...Started application in 2.544 seconds (JVM running for 4.453)
...Tomcat started on port(s): 8080 (http) with context path ''
...Started application in 2.544 seconds (JVM running for 4.453)
```

如上启动日志，启动端口为 8080，就像启动正常的 Java 应用一样，因为 Groovy 本身就是 JVM 系编程语言，在 Java 中创建一个接口可能需要许多项目配置，而 Groovy 只需要一个脚本就够了。

访问 Groovy 脚本中的 /hi 接口，能正常输出结果，如下图所示。

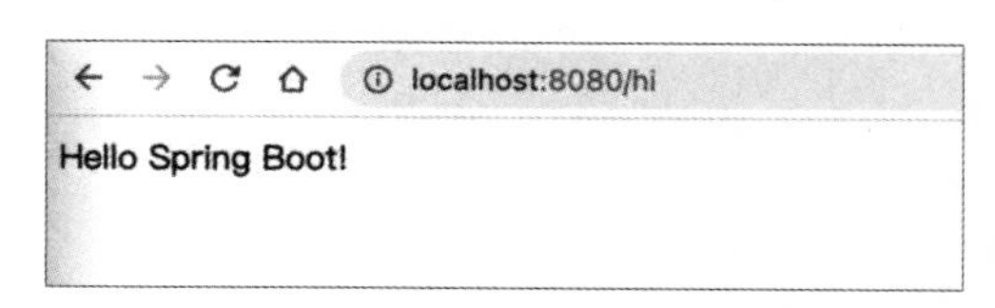

只需一个 Groovy 脚本、一条 Spring Boot CLI 命令就能运行一个 Spring 应用，这太轻松了，省去了 Java 项目工程配置、依赖配置等大量烦琐工作。所以，如果你了解 Groovy 语言，使用 Spring Boot CLI 命令行工具可以比使用传统 Java & Maven 方式更快速地构建 Spring 应用，极大提升生产力。

1.6 开启 Spring Boot 之旅

在前面的章节中，我们了解了 Spring Boot 的背景、基本介绍、集成方式、启动类及其注解的使用，也用了五个小步骤演示了如何从零开始快速开发一个 Spring Boot 应用接口，除了导入指定的依赖，几乎没有任何的配置，甚至 Maven 和 Gradle 都不需要开发人员额外安装。

如果是传统的 Spring应用呢？

我们需要花费更多的时间在工程和 Spring 框架配置上面，还有 Maven、Gradle、Tomcat 容器的安装、配置、部署都需要花费不少时间，而 Spring Boot 从开始到接口运行只要几分钟，这正是 Spring Boot 框架所能解决的问题—极大简化了 Spring 的使用难度，让 Java 开发者能快速上手开发一个 Spring 应用。

另外，使用 Spring Boot CLI & Groovy 集成方式，一个脚本就能运行一个应用，所以，笔者建议开发者尽快上手使用 Spring Boot 框架。

第2章 Spring Boot 配置管理

本章将介绍 Spring Boot 配置相关的一些内容，比如如何自定义配置类、导入配置，以及应用的核心配置文件、配置绑定、外部化配置、多文档配置、基于 Profiles 的配置、配置加密、配置迁移等，只有先理解这些基础配置的用法和原理，打好基础，才能更好地理解 Spring Boot 后续的内容。

2.1 配置类

2.1.1 自定义配置类

第 1 章介绍了 Spring Boot 的特性，其中有一条就是无代码生成、无 XML 配置，虽然 Spring Boot 能做到大部分代码自动配置，但应用还有自定义的组件配置，以及应用自身的配置等，这些都是需要通过 Java 自定义配置类提供的。

Spring 3.0 之前，要使用 Spring 组件就必须要有一个 XML 配置文件，这也是 Spring 的核心配置文件，而 Spring 3.0 之后就不需要 XML 配置文件了，可以通过新的 @Configuration 注解代替 XML 配置文件。

第 1 章的 Spring Boot 集成实战中用到了 Spring Boot 的核心注解 @SpringBootApplication，它其实就包含了 @SpringBootConfiguration 配置类注解，也就是 Spring Boot 应用的专用配置类注解。

> @SpringBootApplication 注解的详细介绍见第 4 章 Spring Boot 启动过程与扩展应用。

@SpringBootConfiguration 注解的源码如下：

```
@Target({ElementType.TYPE})
@Retention(RetentionPolicy.RUNTIME)
@Documented
@Configuration
@Indexed
public @interface SpringBootConfiguration {
    @AliasFor(annotation = Configuration.class)
    boolean proxyBeanMethods() default true;
}
```

如源码所示，它组合了 Spring 框架的原生 @Configuration 注解，用来代替 @Configuration 注解，示例代码如下：

```
@SpringBootConfiguration
public class MainConfig {

    @Bean
    public RestTemplate restTemplate() {
        return new RestTemplate();
    }

}
```

如上，笔者使用 @SpringBootConfiguration 注解替换了 @Configuration 注解，其实使用哪个都一样，一个是 Spring Boot 换了个名字的专用注解，另一个是 Spring 原生注解，它们两个是等效的，所以，在 Spring Boot 中不需要 XML 配置，只需要自定义若干用 @Configuration 或者 @SpringBootConfiguration 注解修饰的 Java 配置类即可，Spring Boot 应用推荐使用 @SpringBootConfiguration 注解。

以上的 @Bean 注解就等于 XML 配置文件中的 <bean ...> 标签，Spring 3.0+ 中有一整套的注解来代替 XML 配置文件中的定义，比如 @Lazy、@Primary、@Scope 等，完全不需要再用 XML 进行配置，并且在 Spring Boot 框架中都可以正常使用，Spring Boot 也会有自己扩展的相关注解，比如上面提到的 @SpringBootConfiguration 注解。

2.1.2　导入配置

1. 导入配置类

我们知道在 Spring Boot 中可以使用一个 @SpringBootConfiguration 或者 @Configuration 注解的配置文件来配置所有 Bean 及其他组件，但其实没必要这么做，我们可以把相关的配置独立出来，放到多个配置类中，这样更好管理。

比如可以有以下参考配置类：

- MainConfiguration：项目主要配置。
- DataSoureceConfiguration：数据源配置。
- RedisConfiguration：Redis 配置。
- MongoDBConfiguration：MongoDB 配置。

……

但实际开发过程中有另外一种情况，可能某些配置类在其他子包中，或者在依赖的 jar 包中，这些配置无法被 Spring Boot 默认扫描注册到，需要额外导入，基于这种情况，Spring 框架提供的 @Import 导入注解就能派上用场了，它的源码如下：

```
@Target(ElementType.TYPE)
@Retention(RetentionPolicy.RUNTIME)
@Documented
public @interface Import {

	/**
	 * {@link Configuration @Configuration}, {@link ImportSelector},
	 * {@link ImportBeanDefinitionRegistrar}, or regular component classes to import.
	 */
	Class<?>[] value();

}
```

从 value 注释可以看出，value 是一个 Class 数组，用于导入多个配置，具体类型可以是以下几种：

- @Configuration。
- ImportSelector。
- ImportBeanDefinitionRegistrar。
- 任何一个 component 组件。

来看下面的示例，笔者创建了两个配置类，并通过 @Import 注解导入：

```
@SpringBootConfiguration
@Import({Configuration1.class, Configuration2.class})
public class MainConfig {
```

```
    @Bean
    public RestTemplate restTemplate() {
        return new RestTemplate();
    }

}
```

其他 @Import 注解用法在后续的章节中会有更多的应用。

当然，如果这些配置类都在类扫描路径下，就不用通过 @Import 导入了，直接用 @ComponentScan 注解添加要扫描的路径即可，因为 @SpringBootApplication 注解包含了 @ComponentScan 注解并扫描所有子包目录，所以不必重复添加。但是，像上面那种情况，即配置类不在默认的类扫描路径下，此时可以用 @ComponentScan 注解指定要扫描的包目录。

2. 导入 XML 配置

Spring 框架还提供了一个 @ImportResource 注解，用于导入额外的 XML 配置文件，它的源码如下：

```
@Retention(RetentionPolicy.RUNTIME)
@Target(ElementType.TYPE)
@Documented
public @interface ImportResource {

    @AliasFor("locations")
    String[] value() default {};

    @AliasFor("value")
    String[] locations() default {};

    Class<? extends BeanDefinitionReader> reader() default BeanDefinitionReader.class;

}
```

使用方式很简单，指定要导入的 XML 配置文件路径即可，比如有些老的技术组件是 XML 配置形式的，如果某些 XML 标签无法找到对应的 Spring 替代注解，或者无法通过 @Configuration 方式配置，那么此时使用 @ImportResource 注解就非常管用。

其实，这两个导入注解也都是 Spring 框架的原生注解，和 Spring Boot 框架本身也并没有什么关系，但在平时开发时要知道如何在 Spring Boot 框架中灵活运用。

2.2 配置文件

2.2.1 application

Spring Boot 应用的配置参数主要在 application 配置文件中，见 StandardConfigDataLocation-Resolver 类源码中的定义，它可以有不同形式的配置方式，该配置文件也不是必需的，可以有多个，也可以一个都没有。

StandardConfigDataLocationResolver 源码中的定义如下：

```
static final String CONFIG_NAME_PROPERTY = "spring.config.name";

static final String[] DEFAULT_CONFIG_NAMES = { "application" };
```

如果不想使用默认的 application 配置文件，则可以在命令行使用 spring.config.name 参数名指定其他的配置文件名称：

```
$ java -jar demo.jar --spring.config.name=app
```

默认的配置文件搜索路径为：

- optional:classpath:/;optional:classpath:/config/。
- optional:file:./;optional:file:./config/;optional:file:./config/*/。

这个在 ConfigDataEnvironment 类静态初始块中可以找到，在后面介绍配置加载机制源码的章节中也会介绍。

默认的配置文件搜索路径可以通过 spring.config.location 命令行参数指定，既可以是目录（需要以“/”结束），也可以是具体文件，比如下面指定了多个配置文件命名：

```
$ java -jar myproject.jar --spring.config.location=\
 optional:classpath:/default.properties,\
 optional:classpath:/override.properties
```

其中的 optional 前缀表示配置是可选的，如果指定的配置文件不存在，默认会报告异常，所以指定 optional 前缀就可以不用管配置文件是否存在。

如果应用想忽略所有的类似异常而正常启动，则可以通过以下参数来设置：

```
spring:
  config:
    on-not-found: ignore
```

另外，这个 spring.config.location 命令行参数会覆盖应用默认的配置文件，如果不想覆盖，则可以使用 spring.config.additional-location 参数追加其他的配置文件。

2.2.2 bootstrap

除了 application 配置文件，还有另外一种 bootstrap 配置文件，但它不能独立于 Spring Boot 应用生效，它属于 Spring Cloud 环境，需要引入 Spring Cloud 依赖。bootstrap 配置文件会优先于 application 配置文件被加载，它主要用于从额外的资源中加载配置参数，通常用来加载外部配置，如 Spring Cloud 用于加载配置中心的配置参数，也可以用来定义系统不会被改变的参数，它们默认不能被本地相同的配置所覆盖。

因此，对比 application 配置文件，bootstrap 配置文件具有以下几个特性：

- bootstrap 由父 ApplicationContext 加载，比 application 优先被加载。
- bootstrap 中的参数不能被覆盖。

一般的 Spring Boot 应用使用 application 配置文件即可，除非用到了 Spring Cloud，并遇到了以下几个应用场景：

- 使用了外部化配置中心。
- 有一些固定的不能被覆盖的参数。

2.2.3 配置文件类型

配置文件类型可以是以下两种：

- .properties：key = value 格式。
- .yml：key: value 树状格式（全称：yaml）。

比如下面的示例，分别用 .properties 和 .yml 两种格式指定了应用的访问端口和访问路径，如果不指定，则默认分别为 8080 和 /。

.properties 格式：

```
server.port = 8080
server.servlet.contextPath= /javastack
```

.yml 格式：

```
server:
  port: 8081
  servlet:
    contextPath: /javastack
```

使用 .yml 配置文件的格式说明如下：

- 键冒号后面需要带一个空格。
- 每个层级的缩进用两个空格。

一般推荐使用 .yml 格式文件，树状的层次结构比较清晰，且节省冗余的参数前缀配置量，一个 .yml 配置文件就能适配多个特定环境，而使用 .properties 格式则需要多个配置文件，这个会在后面的 profiles 章节介绍。

需要注意的是，.yml 配置文件不支持通过 @PropertySource 注解来导入配置，因为这个注解本来就是只针对 .properties 文件的，有用到这个注解的地方需要注意，但也可以有其他解决方案，比如使用 @ConfigurationProperties 注解，它同时支持两种文件格式，这个在接下来的配置绑定一节中会讲到。

虽然 .yml 配置文件不支持 @PropertySource 注解，但它可以通过 YamlPropertiesFactoryBean 类被加载为一个 Properties 参数类，或者通过 YamlMapFactoryBean 类被加载为一个 Map，还可以通过 YamlPropertySourceLoader 被加载为 Spring 环境中的 PropertySource。

2.3 配置绑定

2.3.1 Spring 中的配置绑定

在 Spring 框架中，所有已加载到 Spring 环境中的配置都可以通过注入 Environment 环境 Bean 来获取：

```
@Autowired
private Environment env;

// 获取参数
String getProperty(String key);
```

使用 @Value("${property}") 注解在类成员变量上绑定配置参数，这是最原始的做法，一般不推荐这样做，一方面是太不优雅，另一方面是，如果注入多个参数就会很麻烦，比如以下同时注入多个前缀一样的参数：

```
@Value("${user.username}")
private String username;

@Value("${user.age}")
private int age;

@Value("${user.sex}")
private int sex;

...
```

这样注入还会显得代码太冗余了，也非常不优雅。

另外，Spring 框架也有 @PropertySource 注解绑定配置，但也仅限于 .properties 配置文件，比如绑定资源目录下的 config/db-config.properties 配置参数：

```
@Data
@Component
@PropertySource(value = {"config/db-config.properties"})
public class DbProperties {

    @Value("${db.username}")
    private String username;

    @Value("${db.password}")
    private String password;

}
```

由于 @PropertySource 注解并不支持主流的 .yml 配置文件绑定，其本身也需要结合 @Value 注解使用，所以这也不是推荐的用法。

Spring Boot 提供了多种配置参数注入方法，比如参数绑定、构造器绑定等，支持多种配置文件格式，并且还提供了参数验证，让配置参数注入更加方便、灵活、安全。

2.3.2　参数绑定

本节通过 Java Bean 提供的 setter 方法进行配置参数与 Java Bean 字段的绑定，首先在 application.yml 配置文件中添加以下 javastack.* 配置参数：

```
javastack:
  name: Java 技术栈
  site: www.javastack.cn
  author: 栈长
  users:
    - Jom
    - Lucy
    - Jack
  params:
    tel: 18800008888
    address: China
  security:
    security-key: 123321
    security-code: 666666
```

新建 JavastackProperties 参数类进行参数绑定：

```
@Data
@ConfigurationProperties(prefix = "javastack")
public class JavastackProperties {

    private boolean enabled;

    private String name;

    private String site;

    private String author;

    private List<String> users;

    private Map<String, String> params;

    private Security security;

}

@Data
class Security {

    private String securityKey;

    private String securityCode;

}
```

一般参数类要以 XxxProperties 命名，比如 Spring Boot 中的各种参数类：

- ServerProperties。
- DataSourceProperties。
- JdbcProperties。

……

这样当我们配置任意其他框架时，都可以直接找到其配置参数类，如要配置 Redis，则可以直接打开 RedisProperties 参数类，看看都有什么配置。

通过 @ConfigurationProperties 注解可以将配置参数映射到一个 Java Bean 上，该注解中的 prefix 或者 value 参数用于指定要映射的参数前缀，前缀格式为英文小写，多个前缀以“-”分隔，例如：

```
javastack.user-agent
```

因为 @ConfigurationProperties 注解绑定配置参数需要用到类的 setter 方法，所以这里用到了 Lombok 的 @Data 注解，用于在类编译时生成类的 getter、setter、构造器、toString 等方法，这里不再赘述，具体可以参考笔者的博客文章（链接 5）。

然后在启动类上添加 @EnableConfigurationProperties 注解，用于指定要启用的 @ConfigurationProperties 参数类 JavastackProperties，使其生效：

```
@SpringBootApplication
@RequiredArgsConstructor
@EnableConfigurationProperties(value = {JavastackProperties.class})
@Slf4j
public class Application {

    private final JavastackProperties javastackProperties;

    @Bean
    public CommandLineRunner commandLineRunner() {
        return (args) -> {
            log.info("javastack properties: {}", javastackProperties);
        };
    }

    ....

}
```

然后在要使用参数类的地方直接将其注入，以上 JavastackProperties 参数类使用的是构造器注入方式。

最后在启动类中添加一个 CommandLineRunner 测试一下，这个 CommandLineRunner 是 Spring Boot 内置的命令行运行器，用于在应用启动之后执行一些特定的操作，比如上述程序用于在应用启动时输出一些配置参数。

> 更多关于 CommandLineRunner 的详细介绍请参考第 4 章 Spring Boot 启动过程与扩展应用。

输出结果如下：

```
javastack properties: JavastackProperties(enabled=false, name=Java 技术栈,
site=www.javastack.cn, author= 栈长 , users=[Jom, Lucy, Jack], params={tel=18800008888,
address=China}, security=Security(securityKey=123321, securityCode=666666))
```

配置参数正常输出，下面总结一下 @ConfigurationProperties 注解配置绑定：

- 支持按配置参数的前缀进行绑定，前缀一样的配置参数将被绑定到同一个类上。
- 支持配置参数使用默认值，比如示例中的 enabled 参数并没有配置，但输出了默认值 false。
- 支持松绑定（xx1-xx2-xx3 -> xx1Xx2Xx3），这里的连接符 - 也可以是下画线 _，比如示例中的 Security 嵌套类的 securityKey 字段，对应配置文件中的 security-key 参数，还可以是 XX_XX_XX 的大写格式，这个在绑定环境变量参数时非常有用。
- 支持 Java 集合绑定，比如示例中的 Map、List 字段值正常绑定了。
- 支持嵌套类，比如示例中特别使用了一个 Security 嵌套类，其所有字段值都能正常绑定。
- 支持主要的配置途径，比如 .yml 文件、.properties 文件、环境变量参数等。
- 需要搭配 @EnableConfigurationProperties 注解共同使用。

2.3.3 构造器绑定

本节通过 Java Bean 提供的构造器进行配置参数与 Java Bean 字段的绑定，首先在 application.yml 配置文件中添加以下 member.* 配置参数：

```
member:
  name: Tom
  sex: 1
  age: 18
  birthday: 2000-12-12 12:00:00
```

然后新建一个 MemberProperties 参数类进行参数绑定：

```
@Data
@NoArgsConstructor
@ConfigurationProperties(prefix = "member")
public class MemberProperties {

    private String name;
    private int sex;
    private int age;
    private String country;
    private Date birthday;

    public MemberProperties(String name, int sex, int age) {
        this.name = name;
        this.sex = sex;
        this.age = age;
    }

    @ConstructorBinding
    public MemberProperties(String name,
                            int sex,
                            int age,
                            @DefaultValue("China") String country,
                            @DateTimeFormat(pattern = "yyyy-MM-dd HH:mm:ss") Date birthday) {
        this.name = name;
        this.sex = sex;
        this.age = age;
        this.country = country;
        this.birthday = birthday;
    }

}
```

通过 @ConfigurationProperties 注解指定要绑定的配置参数的前缀，再使用 @ConstructorBinding 注解指定要绑定的构造器方法。

注解说明如下：

- @DefaultValue：提供默认值。
- @DateTimeFormat：指定日期格式。

然后在启动类 @EnableConfigurationProperties 注解中指定要启用的 @ConfigurationProperties 配

置参数类 MemberProperties 并使其生效：

```
@SpringBootApplication
@RequiredArgsConstructor
@EnableConfigurationProperties(value = {JavastackProperties.class, MemberProperties.class})
@Slf4j
public class Application {

    private final JavastackProperties javastackProperties;

    private final MemberProperties memberProperties;

    @Bean
    public CommandLineRunner commandLineRunner() {
        return (args) -> {
            log.info("javastack properties: {}", javastackProperties);
            log.info("member properties: {}", memberProperties);
        };
    }

    ...

}
```

最后注入其参数类，并在 commandLineRunner 方法中添加测试代码。

输出结果如下：

```
member properties: MemberProperties(name=Tom, sex=1, age=18, country=China,
birthday=Tue Dec 12 12:00:00 CST 2000)
```

可以看出，构造器的绑定也很简单，可以共同使用 @ConfigurationProperties 注解和 @EnableConfigurationProperties 注解，如果有多个构造器，则需要通过一个 @ConstructorBinding 注解指定要使用的构造器。

@ConstructorBinding 注解的源码如下：

```
@Target(ElementType.CONSTRUCTOR)
@Retention(RetentionPolicy.RUNTIME)
@Documented
public @interface ConstructorBinding {

}
```

> Spring Boot 3.0 新变化：
>
> （1）@ConstructorBinding 注解由 org.springframework.boot.context.properties 包移到了 org.springframework.boot.context.properties.bind 包，但目前还是 Deprecated 状态，在后续版本中会被彻底删除。
>
> （2）@ConstructorBinding 注解在 Spring Boot 2.x 中是可以同时用于类和构造器方法上的，而在 3.0 版本中只能用在具体某个构造器方法上，如上面的 @ConstructorBinding 注解的源码所示。
>
> （3）如果配置类只有一个参数化的构造器，则无须使用 @ConstructorBinding 注解指定。如果参数类有多个参数化的构造器，那么还是需要使用 @ConstructorBinding 注解指定要绑定的构造器，否则会调用默认的无参构造器并通过 setters 方法注入绑定。

这几点变化能大大简化 @ConfigurationProperties 注解的使用难度，如果通过构造器注入 Bean 时不想使用 @ConstructorBinding 构造器绑定的方式，则可以使用 @Autowired 注解替代。

2.3.4 Bean 配置绑定

除了使用 @ConfigurationProperties 注解在一个类上，还可以在 @Bean 方法上使用它，比如将参数绑定到一个第三方类上时，因为它们可能在 jar 包中，所以无法在其类、字段上添加注解等设置。

比如新建一个 OtherMember 其他成员类：

```
@Data
@NoArgsConstructor
public class OtherMember {

    private String name;
    private int sex;
    private int age;

}
```

再新建一个参数类，并添加一个 OtherMember 类的 Bean：

```
@SpringBootConfiguration
public class MainConfig {

    ...

    @Bean
```

```
    @ConfigurationProperties(prefix = "member")
    public OtherMember otherMember() {
        return new OtherMember();
    }

}
```

可以看到，还是用 @ConfigurationProperties 注解绑定的配置参数，只是把该注解从参数类中移到了 @Bean 方法之上。因为使用 @Bean 方式，所以不需要使用 @EnableConfigurationProperties 注解来使 @ConfigurationProperties 配置类生效，直接注入就能使用，示例如下：

```
@SpringBootApplication
@RequiredArgsConstructor
@EnableConfigurationProperties(value = {JavastackProperties.class, MemberProperties.class})
@Slf4j
public class Application {

    private final JavastackProperties javastackProperties;

    private final MemberProperties memberProperties;

    private final OtherMember otherMember;

    @Bean
    public CommandLineRunner commandLineRunner() {
        return (args) -> {
            log.info("javastack properties: {}", javastackProperties);
            log.info("member properties: {}", memberProperties);
            log.info("other member: {}", otherMember);
        };
    }

    ...

}
```

最后注入其参数类，并在 commandLineRunner 方法中添加测试代码。

输出结果：

```
other member: OtherMember(name=Tom, sex=1, age=18)
```

配置参数成功绑定，在不能修改配置参数类的情况下确实有用，但这种方式只能绑定简单的数据类型，像 Date 这种数据类型就没有对应的解决方案。

2.3.5　参数类扫描

前面章节中的参数绑定和构造器绑定需要将 @ConfigurationProperties 注解搭配 @EnableConfigurationProperties 注解使用，不能自行使用 @Component 等注解进行注册，如果后面新增了别的参数类，而忘了在 @EnableConfigurationProperties 注解中指定，则参数类不会生效，比较麻烦。

为此，Spring Boot 提供了一种参数类扫描方式：

```
@SpringBootApplication
@RequiredArgsConstructor
//@EnableConfigurationProperties(value = {JavastackProperties.class, MemberProperties.class})
@ConfigurationPropertiesScan
@Slf4j
public class Application {
```

在启动类上使用 @ConfigurationPropertiesScan 注解，就可以扫描所有包目录下的参数类，如果要扫描指定包目录，则在注解的 basePackages 参数中指定具体的包即可。

2.3.6　配置验证

使用 @ConfigurationProperties 注解还能有效地进行参数验证，Spring Boot 支持 JSR-303 javax.validation 规范注解，首先添加一个遵循 JSR-303 标准实现的依赖：

```
<dependency>
  <groupId>org.hibernate.validator</groupId>
  <artifactId>hibernate-validator</artifactId>
</dependency>
```

然后在字段 name 上面使用一个 @NotNull 不为空注解：

```
@Data
@Validated
@ConfigurationProperties(prefix = "javastack")
public class JavastackProperties {

    private boolean enabled;

    @NotNull
    private String name;
```

```
    ...
}
```

在配置文件中把 name 改为 names，重启应用进行测试。结果应用启动失败，并输出了以下错误信息：

```
***************************
APPLICATION FAILED TO START
***************************
Description:
Binding to target org.springframework.boot.context.properties.bind.BindException: Failed
to bind properties under 'javastack' to cn.javastack.springboot.properties.props.
JavastackProperties failed:
    Property: javastack.name
    Value: null
    Reason: 不能为 null
Action:
Update your application's configuration
```

提示 javastack.name 参数不能为 null，并提示更新应用的配置，所以这个配置验证也很有用，可避免配置参数被误删除等操作导致的配置缺失。

2.4 外部化配置

2.4.1 配置源

Spring Boot 还可以将配置外部化，即除了把配置放在应用配置文件中，还可以使用各种外部配置源，主要包含以下几种：

- properties 文件。
- YAML 文件。
- 环境变量。
- 命令行参数。

2.4.2 配置优先级

Spring Boot 使用了一个非常特殊的 PropertySource 顺序，该顺序可以按优先级自动覆盖参数值，如果多个配置源的参数值相同，则优先级高的参数值最后会覆盖优先级低的参数值。

配置源优先级从低到高如下：

（1）默认参数（SpringApplication.setDefaultProperties）。

（2）使用 @PropertySource 注解绑定的配置。

（3）应用配置文件中的参数（application）。

（4）配置了 random.* 随机数的参数。

（5）系统环境变量。

（6）Java System properties。

（7）java:comp/env 的 JNDI 参数。

（8）ServletContext 初始化参数。

（9）ServletConfig 初始化参数。

（10）来自 SPRING_APPLICATION_JSON 的参数。

（11）命令行参数。

（12）单元测试上的参数。

（13）使用 @TestPropertySource 注解绑定的配置。

（14）Devtools 全局设置参数（来自 $HOME/.config/spring-boot）。

多个应用配置文件（application）的优先级从低到高如下：

（1）应用配置文件（jar 包内）。

（2）指定了 profile 的配置文件，如 application-dev.properties（jar 包内）。

（3）应用配置文件（jar 包外）。

（4）指定了 profile 的配置文件，如 application-dev.properties（jar 包外）。

举个例子，比如注入一个 user.name 参数：

```
@Value("${user.name}")
private String username;
```

如果 application.yml 配置文件配置了该参数值，但系统环境变量中也有此参数值，因为系统环境变量的优先级高于应用配置文件，所以最后会使用系统环境变量中的参数值。

建议在整个应用中统一使用一种格式，但如果同一位置同时有 .properties 和 .yml 两种格式的配置文件，则 Spring Boot 会优先使用 .properties 配置文件。

2.4.3 命令行参数

命令行参数指的是使用 java -jar --key=value 命令启动应用时指定的参数，即以 "--" 开头的参数，比如：

```
java -jar demo.jar --server.port=9090
```

从配置优先级可以看到，命令行参数的优先级始终是高于配置文件的，所以以上命令会使用 9090 端口代替配置文件中的端口，或者默认的 8080 端口。

默认情况下，Spring Boot 会将所有命令行参数转换并添加到 Spring 环境中，如果不想添加到 Spring 环境中，则可以将其禁用：

```
public static void main(String[] args) {
    SpringApplication.setAddCommandLineProperties(false);
    SpringApplication.run(Application.class, args);
}
```

测试以上设置，先注入 server.port 参数并把它打印出来：

```
@Value("${server.port}")
private int serverPort;

@Bean
public CommandLineRunner commandLineRunner() {
    return (args) -> {
        log.info("server.port: {}", serverPort);
    };
}
```

如果是在 IDE 中使用 Maven 启动应用，则可以参考以下命令使用方法，具体如下图所示。

```
spring-boot:run -Dspring-boot.run.jvmArguments='-Dserver.port=9090'
```

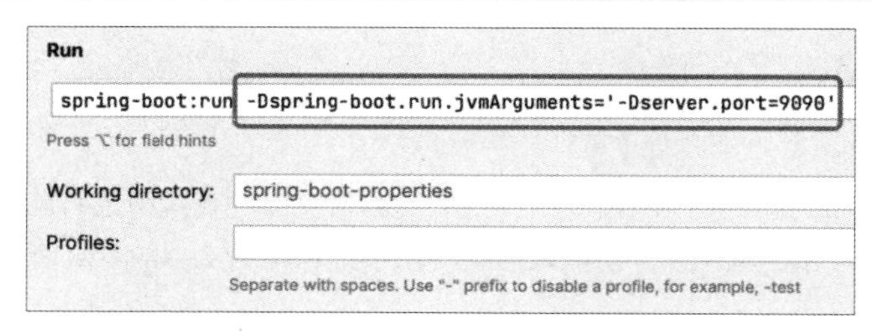

应用重启后输出结果如下：

```
...
Tomcat initialized with port(s): 9090 (http)
...
server.port: 9090
```

端口绑定正确，Spring 环境中的参数值取的也是命令行参数的值。所以，有时候想临时改变应用的配置参数，可以使用命令行参数，因为它的优先级够高，而且使用也很方便。

2.5 导入配置

可以使用 spring.config.import 参数来指定要导入的配置文件的路径，比如下面指定了要导入一个 app.yml 配置文件：

```
spring:
  config:
    import:
      - optional:classpath:/config/app.yml
```

然后新建该配置文件：

```
javastack:
  enabled: true
  site: app.javastack.cn
```

使用之前的示例重启测试，输出结果：

```
javastack properties: JavastackProperties(enabled=true, name=Java 技术栈 ,
site=app.javastack.cn, author= 栈长 , users=[Jom, Lucy, Jack], params={tel=18800008888,
address=China}, security=Security(securityKey=123321, securityCode=666666))
```

字段 enabled 有了指定的 true 值，site 值也被覆盖了，这是因为导入的其他位置的配置文件的优先级要高于其触发导入配置文件的优先级。

这里的 optional 前缀及配置文件不存在时抛出异常的处理和 spring.config.location 参数的处理一致，具体可参考前面的配置文件章节。

2.6 随机值配置

Spring Boot 提供的 RandomValuePropertySource 类可用于注入随机值，它可以生成整数（int）、长整数（long）、UUID 及字符串。

比如下面的示例，修改了前面章节的 .yml 配置文件：

```
javastack:
  name: Java 技术栈
  site: www.javastack.cn
  author: 栈长
  users:
    - Jom
    - Lucy
    - Jack
  params:
    tel: 18800008888
    address: China
  security:
    # 生成随机 32 位 MD5 字符串
    security-key: ${random.value}
    security-code: ${random.uuid}

member:
  name: Tom
  sex: 1
  age: ${random.int[18,100]}
  birthday: 2000-12-12 12:00:00
```

上面以 ${random.*} 标识的就是随机数，配置了随机字符串（value）、UUID，以及带范围的随机整数。

重启应用来看一下效果：

```
javastack properties: JavastackProperties(enabled=true, name=Java 技术栈,
site=app.javastack.cn, author= 栈长 , users=[Jom, Lucy, Jack], params={tel=18800008888,
address=China}, security=Security(securityKey=d2b8c180f07844c733a1fcbc8561164d,
securityCode=58ec8e15-4604-4708-8662-0cb1a3ae7d4b))

member properties: MemberProperties(name=Tom, sex=0, age=61, country=China,
birthday=Tue Dec 12 12:00:00 CST 2000)
```

输出结果达到了预期，下面看一下获取指定随机值的方法的源码：

```
private Object getRandomValue(String type) {
    if (type.equals("int")) {
        return getSource().nextInt();
    }
    if (type.equals("long")) {
        return getSource().nextLong();
    }
    String range = getRange(type, "int");
    if (range != null) {
        return getNextIntInRange(Range.of(range, Integer::parseInt));
    }
    range = getRange(type, "long");
    if (range != null) {
        return getNextLongInRange(Range.of(range, Long::parseLong));
    }
    if (type.equals("uuid")) {
        return UUID.randomUUID().toString();
    }
    return getRandomBytes();
}
```

实现很简单，其实就是使用 Java 自带的 java.util.Random 和 java.util.UUID 等工具类，这里就不再详细解析了，感兴趣的读者可以自行查看 RandomValuePropertySource 类的实现源码。

实际工作中很少使用随时数配置，但不能不知道它的存在，笔者知道的一个典型应用场景就是微服务随机端口，这样就能快速地进行扩容，当然也需要处理端口重复占用的情况。

2.7　多文档配置

Spring Boot 可以将单个物理的配置文件拆分为多个独立的逻辑配置文档，配置参数从上到下按顺序处理，后面的配置参数值可以覆盖前面的配置参数值。

多文档配置一般用于指定多套不同的配置，比如开发、测试、生产环境，它们的环境配置可能是不同的，比如端口，所以多文档配置中配置只在指定激活的环境中才会生效。

2.7.1　配置格式

.yaml 配置文件多文档之间用三个连接符（---）进行分隔，.properties 配置文件多个文档之间用 #---

进行分隔，如下面示例所示。

.yaml 配置文件：

```
spring:
  application:
    name: "javastack"
---
spring:
  application:
    name: "javastack-cloud"
  config:
    activate:
      on-cloud-platform: "kubernetes"
```

.properties 配置文件：

```
spring.application.name=javastack
#---
spring.application.name=javastack-cloud
spring.config.activate.on-cloud-platform=kubernetes
```

以上代码在一个物理配置文件中定义了 2 个逻辑文档。

使用多文档配置需要注意以下几点：

- 多文档配置的分隔符前面不能有空格，并且分隔符要连续。
- 多文档配置的分隔符前、后一行不能是注释。
- 多文档配置不能被 @PropertySource 和 TestPropertySource 注解加载。

2.7.2 激活多文档配置

多文档配置一般要配合指定的条件激活，比如在指定的环境中才激活生效，Spring Boot 支持根据指定的 Profile 和云平台进行激活，激活参数及说明如下表所示。

激活参数	说明
spring.config.activate.on-profile	根据指定的 Profile 激活
spring.config.activate.on-cloud-platform	根据指定的云平台激活

下面新建了两个配置文档，而第二个配置文档只在 dev 和 test 环境中生效：

```
spring:
```

```
  profiles:
    active: dev
  config:
    import:
      - optional:classpath:/config/app.yml

javastack:
  name: Java 技术栈
  site: www.javastack.cn
  author: 栈长
  users:
    - Jom
    - Lucy
    - Jack
  params:
    tel: 18800008888
    address: China
  security:
    # 生成随机 32 位 MD5 字符串
    security-key: ${random.value}
    security-code: ${random.uuid}

member:
  name: Tom
  sex: 1
  age: ${random.int[18,100]}
  birthday: 2000-12-12 12:00:00

---
member:
  name: Jack
  sex: 1
  age: 20
  birthday: 2000-01-01 12:00:00

spring:
  config:
    activate:
      on-profile: "dev | test"
```

然后通过 spring.profiles.active 激活了 dev 环境。

重启应用再测试一下：

```
...

The following 1 profile is active: "dev"

...

member properties: MemberProperties(name=Jack, sex=1, age=20, country=China,
birthday=Sat Jan 01 12:00:00 CST 2000)
```

从启动日志可以看到，dev 环境被激活了，member 参数的系列值被覆盖为 Jack 的了，即后面激活的文档配置覆盖了前面的文档配置。

2.8 Profile

基于 Profile 的配置也是 Spring Boot 的重要特性，在前面的章节中也基本介绍了 Profile 的功能，Profile 提供了一种分离应用配置的方法，让配置仅在某些具体的环境中才被激活生效，但在 Spring Boot 系列版本中也发生了不少变更，使用时特别需要注意。

2.8.1 默认的 Profile

如果不指定要激活的 Profile，Spring Boot 就会使用默认的 Profiledefault，这个可以从一开始的启动日志中看到：

```
No active profile set, falling back to 1 default profile: "default"
```

也可以调整默认的 Profile，如下所示。

```
spring:
  profiles:
    default: dev
```

以上表示如果没有指定 Profile，则默认使用 dev 环境。

2.8.2 激活 Profile

可以使用 spring.profiles.active 参数指定要激活的 Profile，比如在 application.yml 配置文件中指定激活 dev 和 test 环境：

```
spring:
  profiles:
    active: dev, test
```

还可以在应用启动方法上进行激活：

```
public static void main(String[] args) {
    SpringApplication springApplication = new SpringApplication(Application.class);
    springApplication.setAdditionalProfiles("dev", "main");
    springApplication.run(args);
}
```

Profile 不仅可激活配置文件，还可以激活 Java 配置类，比如可以用在 @Component、@Configuration 或 @ConfigurationProperties 注解的类上。

下面是一个 @Configuration 使用示例：

```
@Profile("main")
@SpringBootConfiguration
@Import({Configuration1.class, Configuration2.class})
public class MainConfig {

    ...

}
```

这样只有在激活 main 环境时，这个配置类中的配置才会被加载，然后通过配置文件被激活：

```
spring:
  profiles:
    default: dev
    active: dev, main
```

在使用 @Component、@Configuration 注解时，可以直接把 @Profile 注解用在该注解所在的类上，但 @ConfigurationProperties 的使用需要说明一下：

- 如果使用的是直接扫描的注册方式，则 @Profile 注解可以直接用在 @ConfigurationProperties 类上。
- 如果使用的是 @EnableConfigurationProperties 注册方式，则需要把 @Profile 注解用在 @EnableConfigurationProperties 配置类上。

2.8.3 切换 Profile

这个 spring.profiles.active 参数和其他参数遵循一样的顺序加载规则，比如即使在配置文件这里激活了，还是可以通过命令行参数临时替换 Profile：

```
$ java -jar demo.jar --spring.profiles.active=test
```

如果使用 IDE 启动：

```
$ mvn spring-boot:run -Dapp.profiles=test
```

因为命令行参数的优先级比配置文件的优先级高，所以这样就使用了 test 环境，配置文件中激活的 dev 在 main 环境中不会被用到。

所以，如果不想直接被优先级更高的配置源替换，也可以使用 spring.profiles.include 附加模式：

```
spring:
  profiles:
    default: dev
    # 可以被命令行替换
    active: dev, main
    # 不会被替换
    include:
      - dev
      - main
```

这样即使在命令行用 --spring.profiles.active 参数切换了 Profile，也不会影响 dev 和 main 环境的激活。

2.8.4 Profile 分组

有时候，除了 dev、test、main 这些 Profile，可能还会有一些独立存在的 Profile，比如以下示例：

```
...

---
member:
  name: Jack1

spring:
  config:
    activate:
```

```
      on-profile: main1

---
member:
  name: Jack2

spring:
  config:
    activate:
      on-profile: main2
```

这些 main1、main2 等 Profile 只有在激活 main 环境时才被激活，因为使用之前的参数配置就会比较麻烦，所以 Spring Boot 提供了 Profile 分组功能。

以下代码定义了一个 main 组，其中包含上面所说的两个 Profile：

```
spring:
  profiles:
    default: dev
    # 可以被命令行切换
    active: dev, main
    # 不会被切换
    include:
      - dev
      - main
    group:
      main:
        - main1
        - main2
```

只要 main 环境激活，main1 和 main2 也会一起被激活。

重启应用查看输出日志：

```
...

The following 4 profiles are active: "dev", "main", "main1", "main2"

...

member properties: MemberProperties(name=Jack2, sex=1, age=20, country=China,
birthday=Sat Jan 01 12:00:00 CST 2000)
```

日志显示输出了被激活的几个 Profile，Member 参数的 name 值也变成了 Jack2，这是因为最后的 main2 覆盖了所有之前的 Profile 的参数值。

2.8.5 指定 Profile 配置文件

除了之前章节介绍的多文档配置，现在有了 Profile 功能，应用就可以有多个基于 Profile 的配置文件，格式为：

```
application-${profile}
```

比如可以有以下配置文件：

- application.yml。
- application-dev.yml。
- application-test.yml。
- application-prod.yml。
- application-main.yml。

……

这样只有激活指定的 Profile，对应 Profile 的配置文件才会被激活。

从之前的配置优先级章节中可以知道，基于 Profile 的配置文件的优先级要高于默认的配置文件的优先级，所以如果按顺序同时激活了 dev 和 main 环境，则配置文件的优先级为：

```
默认＞ dev ＞ main
```

优先级高的配置参数会覆盖优先级低的配置参数，所以如果以上三个配置文件都有相同的值，那么最后会取 main 中的配置参数。

下面新建两个配置文件：

- application-dev.yml。
- application-main.yml。

分别在这两个配置文件中把 member.name 参数的值设置为 Lucy1 和 Lucy2，再重启应用查看日志：

```
...
The following 4 profiles are active: "dev", "main", "main1", "main2"
```

```
...

member properties: MemberProperties(name=Lucy2, sex=1, age=20, country=China,
birthday=Sat Jan 01 12:00:00 CST 2000)
```

日志显示输出了被激活的几个 Profile，Member 参数的 name 值也变成了 Luck2，这样就看出基于 Profile 配置文件的优先级了。

2.8.6 使用限制

Profile 中常用的几个参数如下表所示。

参数	功能	说明
spring.profiles.default	指定默认的 Profile	当不指定 Profile 时，默认生效的 Profile
spring.profiles.active	激活指定的 Profile	可以被优先级高的配置源替换
spring.profiles.include	激活指定要包含的 Profile	不会被其他配置源替换
spring.profiles.group	指定 Profile 分组	激活一个分组，就会激活组下所有 Profile

注意，Profile 有一些使用上的限制，不能随意搭配使用，以上几个参数不能用于多文档配置中和指定 Profile 配置文件。

比如在 application.yml 多文档配置中使用 spring.profiles.active 参数激活：

```
...

---
member:
  name: Jack2

spring:
  config:
    activate:
      on-profile: main2
  profiles:
    active: main
```

重启应用就会抛出异常：

```
ERROR org.springframework.boot.SpringApplication - Application run failed
org.springframework.boot.context.config.InactiveConfigDataAccessException: Inactive
property source 'Config resource 'class path resource [application.yml]' via location
'optional:classpath:/' (document #3)' imported from location 'class path resource
```

```
[application.yml]' cannot contain property 'spring.profiles.active' [origin: class
path resource [application.yml] - 70:13]
	at org.springframework.boot.context.config.InactiveConfigDataAccessException.
throwIfPropertyFound(InactiveConfigDataAccessException.java:126)
	at org.springframework.boot.context.config.ConfigDataEnvironmentContributors
$InactiveSourceChecker.onSuccess(ConfigDataEnvironmentContributors.java:310)
```

比如在 application-dev.yml 配置文件中使用 spring.profiles.active 参数激活：

```
spring:
  profiles:
    active: main
```

重启应用后也会抛出异常：

```
ERROR org.springframework.boot.SpringApplication - Application run failed
org.springframework.boot.context.config.InvalidConfigDataPropertyException: Property
'spring.profiles.active' imported from location 'class path resource [application-dev.
yml]' is invalid in a profile specific resource [origin: class path resource [application-dev.
yml] - 3:13]
    at org.springframework.boot.context.config.InvalidConfigDataPropertyException.
lambda$throwIfPropertyFound$1(InvalidConfigDataPropertyException.java:121)
    at java.base/java.lang.Iterable.forEach(Iterable.java:75)
    at java.base/java.util.Collections$UnmodifiableCollection.forEach (Collections.
java:1092)
    at org.springframework.boot.context.config.InvalidConfigDataPropertyException.
throwIfPropertyFound
```

所以，如果要将这几个参数放在配置文件中，则只能放在默认配置文件且没有多文档标识的部分。

2.9 加载机制

Spring Boot 中的配置文件是通过 PropertySourceLoader 接口实现的：

```
public interface PropertySourceLoader {

	/**
	 * Returns the file extensions that the loader supports (excluding the '.').
	 * @return the file extensions
	 */
```

```
    String[] getFileExtensions();

    /**
     * Load the resource into one or more property sources. Implementations may either
     * return a list containing a single source, or in the case of a multi-document format
     * such as yaml a source for each document in the resource.
     * @param name the root name of the property source. If multiple documents are loaded
     * an additional suffix should be added to the name for each source loaded.
     * @param resource the resource to load
     * @return a list property sources
     * @throws IOException if the source cannot be loaded
     */
    List<PropertySource<?>> load(String name, Resource resource) throws IOException;

}
```

该接口提供了两个抽象方法：

- getFileExtensions：获取文件后缀。
- load：加载配置。

前面的章节中提到的 YamlPropertySourceLoader 就实现了该 PropertySourceLoader 接口，包括它在内，该接口一共有以下两个实现类：

- YamlPropertySourceLoader。
- PropertiesPropertySourceLoader。

类结构如下图所示。

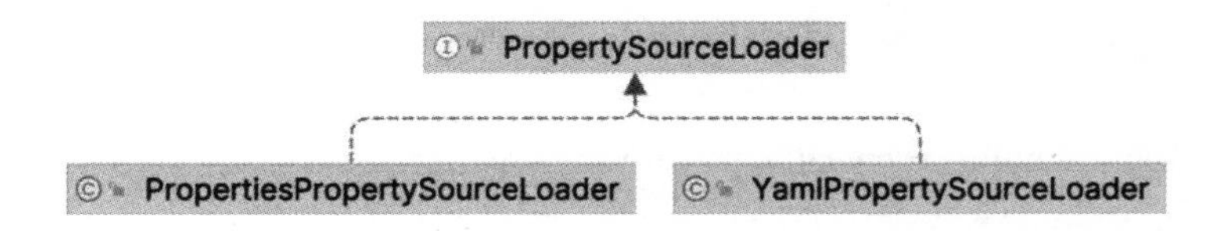

它们分别用来加载 .yaml 和 .properties 配置文件，然后在 Spring Boot 主包 spring-boot-3.0.0.jar 的 /META-INF/spring.factories 自动配置文件中定义了一系列的应用监听器，如下图所示。

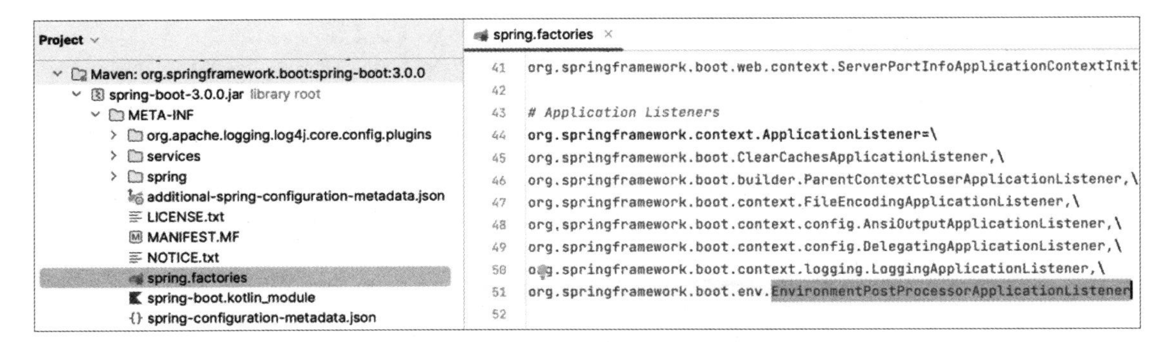

其中就有一个 EnvironmentPostProcessorApplicationListener 监听器，它监听了以下三个事件：

- ApplicationEnvironmentPreparedEvent。
- ApplicationPreparedEvent。
- ApplicationFailedEvent。

源码如下所示。

```
@Override
public void onApplicationEvent(ApplicationEvent event) {
    if (event instanceof ApplicationEnvironmentPreparedEvent) {
        onApplicationEnvironmentPreparedEvent((ApplicationEnvironmentPreparedEvent) event);
    }
    if (event instanceof ApplicationPreparedEvent) {
        onApplicationPreparedEvent();
    }
    if (event instanceof ApplicationFailedEvent) {
        onApplicationFailedEvent();
    }
}
```

其中监听的 ApplicationEnvironmentPreparedEvent 事件用于加载配置，具体可以跟踪到 ConfigDataEnvironment#processAndApply 类的方法：

```
static {
    List<ConfigDataLocation> locations = new ArrayList<>();
    locations.add(ConfigDataLocation.of("optional:classpath:/;optional:classpath:
/config/"));
    locations.add(ConfigDataLocation.of("optional:file:./;optional:file:./config/;
optional:file:./config/*/"));
    DEFAULT_SEARCH_LOCATIONS = locations.toArray(new ConfigDataLocation[0]);
```

```
}

void processAndApply() {
    ConfigDataImporter importer = new ConfigDataImporter(this.logFactory,
this.notFoundAction, this.resolvers,
            this.loaders);
    registerBootstrapBinder(this.contributors, null, DENY_INACTIVE_BINDING);

    // 初始化配置处理
    ConfigDataEnvironmentContributors contributors = processInitial(this.contributors, importer);
    ConfigDataActivationContext activationContext = createActivationContext(
            contributors.getBinder(null, BinderOption.FAIL_ON_BIND_TO_INACTIVE_SOURCE));

    // 处理没有 Profile 的配置
    contributors = processWithoutProfiles(contributors, importer, activationContext);
    activationContext = withProfiles(contributors, activationContext);

    // 最后处理带有 Profile 的配置
    contributors = processWithProfiles(contributors, importer, activationContext);
    applyToEnvironment(contributors, activationContext, importer.getLoadedLocations(),
            importer.getOptionalLocations());
}
```

在类源码上方看到了静态初始块中的配置文件的默认搜索路径，然后在 processAndApply 方法中看到这个 contributors 变量被先后定义了三次：

- 初始化配置处理。
- 处理没有 Profile 的配置。
- 最后处理带有 Profile 的配置。

从这个加载顺序就可以看到基于 Profile 配置的优先级了，然后在获取 contributors 的方法中可以一路追踪到调用 PropertySourceLoader 加载配置文件的地方。

既然有监听事件，就有发布事件的地方，可以在 Spring Boot 启动的 run 方法中找到：

```
public ConfigurableApplicationContext run(String... args) {
    long startTime = System.nanoTime();
    DefaultBootstrapContext bootstrapContext = createBootstrapContext();
    ConfigurableApplicationContext context = null;
    configureHeadlessProperty();
    SpringApplicationRunListeners listeners = getRunListeners(args);
    listeners.starting(bootstrapContext, this.mainApplicationClass);
```

```
    try {
        ApplicationArguments applicationArguments = new DefaultApplicationArguments(args);

        // 准备环境
        ConfigurableEnvironment environment = prepareEnvironment(listeners,
bootstrapContext, applicationArguments);
```

根据准备环境 prepareEnvironment 方法可以一路找到 EventPublishingRunListener#environmentPrepared 方法：

```
public void environmentPrepared(ConfigurableBootstrapContext bootstrapContext,
        ConfigurableEnvironment environment) {
    this.initialMulticaster.multicastEvent(
            new ApplicationEnvironmentPreparedEvent(bootstrapContext, this.application,
this.args, environment));
}
```

事件就是 Spring Boot 启动过程中发出的，所以整个流程就是 Spring Boot 启动时发布事件，然后通过注册的监听器进行监听，监听器监听到事件后就马上加载指定搜索路径的配置文件。

> **Spring Boot 3.0 新变化：**
> 在 Spring Boot 2.4 之前使用的是 ConfigFileApplicationListener 监听器，而在 Spring Boot 2.4 中将它标识为 @Deprecated 状态了，并且在 Spring Boot 3.0 中正式删除了，如下图所示。
>
> 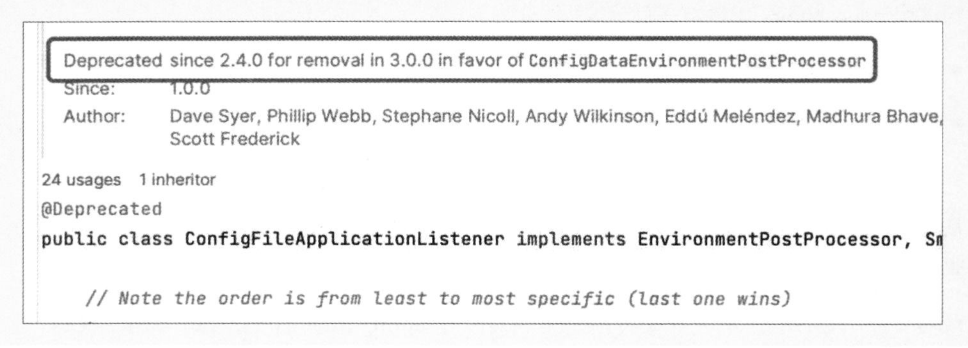
>

关于自动配置文件的变更：

> 自动配置文件 /META-INF/spring.factories 在 Spring Boot 2.7 中有了重大变更，这个会在第 3 章 Spring Boot Starter 与自动配置章节中详细介绍。

2.10 配置加密

2.10.1 概述

项目中会涉及各种敏感配置信息，比如各种中间件的连接用户名、密码信息，以及各种第三方的 Key、密钥等，它们有些存储在项目中，有些存储在项目之外，比如配置中心，一般主流的配置中心都支持自动加 / 解密。

如果这些信息存储在项目中，比如基于 Spring Boot 的传统 MVC 项目，所有的代码、配置都几乎在同一个项目中，那么 Spring Boot 项目的配置自然就是 application 配置文件了，如果这些敏感信息直接放在配置文件中，那么肯定是不安全的，甚至在很多行业及领域（比如支付领域）都是不合规的，所以需要保护 Spring Boot 中的敏感配置信息，这也是对一个项目最基本的安全要求。

Spring Boot 并没有对配置的加 / 解密这一块提供标准的解决方案，但是提供了一个 EnvironmentPostProcessor 接口，可用于在应用启动前控制 Spring 环境，所以通过它可以实现对配置在应用启动前进行解密，但是一般很少使用。对于 Spring Boot 应用的配置的加 / 解密可以有以下几种解决方案：

- 使用配置中心（支持自动加 / 解密）。
- 使用数据库机制。
- 使用自定义加 / 解密机制。
- 使用 Jasypt Spring Boot（第三方加 / 解密方案）。

下面的章节会分别介绍这几种方案的使用场景及处理方式，读者可以根据项目不同的需要进行选择。

2.10.2 使用配置中心（支持自动解密）

如果使用了外置的第三方配置中心，那种支持自动解密的中间件，就可以把所有的动态配置和敏感配置信息都存储在配置中心，比如 Spring Cloud 微服务生态中的 Spring Cloud Config 配置中心，使用其自带的加 / 解密机制来保护敏感信息：

```
spring:
  datasource:
  username: '{cipher}t1s293294187a31f35dea15e8bafaf7774532xxcc20d6d6dd0dfa5ae753d6836'
```

这里的密文由 Spring Cloud Config Server 生成，需要实现自动加 / 解密的内容以 {cipher} 开头标识，注意要使用单引号包起来，Spring Boot 配置文件中可以只存储一些无关紧要的应用配置。

2.10.3 使用数据库机制

一种传统的做法是把所有配置信息存储到数据库中，敏感信息以对称加密算法进行加密存储，当 Spring Boot 应用启动的时候，查询数据库配置并加载到内存中，如果是密文则解密后再加载到内存中。

这是最传统的配置管理方法，其实也可以理解为一个原始的、简易的配置中心，只是功能不那么强大而已，因为现在很多配置中心也都是把配置持久化到数据库中进行存储，然后提供一系列的配置管理功能而已。这里的数据库可以是关系型数据库（MySQL、Oracle）、内存数据库（Redis、ZooKeeper）等，这是用得比较多的中间件技术。

2.10.4 使用自定义加 / 解密机制

如果应用只有极少的敏感配置，比如只有数据库连接信息，那么可以考虑使用现有系统中的对称加密算法，再自定义一个数据源类实现自定义的加 / 解密机制，比如我们可以模仿 Spring Cloud 加密机制，先使用系统已有的对称加密算法对数据库连接信息加密：

```
spring:
  datasource:
  username: '{cipher}t1s293294187a31f35dea15e8bafaf7774532xxcc20d6d6dd0dfa5ae753d6836'
```

先排除 Spring Boot 系统自带的数据源自动配置，然后自行组装数据源 Spring Bean，判断获取的配置值是否以 {cipher} 这个标识开头，如果是，则使用系统约定的对称加密算法进行解密，然后设置数据源，比如以下示例代码：

```
@Bean
public DataSource dataSource(){
        DataSource dataSource = new DruidDataSource();

        // 解密
        String username = this.getUsername();
        if (username.startWith('{cipher}')){
                username = Encrypt.decrypt(username, this.getKey()))
        }
        dataSource.setUsername(username);
```

```
        ...

        return dataSource;
}
```

这种手动加 / 解密机制使用简单，不用额外引入任何第三方包，可以满足只有少量敏感配置的应用场景。但如果是 Spring Boot 自动配置的场景，比如数据源自动配置、Redis 自动配置等，在系统启动的时候就会默认加载配置并自动配置，所以需要在 Spring Boot 配置装载之后自动配置之前对配置进行后置的解密处理。

2.10.5 Jasypt Spring Boot

1. 加 / 解密实战

Jasypt Spring Boot 是一个专门为 Spring Boot 项目中的属性提供加密支持的框架，笔者在撰写本章时已经有 2.3K+ 的 Star 数了，还是挺受欢迎的，是现在市面上使用最广泛的 Spring Boot 第三方加 / 解密框架，也是事实上的 Spring Boot 应用加 / 解密标准。

Jasypt Spring Boot 的开源地址为链接 6。

在笔者撰写本章期间，Jasypt Spring Boot 最新只支持 Spring boot 2.x，最新的 Spring Boot 3.0 可能需要等一段时间才能适配，本章也是基于 Spring Boot 2.7.0 进行实战演示的。

Jasypt Spring Boot 有 3 种集成方法：

- 如果开启了 Spring Boot 的自动配置（使用了 @SpringBootApplication 或者 @EnableAuto-Configuration 注解），则只需要添加 jasypt-spring-boot-starter 依赖，这种会在整个 Spring Environment 中启用可加密属性。
- 添加 jasypt-spring-boot 依赖，同时在 Spring 主要配置类上添加 @EnableEncryptable-Properties 注解，这种会在整个 Spring Environment 中启用可加密属性。
- 添加jasypt-spring-boot依赖，使用 @EncrytablePropertySource 注解声明各个可加密的参数上，这种只适用于独立配置参数的加 / 解密。

一般正常的 Spring Boot 应用都会开启自动配置，再排除个别的自动配置，所以很少会有全部禁用自动配置的情况，不然使用 Spring Boot 的意义不大，这里我们使用第 1 种集成方式进行演示。

首先引入 Jasypt Spring Boot 的依赖和插件：

```
<dependency>
```

```
    <groupId>com.github.ulisesbocchio</groupId>
    <artifactId>jasypt-spring-boot-starter</artifactId>
    <version>3.0.4</version>
</dependency>

<build>
   <plugins>
       <plugin>
           <groupId>com.github.ulisesbocchio</groupId>
           <artifactId>jasypt-maven-plugin</artifactId>
           <version>${jasypt-spring-boot.version}</version>
       </plugin>
   </plugins>
</build>
```

然后添加 jasypt 加密使用的密钥：

```
jasypt:
  encryptor:
  password: G9w0BAQEFAASCBKYwggSiAgEAAoIBAQC
  property:
  prefix: "ENC@["
  suffix: "]"
```

这个 jasypt.encryptor.password 参数是必需的，相当于 Salt（盐），以保证密码的安全性，prefix 和 prefix 是自定义的密码串标识，不配置则默认为 ENC(...)。实现原理可以参考 Jasypt Spring Boot 的自动配置类：

> com.ulisesbocchio.jasyptspringbootstarter.JasyptSpringBootAutoConfiguration

比如在自动配置类中注册了一个 StringEncryptor 字符串加密器，然后笔者基于此提供了一个敏感信息加密测试类：

```
@Slf4j
@SpringBootTest
public class JasyptTest {

  @Autowired
  private StringEncryptor stringEncryptor;

  @Test
  public void encrypt() {
  String usernameEnc = stringEncryptor.encrypt("javastack");
```

```
    String passwordEnc = stringEncryptor.encrypt("javastack.cn");

    log.info("test username encrypt is {}", usernameEnc);
    log.info("test password encrypt is {}", passwordEnc);

    log.info("test username is {}", stringEncryptor.decrypt(usernameEnc));
    log.info("test password is {}", stringEncryptor.decrypt(passwordEnc));
  }
}
```

> Spring Boot 单元测试详解请参考第 10 章 Spring Boot 调试与单元测试。

笔者在以上单元测试类中注入了一个 StringEncryptor，其类结构如下图所示。

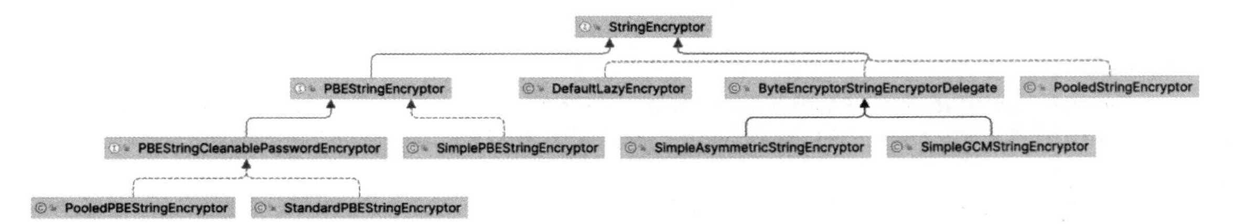

如果当前没有自定义 StringEncryptor，那么 Jasypt Spring Boot 的自动配置会默认创建一个 StringEncryptor 实例，直接使用即可，其构造器的默认值如下表所示。

Key	Required	Default Value
jasypt.encryptor.password	True	-
jasypt.encryptor.algorithm	False	PBEWITHHMACSHA512ANDAES_256
jasypt.encryptor.key-obtention-iterations	False	1000
jasypt.encryptor.pool-size	False	1
jasypt.encryptor.provider-name	False	SunJCE
jasypt.encryptor.provider-class-name	False	null
jasypt.encryptor.salt-generator-classname	False	org.jasypt.salt.RandomSaltGenerator
jasypt.encryptor.iv-generator-classname	False	org.jasypt.iv.RandomIvGenerator
jasypt.encryptor.string-output-type	False	base64
jasypt.encryptor.proxy-property-sources	False	false
jasypt.encryptor.skip-property-sources	False	empty list

然后运行测试用例，测试结果如下图所示。

```
springboot.jasypt.JasyptTest          : Started JasyptTest in 1.004 seconds (JVM running for 2.257)
springboot.jasypt.Application         : javastack.username = javastack
springboot.jasypt.Application         : javastack.password = javastack.cn
springboot.jasypt.JasyptTest          : test username encrypt is VA058GV86AaUx4wSoCKekKrDJv6tP4eTwWkoCG
springboot.jasypt.JasyptTest          : test password encrypt is 7Y7FVX05qS4lAx3o5B4atyHW8kGUv0o6903xL2
springboot.jasypt.JasyptTest          : test username is javastack
springboot.jasypt.JasyptTest          : test password is javastack.cn
```

加 / 解密单元测试成功了，如果不想使用测试这种方法生成密文，则可以使用 Maven 插件，这就是前面为什么要加 Maven 插件的原因，使用方式如下：

```
mvn jasypt:encrypt-value -Djasypt.encryptor.password="G9w0BAQEFAASCBKYwggSiAgEAAoIBA
QC" -Djasypt.plugin.value="javastack"
```

然后将生成的密文填充到 application 配置文件中：

```
javastack:
  username: ENC@[VA058GV86AaUx4wSoCKekKrDJv6tP4eTwWkoCGHnt6Tb3F7ETJnKAs3QiRkZFD5k]
  password: ENC@[7Y7FVX05qS4lAx3o5B4atyHW8kGUv0o6903xL2ofTOicdcJw71/m+UeFUVfX0VPM]
```

注意 ENC@[] 这个占位符是上面自定义配置的，写一个程序尝试打印出来：

```
@Slf4j
@SpringBootApplication
public class Application {

    @Value("${javastack.username}")
    private String username;

    @Value("${javastack.password}")
    private String password;

    public static void main(String[] args) {
        SpringApplication.run(Application.class);
    }

    @Bean
    public CommandLineRunner commandLineRunner() {
        return (args) -> {
            log.info("javastack.username = {}", username);
            log.info("javastack.password = {}", password);
        };
    }

}
```

笔者这里写了一个 CommandLineRunner，在系统启动之后将密文的原文打印出来，不需要做任何处理，直接注入、打印就行。应用启动之后查看启动日志，看一下配置是否为明文，如下图所示。

```
c.u.j.c.StringEncryptorBuilder          : Encryptor config not found for property jasypt.
c.u.j.c.StringEncryptorBuilder          : Encryptor config not found for property jasypt.
c.u.j.c.StringEncryptorBuilder          : Encryptor config not found for property jasypt.
c.u.j.c.StringEncryptorBuilder          : Encryptor config not found for property jasypt.
c.j.springboot.jasypt.Application       : Started Application in 1.308 seconds (JVM runni
c.j.springboot.jasypt.Application       : javastack.username = javastack
c.j.springboot.jasypt.Application       : javastack.password = javastack.cn
```

结果正常，配置显示明文了，配置自动解密成功。

> CommandLineRunner 的详解介绍请参考第 3 章 Spring Boot Starter 与自动配置。

2. 密钥安全性

前面的章节中笔者把 Jasypt 密钥（password）存放在了 application 配置文件中，这样敏感信息还是在项目代码中，理论上也不是很安全，建议通过命令行参数的方式传入，比如在 IDEA 中可以这样设置 JVM 启动参数，如下图所示。

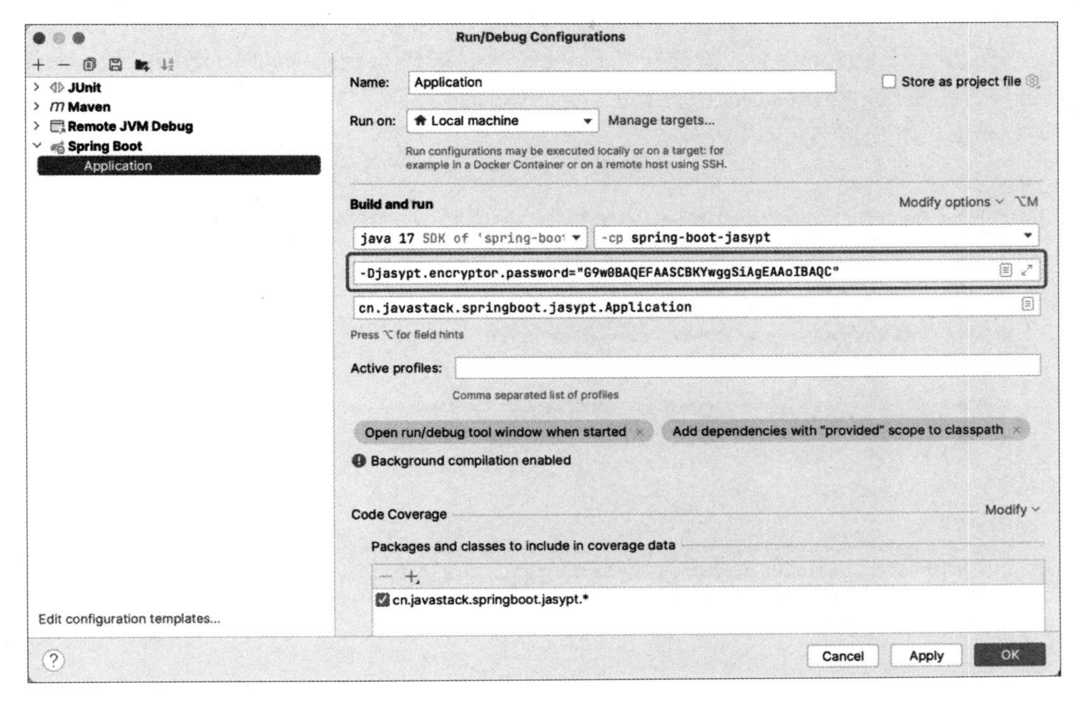

如果是生产环境，则可以通过命令的方式传入：

```
java -Djasypt.encryptor.password=password -jar xx.jar
```

甚至可以配置在服务器环境变量中，StringEncryptor可以通过系统参数、配置文件、命令行参数、环境变量等方式进行构造，这样Spring Boot中的配置信息就彻底安全了！

当然，Jasypt Spring Boot的功能远不止如此，实际功能要更强大和复杂，笔者这里只是进行基本介绍及简单的加 / 解密应用，更多的自定义加 / 解密需求可以参考官方文档。

3. 实现原理

Jasypt Spring Boot自动配置类的源码如下图所示。

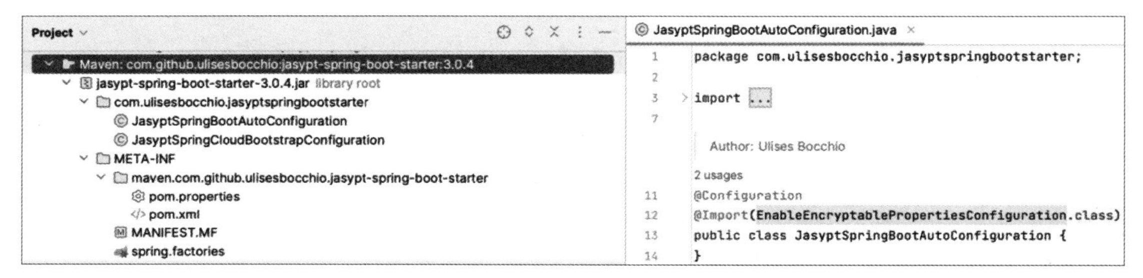

从自动配置类引入的EnableEncryptablePropertiesConfiguration配置类中可以找到Jasypt Spring Boot，它注册了一个Spring后置处理器类EnableEncryptablePropertiesBeanFactoryPost- Processor，它修饰包含在Spring Environment中的所有PropertySource对象，并按照Jasypt的配置约定对属性进行加 / 解密，如下图所示。

```
@Slf4j
public class EnableEncryptablePropertiesBeanFactoryPostProcessor implements BeanFactoryPostProcessor, Ordered {

    2 usages
    private final ConfigurableEnvironment environment;
    2 usages
    private final EncryptablePropertySourceConverter converter;

    1 usage
    public EnableEncryptablePropertiesBeanFactoryPostProcessor(ConfigurableEnvironment environment, EncryptableP
        this.environment = environment;
        this.converter = converter;
    }

    @Override
    public void postProcessBeanFactory(ConfigurableListableBeanFactory beanFactory) throws BeansException {
        log.info("Post-processing PropertySource instances");
        MutablePropertySources propSources = environment.getPropertySources();
        converter.convertPropertySources(propSources);
    }

    @Override
    public int getOrder() { return Ordered.LOWEST_PRECEDENCE - 100; }
}
```

在 EncryptablePropertySourceConverter 转换类的初始化参数方法中，根据不同的参数类型定义了不同类型的加密参数包装类，如下图所示。

```
/unchecked, rawtypes/
private <T> PropertySource<T> instantiatePropertySource(PropertySource<T> propertySource) {
    PropertySource<T> encryptablePropertySource;
    if (needsProxyAnyway(propertySource)) {
        encryptablePropertySource = proxyPropertySource(propertySource);
    } else if (propertySource instanceof SystemEnvironmentPropertySource) {
        encryptablePropertySource = (PropertySource<T>) new EncryptableSystemEnvironmentPropertySourceW
    } else if (propertySource instanceof MapPropertySource) {
        encryptablePropertySource = (PropertySource<T>) new EncryptableMapPropertySourceWrapper((MapPro
    } else if (propertySource instanceof EnumerablePropertySource) {
        encryptablePropertySource = new EncryptableEnumerablePropertySourceWrapper<>((EnumerablePropert
    } else {
        encryptablePropertySource = new EncryptablePropertySourceWrapper<>(propertySource, propertyRes
    }
    return encryptablePropertySource;
}
```

源码有点复杂，根据 *Wrapper 包装类中的获取参数 getProperty 方法，再一路找到了 DefaultPropertyResolver 这个参数解析处理器，如下图所示。

```
public class DefaultPropertyResolver implements EncryptablePropertyResolver {                Reader Mode

    3 usages
    private final Environment environment;
    2 usages
    private StringEncryptor encryptor;
    3 usages
    private EncryptablePropertyDetector detector;

    public DefaultPropertyResolver(StringEncryptor encryptor, Environment environment) {
        this(encryptor, new DefaultPropertyDetector(), environment);
    }

    2 usages
    public DefaultPropertyResolver(StringEncryptor encryptor, EncryptablePropertyDetector detector, Envi
        this.environment = environment;
        Assert.notNull(encryptor,  message: "String encryptor can't be null");
        Assert.notNull(detector,  message: "Encryptable Property detector can't be null");
        this.encryptor = encryptor;
        this.detector = detector;
    }

    2 usages
    @Override
    public String resolvePropertyValue(String value) {
        return Optional.ofNullable(value)
                .map(environment::resolvePlaceholders)
                .filter(detector::isEncrypted)
                .map(resolvedValue -> {
                    try {
                        String unwrappedProperty = detector.unwrapEncryptedValue(resolvedValue.trim());
                        String resolvedProperty = environment.resolvePlaceholders(unwrappedProperty);
                        return encryptor.decrypt(resolvedProperty);
                    } catch (EncryptionOperationNotPossibleException e) {
```

它也是注入了 StringEncryptor 这个实例，获取配置时，会在解密后再返回。另外，这个 Resolver 解析处理器也是支持自定义的，感兴趣的读者可以再深入研究。

另外，笔者通过在单元测试类中进行 Debug 调试，可以看到 StringEncryptor 其实是一个 DefaultLazyEncryptor 实例，如下图所示。

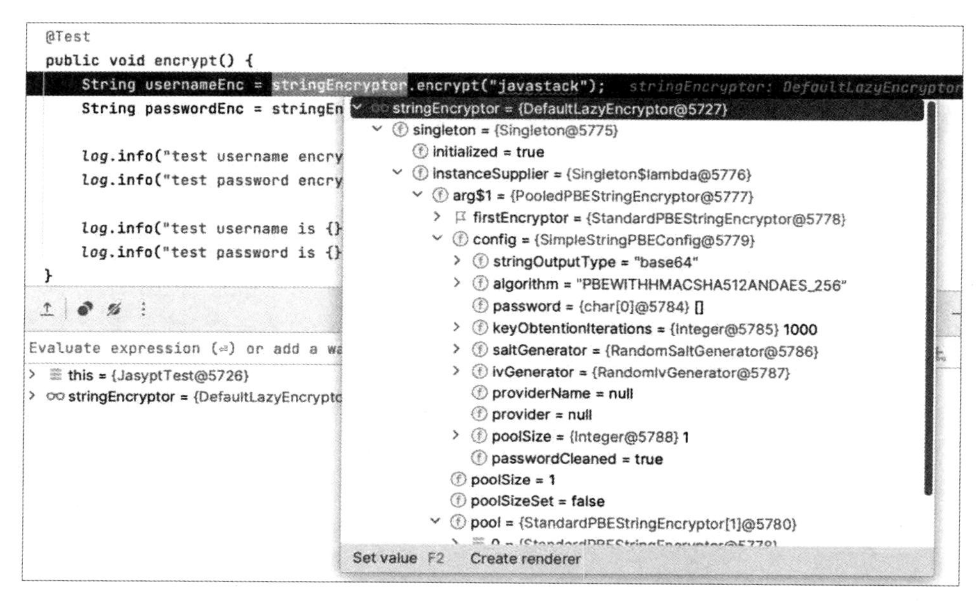

这一点在自动配置类中也可以找到，当然也支持自定义的 Encryptor，有需要的读者可以自行定制。

2.11 配置迁移

2.11.1 迁移方案

在升级 Spring Boot 新版本时，某些配置参数可能已经被改名或者删除了，Spring Boot 提供了一个配置迁移依赖，方便开发者快速进行配置迁移，如下所示。

```
<dependency>
    <groupId>org.springframework.boot</groupId>
    <artifactId>spring-boot-properties-migrator</artifactId>
    <scope>runtime</scope>
</dependency>
```

在应用中加入此依赖，就可以分析应用的环境并在启动时打印诊断信息，还可以在运行时进行临时的配置迁移。

笔者使用 Spring Boot 3.0 中的一个变更参数测试了一下，效果如下图所示。

```
The use of configuration keys that have been renamed was found in the environment:

Property source 'Config resource 'class path resource [application.yml]' via location 'optional:classpath:/'':
    Key: management.metrics.export.prometheus.pushgateway.enabled
        Line: 34
        Replacement: management.prometheus.metrics.export.pushgateway.enabled


Each configuration key has been temporarily mapped to its replacement for your convenience. To silence this warning,
```

Spring Boot 会在启动日志中打印需要替换的旧的配置参数和新的配置参数，替换后再重新启动测试即可。

需要注意的是：

（1）在应用启动后再加入 Spring 环境中的配置参数不在配置迁移支持范围之内，比如通过 @PropertySource 注解加载的配置。

（2）应用完成迁移后，建议删除该配置迁移依赖，以免部署到生产环境后影响性能或者出现潜在的问题。

2.11.2 实现原理

spring-boot-properties-migrator 配置迁移依赖包的所有内容如下图所示。

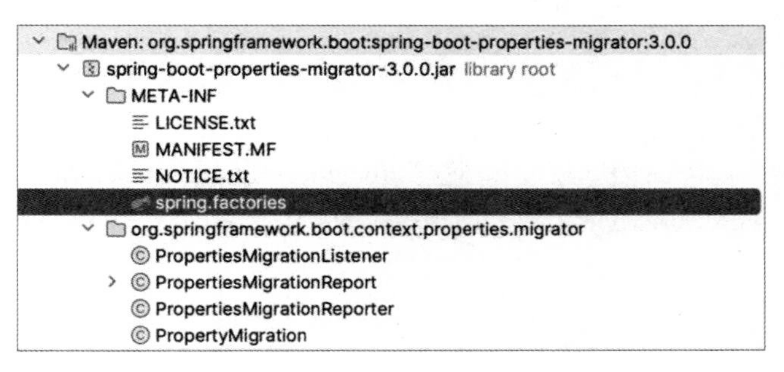

在 spring.factories 自动配置文件中注册了一个 PropertiesMigrationListener 监听器：

```
org.springframework.context.ApplicationListener=\
org.springframework.boot.context.properties.migrator.PropertiesMigrationListener
```

监听器的部分源码如下所示。

```
class PropertiesMigrationListener implements ApplicationListener<SpringApplicationEvent> {

    private static final Log logger = LogFactory.getLog(PropertiesMigrationListener.class);

    private PropertiesMigrationReport report;

    private boolean reported;

    @Override
    public void onApplicationEvent(SpringApplicationEvent event) {
        if (event instanceof ApplicationPreparedEvent preparedEvent) {
            onApplicationPreparedEvent(preparedEvent);
        }
        if (event instanceof ApplicationReadyEvent || event instanceof
ApplicationFailedEvent) {
            logLegacyPropertiesReport();
        }
    }

    ...

}
```

它实现了 ApplicationListener 接口，并监听 SpringApplicationEvent 事件，事件是 Spring Boot 启动过程中发出的。

> Spring Boot 启动事件和监听器机制可以参考第 4 章 Spring Boot 启动过程与扩展应用中的启动事件和监听器一节。

在第一个应用准备事件发出后，监听器先生成配置参数迁移报告 PropertiesMigrationReport，然后在第二个应用就绪 / 失败事件发出后，再通过 PropertiesMigrationReport#getWarningReport 和 PropertiesMigrationReport#getErrorReport 方法获取相对应的报告并输出日志。

第3章 Spring Boot Starter 与自动配置

Spring Boot Starter 是 Spring Boot 框架能快速集成并使用其他技术框架的首要条件，它涉及大量的核心技术，本章会介绍 Spring Boot Starter 的方方面面，包括 Starter 的基本介绍、命名规范、分类、自动配置、自定义 Starter 等，这些都是掌握 Spring Boot 框架必不可少的技术要点。

3.1 概述

Starter 可以理解为 Spring Boot 中的一站式集成启动器，它包含了一系列可以集成到应用中的依赖项（dependencies），可以快速一站式集成 Spring 组件及其他框架，而不需要到处找示例代码、依赖包及复杂的组件配置，我们只需要通过简单的参数配置就可以实现“开箱即用”。

一个完整的 Spring Boot Starter 一般需要包含以下组件：

- 完成自动配置的自动配置模块。
- 为自动配置模块提供的所有依赖项。

简而言之，就是添加一个 Starter 应该提供使用该 Starter 所需的一切。比如，我们想使用 Spring JPA 访问数据库，只要加入官方提供的 spring-boot-starter-data-jpa 启动器依赖即可：

```
<dependency>
  <groupId>org.springframework.boot</groupId>
  <artifactId>spring-boot-starter-data-jpa</artifactId>
</dependency>
```

加入依赖后就能完成默认的自动配置并能直接使用，如果有扩展的需要，也可以通过一系列

spring.jpa.* 参数进行额外的自定义配置，下载下来的 spring-boot-starter-data-jpa 依赖坐标内容如下图所示。

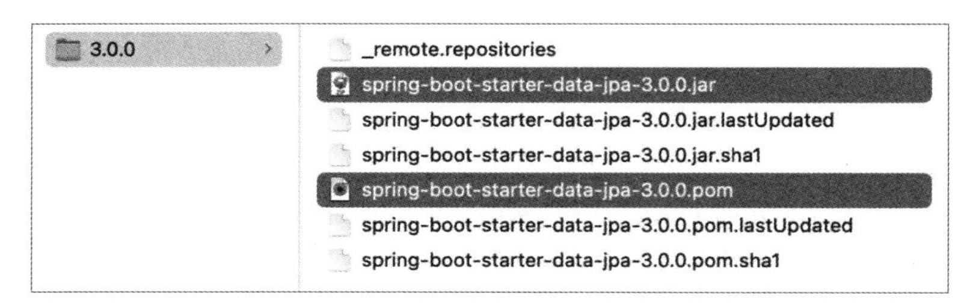

一个 Spring Boot Starter 包含一个 jar 包和一个依赖配置文件，官方 jar 包中没有任何实质内容，官方提供的自动配置类和自动配置逻辑全部存储在另外一个自动配置模块中（spring-boot-autoconfigure），具体内容可见 3.4 节，Starter 存在的意思是通过依赖配置文件导入所有需要的依赖。

3.2 Starter 的命名规范

Spring Boot 官方的 Starter 启动器都是以 spring-boot-starter-* 命名的，* 代表任意一个特定的应用类型，比如 Redis 的 spring-boot-starter-data-redis。而第三方的启动器不能以 spring-boot 开头命名，它们都被 Spring Boot 官方保留，一般一个第三方的 Starter 应该以 *-spring-boot-starter 这样的形式命名，即和官方的命名反过来，比如 druid 连接池：druid-spring-boot-starter。

有了这样的命名规范，就能帮助我们快速找到想要的 Starter 启动器，比如在开发工具编辑器中配置 Maven 依赖时，输入前缀就会弹出各种已有的依赖提示，如下图所示。

```
    <dependency>
        <groupId></groupId>
        <artifactId>spring-boot-starter-</artifactId>
    </dependency>          spring-boot-starter-hdiv
</dependencies>            spring-boot-starter-hdiv-commons
                           spring-boot-starter-hdiv-jsp
<build>                    spring-boot-starter-hdiv-thymeleaf
    <plugins>              spring-boot-starter-ip2region
        <plugin>           spring-boot-starter-mysql-lock
            <groupI        spring-boot-starter-rxtx
            <artifa        spring-boot-starter-swagger
            <execut        spring-boot-admin-starter-client
                <ex
```

这样也能规范第三方的 Starter 启动器的开发和使用。

3.3 Starter 的分类

Spring Boot 官方提供了各种各样的 Starter，几乎涉及了市面上所有主流的技术，不需要我们额外开发，也不需要第三方技术厂商去定制，官方把它们分成了以下三个类别。

3.3.1 application starter

Spring Boot 所有应用级的 Starter 如下表所示。

编号	Starter 名称	Starter 描述
1	spring-boot-starter	核心 Starter，包括自动配置、日志及 YAML 支持等
2	spring-boot-starter-amqp	集成 Spring AMQP 和 Rabbit MQ 的消息队列
3	spring-boot-starter-aop	集成 Spring AOP 和 AspectJ 面向切面编程
4	spring-boot-starter-artemis	集成 Apache Artemis，基于 JMS 的消息队列
5	spring-boot-starter-batch	集成 Spring Batch（批处理）
6	spring-boot-starter-cache	集成 Spring Cache（缓存）
7	spring-boot-starter-data-cassandra	集成 Cassandra（分布式数据库）和 Spring Data Cassandra
8	spring-boot-starter-data-cassandra-reactive	集成 Cassandra（分布式数据库）和 Spring Data Cassandra Reactive
9	spring-boot-starter-data-couchbase	集成 Couchbase（文档型数据库）和 Spring Data Couchbase
10	spring-boot-starter-data-couchbase-reactive	集成 Couchbase（文档型数据库）和 Spring Data Couchbase Reactive
11	spring-boot-starter-data-elasticsearch	集成 Elasticsearch（搜索引擎）和 Spring Data Elasticsearch
12	spring-boot-starter-data-jdbc	集成 Spring Data JDBC
13	spring-boot-starter-data-jpa	集成 Spring Data JPA 结合 Hibernate
14	spring-boot-starter-data-ldap	集成 Spring Data LDAP
15	spring-boot-starter-data-mongodb	集成 MongoDB（文档型数据库）和 Spring Data MongoDB
16	spring-boot-starter-data-mongodb-reactive	集成 MongoDB（文档型数据库）和 Spring Data MongoDB Reactive
17	spring-boot-starter-data-neo4j	集成 Neo4j（图形数据库）和 Spring Data Neo4j
18	spring-boot-starter-data-r2dbc	集成 Spring Data R2DBC

续表

编号	Starter 名称	Starter 描述
19	spring-boot-starter-data-redis	集成 Redis（内存数据库）结合 Spring Data Redis 和 Lettuce 客户端
20	spring-boot-starter-data-redis-reactive	集成 Redis（内存数据库）结合 Spring Data Redis reactive 和 Lettuce 客户端
21	spring-boot-starter-data-rest	集成 Spring Data REST 暴露 Spring Data repositories 并输出 REST 资源
22	spring-boot-starter-freemarker	集成 FreeMarker 视图构建 MVC Web 应用
23	spring-boot-starter-graphql	集成 Spring GraphQL 构建 GraphQL 应用
24	spring-boot-starter-groovy-templates	集成 Groovy 模板视图构建 MVC Web 应用
25	spring-boot-starter-hateoas	集成 Spring MVC 和 Spring HATEOAS 构建超媒体 RESTful Web 应用
26	spring-boot-starter-integration	集成 Spring Integration
27	spring-boot-starter-jdbc	集成 JDBC 结合 HikariCP 连接池
28	spring-boot-starter-jersey	集成 JAX-RS 和 Jersey 构建 RESTful Web 应用，是 spring-boot-starter-web 的一种替代方案
29	spring-boot-starter-jooq	集成 jOOQ 访问 SQL 数据库，是 spring-boot-starter-data-jpa 或者 spring-boot-starter-jdbc 的替代 Starter
30	spring-boot-starter-json	用于读写 JSON
31	spring-boot-starter-mail	集成 Java Mail 和 Spring 框架的邮件发送功能
32	spring-boot-starter-mustache	集成 Mustache 视图构建 Web 应用
33	spring-boot-starter-oauth2-client	集成 Spring Security’s OAuth2/OpenID 连接客户端功能
34	spring-boot-starter-oauth2-resource-server	集成 Spring Security’s OAuth2 资源服务器功能
35	spring-boot-starter-quartz	集成 Quartz 任务调度
36	spring-boot-starter-rsocket	构建 RSocket 客户端和服务端
37	spring-boot-starter-security	集成 Spring Security
38	spring-boot-starter-test	集成 JUnit Jupiter、Hamcrest 和 Mockito 测试 Spring Boot 应用和类库
39	spring-boot-starter-thymeleaf	集成 Thymeleaf 视图构建 MVC Web 应用
40	spring-boot-starter-validation	集成 Java Bean Validation 结合 Hibernate Validator
41	spring-boot-starter-web	集成 Spring MVC 构建 RESTful Web 应用，使用 xa0;Tomcat 作为默认内嵌容器
42	spring-boot-starter-web-services	集成 Spring Web Services

续表

编号	Starter 名称	Starter 描述
43	spring-boot-starter-webflux	集成 Spring Reactive Web 构建 WebFlux 应用
44	spring-boot-starter-websocket	集成 Spring WebSocket 构建 WebSocket 应用

Spring Boot 3.0.0 共有 44 个应用级 Starter，较之前的版本略有变动。Spring Boot 官方自带的 Starter 可以直接拿来使用，用到哪个技术就引用哪个技术的 Spring Boot Starter，没有必要重复造轮子，也不用再手动配置复杂的技术组件了。

如果 Spring Boot 官方没有自带的应用级 Starter，一般第三方的应用框架也会提供自制的 Spring Boot Starter，如 Dubbo、ZooKeeper、MyBatis 等，这样只需几个 Starter 依赖、几行配置参数就能轻松集成并使用各种技术。

当然，除了 Spring Boot 官方和第三方技术的 Starter，使用 Spring Boot 框架的公司一般也会有内部私有定制的 Starter，用于在内部各业务部门快速集成和使用，避免各团队重复造轮子。

3.3.2 production starter

除了上面的应用级 Starter，还有生产级 Starter，如下表所示。

编号	Starter 名称	Starter 描述
1	spring-boot-starter-actuator	集成 Spring Boot Actuator，提供生产功能以帮助开发者监控和管理应用

截至 Spring Boot 3.0.0，Spring Boot 只有这一个生产级 Starter。生产级 Starter 意味着和任何技术、业务没有关系，只要用了 Spring Boot 框架，在生产环境中就能使用，也不是只有在生产环境中才能使用，只是在生产环境中使用才体现了它的意义，毕竟本地和测试环境没有增加监控功能的必要。

3.3.3 technical starter

除了应用级别和生产级别的 Starter，Spring Boot 还包括技术级别的 Starter，用于帮助开发者排除或者替换 Spring Boot 框架内部默认的技术组件，如下表所示。

编号	Starter 名称	Starter 描述
1	spring-boot-starter-jetty	集成 Jetty 作为内嵌的 Servlet 容器，可用于替代 spring-boot- starter-tomcat
2	spring-boot-starter-log4j2	集成 Log4j2 日志框架，可用于替代 spring-boot-starter-logging

续表

编号	Starter 名称	Starter 描述
3	spring-boot-starter-logging	集成 Logback 日志框架，这个也是默认的日志 Starter
4	spring-boot-starter-reactor-netty	集成 Netty 作为内嵌的响应式 HTTP 服务器
5	spring-boot-starter-tomcat	集成 Tomcat 作为内嵌的 Servlet 容器，这个也是默认内嵌的 Servlet 容器，它也被集成 spring-boot-starter-web 启动器中
6	spring-boot-starter-undertow	集成 Undertow 作为内嵌的 Servlet 容器，可用于替代 spring-boot-starter-tomcat

Spring Boot 默认内嵌的 Servlet 容器为 Tomcat，如果想换成 Jetty、Undertow 或者其他容器，又或者想换成其他的日志框架，都可以在上面的这个表格中找到，怎么换可以见后续的章节。

Spring Boot 3.0 共收集了 51 个官方 Spring Boot Starter，当然也不限于这 51 个，随着 Spring Boot 版本的不断升级，后续可能会增加更多的 Starter，当然也有少数 Starter 可能会被删除，相比之前的 2.x 版本，3.0.0 版本中的 Starter 是有减少和调整的，不同版本的 Starter 略有不同。

3.4 自动配置

3.4.1 概述

Spring Boot Starter 的核心原理就是自动配置，这也是整个 Spring Boot 框架的核心，开发者只需要按约定提供些许配置参数就能完成各种技术复杂组件的自动组装配置，这正是 Spring Boot 框架能迅速上手使用的原因。

开发者必须搞懂自动配置的原理，以便出现问题时知道怎么去迅速解决，也方便自己在 Spring Boot 提供的默认自动配置不满足要求的情况下定制自己的需求。

Spring Boot 官方所有的自动配置类都是由 spring-boot-autoconfigure 这个模块提供的，如下图所示。

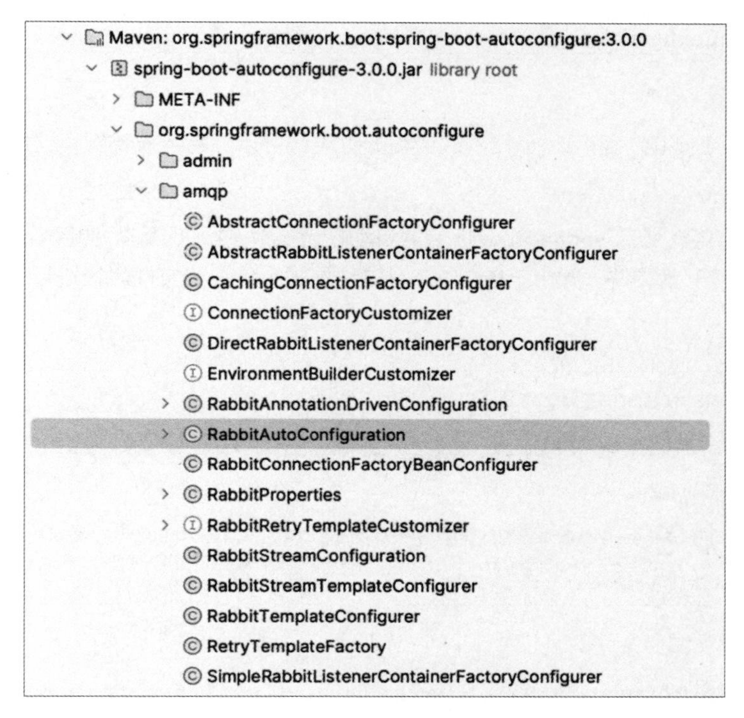

比如在 amqp 子包中定义了 RabbitMQ 自动配置相关的类，其他自带技术的自动配置也都分布在该模块不同的包中，而底层依赖的具体的技术依赖项则需要再通过 Starter 引入，比如想使用 RabbitMQ，则需要引入 RabbitMQ 的启动器：

```
<dependency>
    <groupId>org.springframework.boot</groupId>
    <artifactId>spring-boot-starter-amqp</artifactId>
</dependency>
```

只有引入了对应组件的依赖，Spring Boot 扫描到相关类时才会自动配置，没有引入是不会启用自动配置的，这也是自动配置的魅力，具体内容可见自动配置原理一节。

3.4.2 命名规范

Spring Boot 自动配置类一般以 XxxAutoConfiguration 命名，比如以下几个组件：

- RabbitAutoConfiguration。
- FlywayAutoConfiguration。

- RedisAutoConfiguration。

……

有了这样的命名规范，任何一个技术的自动配置类都可以轻易地被找到，这也是约定大于配置思想的体现。

自动配置类需要注册到 Spring Boot 指定的自动配置文件中，低版本为类路径下的 META-INF/spring.factories 自动配置文件，Spring Boot 2.7 对自动配置类的注册文件路径进行了变更，如下图中第一条所示。

Deprecations in Spring Boot 2.7

- Loading auto-configurations from `spring.factories` is deprecated. See above for more details.
- `DatabaseDriver.GAE`
- Properties under `spring.security.saml2.relyingparty.registration.{id}.identityprovider` have been moved to `spring.security.saml2.relyingparty.registration.{id}.assertingparty`. Using the old property names result in log messages on WARN level on startup.

新的自动配置类注册文件的路径如下：

```
META-INF/spring/org.springframework.boot.autoconfigure.AutoConfiguration.imports
```

以 Spring Boot 3.0.0 为例，如下图所示。

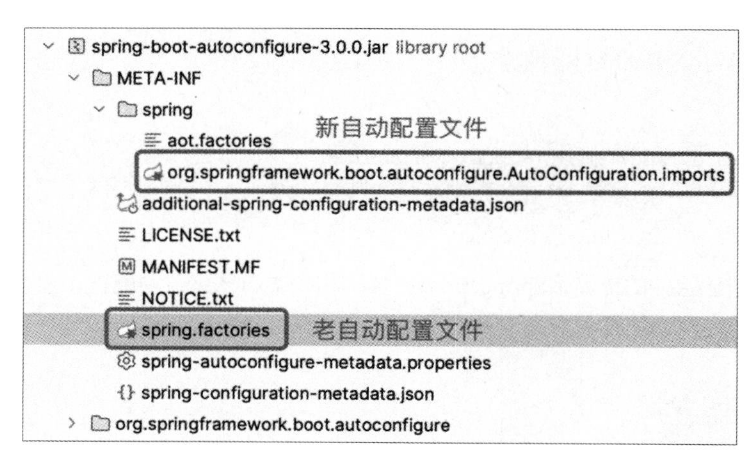

老的自动配置文件 META-INF/spring.factories 虽然在 Spring Boot 2.7 中被宣布废弃了，但是为了向后兼容，Spring Boot 2.7.x 系列版本的自动配置文件仍然可以使用。

另外，自动配置项的编写格式也变了，Spring Boot 2.7+ 中的每一行是一个自动配置类，如下图所示。

org.springframework.boot.autoconfigure.AutoConfiguration.imports

Spring Boot 2.7+

```
1  org.springframework.boot.autoconfigure.admin.SpringApplicationAdminJmxAutoConfiguration
2  org.springframework.boot.autoconfigure.aop.AopAutoConfiguration
3  org.springframework.boot.autoconfigure.amqp.RabbitAutoConfiguration
4  org.springframework.boot.autoconfigure.batch.BatchAutoConfiguration
5  org.springframework.boot.autoconfigure.cache.CacheAutoConfiguration
6  org.springframework.boot.autoconfigure.cassandra.CassandraAutoConfiguration
7  org.springframework.boot.autoconfigure.context.ConfigurationPropertiesAutoConfiguration
8  org.springframework.boot.autoconfigure.context.LifecycleAutoConfiguration
9  org.springframework.boot.autoconfigure.context.MessageSourceAutoConfiguration
10 org.springframework.boot.autoconfigure.context.PropertyPlaceholderAutoConfiguration
11 org.springframework.boot.autoconfigure.couchbase.CouchbaseAutoConfiguration
12 org.springframework.boot.autoconfigure.dao.PersistenceExceptionTranslationAutoConfiguration
13 org.springframework.boot.autoconfigure.data.cassandra.CassandraDataAutoConfiguration
14 org.springframework.boot.autoconfigure.data.cassandra.CassandraReactiveDataAutoConfiguration
15 org.springframework.boot.autoconfigure.data.cassandra.CassandraReactiveRepositoriesAutoConfiguration
16 org.springframework.boot.autoconfigure.data.cassandra.CassandraRepositoriesAutoConfiguration
17 org.springframework.boot.autoconfigure.data.couchbase.CouchbaseDataAutoConfiguration
```

spring.factories

Spring Boot 2.6-

```
1  # Initializers
2  org.springframework.context.ApplicationContextInitializer=\
3  org.springframework.boot.autoconfigure.SharedMetadataReaderFactoryContextInitializer,\
4  org.springframework.boot.autoconfigure.logging.ConditionEvaluationReportLoggingListener
5
6  # Application Listeners
7  org.springframework.context.ApplicationListener=\
8  org.springframework.boot.autoconfigure.BackgroundPreinitializer
9
10 # Environment Post Processors
11 org.springframework.boot.env.EnvironmentPostProcessor=\
12 org.springframework.boot.autoconfigure.integration.IntegrationPropertiesEnvironmentPostProcessor
13
14 # Auto Configuration Import Listeners
15 org.springframework.boot.autoconfigure.AutoConfigurationImportListener=\
16 org.springframework.boot.autoconfigure.condition.ConditionEvaluationReportAutoConfigurationImportListener
```

新的自动配置文件的编写格式确实比之前方便多了，很大程度上能防止配置出错，但同时也有缺点，新的自动配置的文件名也太长了，比较难记，也难以手动编写。

> **Spring Boot 3.0.0 的新变化：**
> 老的自动配置文件 META-INF/spring.factories 已经正式被废除了，虽然配置文件还在，但仅保留系统级别的组件注册，不再作为应用级别的自动配置类的注册文件了。也就是说，它已经不能再注册自定义的自动配置类了，必须使用新的自动配置文件规范进行注册，具体见下面的源码分析。

3.4.3 自动配置文件的加载原理

在之前的章节中介绍了 Spring Boot 开启自动配置使用的是 @EnableAutoConfiguration 注解，一

般使用组合了它的 @SpringBootApplication 主注解即可，@EnableAutoConfiguration 注解的源码如下：

```
@Target(ElementType.TYPE)
@Retention(RetentionPolicy.RUNTIME)
@Documented
@Inherited
@AutoConfigurationPackage
@Import(AutoConfigurationImportSelector.class)
public @interface EnableAutoConfiguration {

	/**
	 * Environment property that can be used to override when auto-configuration is
	 * enabled.
	 */
	String ENABLED_OVERRIDE_PROPERTY = "spring.boot.enableautoconfiguration";

	/**
	 * Exclude specific auto-configuration classes such that they will never be applied.
	 * @return the classes to exclude
	 */
	Class<?>[] exclude() default {};

	/**
	 * Exclude specific auto-configuration class names such that they will never be
	 * applied.
	 * @return the class names to exclude
	 * @since 1.3.0
	 */
	String[] excludeName() default {};

}
```

上面最关键的两个注解是：

- **@AutoConfigurationPackage**：注册需要自动配置的包，如果不指定就是当前注解所在类的包。
- **@Import**：导入配置类，这个注解在之前的章节介绍过，这里导入的就是其中的 ImportSelector 接口类。

自动配置 @Import 注解导入的是 AutoConfigurationImportSelector.class 类，这也是这个注解的关键所在，它实现了 ImportSelector 接口：

```
public interface ImportSelector {

        /**
         * Select and return the names of which class(es) should be imported based on
         * the {@link AnnotationMetadata} of the importing @{@link Configuration} class.
         * @return the class names, or an empty array if none
         */
        String[] selectImports(AnnotationMetadata importingClassMetadata);

        /**
         * Return a predicate for excluding classes from the import candidates, to be
         * transitively applied to all classes found through this selector's imports.
         * <p>If this predicate returns {@code true} for a given fully-qualified
         * class name, said class will not be considered as an imported configuration
         * class, bypassing class file loading as well as metadata introspection.
         * @return the filter predicate for fully-qualified candidate class names
         * of transitively imported configuration classes, or {@code null} if none
         * @since 5.2.4
         */
        @Nullable
        default Predicate<String> getExclusionFilter() {
                return null;
        }

}
```

这个注解只有两个方法：

- selectImports（**抽象方法**）：选择要导入配置类。
- Predicate（**接口默认方法**）：用于返回要排除的类。

然后找到 AutoConfigurationImportSelector 类的 selectImports 方法的源码：

```
@Override
public String[] selectImports(AnnotationMetadata annotationMetadata) {
    if (!isEnabled(annotationMetadata)) {
        return NO_IMPORTS;
    }
    AutoConfigurationEntry autoConfigurationEntry=getAutoConfigurationEntry(annotationMetadata);
    return StringUtils.toStringArray(autoConfigurationEntry.getConfigurations());
}
```

再根据 getAutoConfigurationEntry 方法进入 getCandidateConfigurations 方法。

以下是 Spring Boot 2.7 的获取自动配置的源码：

```
protected List<String> getCandidateConfigurations(AnnotationMetadata metadata,
AnnotationAttributes attributes) {
    // 1. 获取自动配置列表（Spring Boot 2.7）
    List<String> configurations = new ArrayList<>(
            SpringFactoriesLoader.loadFactoryNames(getSpringFactoriesLoaderFactoryClass(),
getBeanClassLoader()));

    // 2. 导入候选自动配置列表
    ImportCandidates.load(AutoConfiguration.class, getBeanClassLoader()).forEach
(configurations::add);
    Assert.notEmpty(configurations,
            "No auto configuration classes found in META-INF/spring.factories nor in
META-INF/spring/org.springframework.boot.autoconfigure.AutoConfiguration.imports. If you"
                    + "are using a custom packaging, make sure that file is correct.");
    return configurations;
}
```

代码注释的1、2处就是获取自动配置列表的关键位置，第1处是Spring Boot 2.7之前的保留逻辑，第2处是导入候选自动配置列表，这是Spring Boot 2.7变更的地方。现在第1处已经在Spring Boot 3.0.0中被彻底移除，第2处正式“转正”替换第1处。以下是Spring Boot 3.0.0的获取自动配置的源码：

```
protected List<String> getCandidateConfigurations(AnnotationMetadata metadata,
AnnotationAttributes attributes) {
    // 导入候选自动配置列表
    List<String> configurations = ImportCandidates.load(AutoConfiguration.class,
getBeanClassLoader())
          .getCandidates();

    Assert.notEmpty(configurations,
                    "No auto configuration classes found in "
                    + "META-INF/spring/org.springframework.boot.autoconfigure.
AutoConfiguration.imports. If you "
                    + "are using a custom packaging, make sure that file is correct.");
    return configurations;
}
```

下面分析获取老的自动配置列表（spring.factories）的逻辑，先看第1处获取自动配置列表的逻辑，进入关键的 loadFactoryNames 方法：

```
public static List<String> loadFactoryNames(Class<?> factoryType, @Nullable ClassLoader
classLoader) {
```

```
    return forDefaultResourceLocation(classLoader).loadFactoryNames(factoryType);
}

public static SpringFactoriesLoader forDefaultResourceLocation(@Nullable ClassLoader
classLoader) {
    return forResourceLocation(classLoader, FACTORIES_RESOURCE_LOCATION);
}
```

这里可以看到加载自动配置文件的地方，常量 FACTORIES_RESOURCE_LOCATION 的定义如下：

```
/**
* The location to look for factories.
* <p>Can be present in multiple JAR files.
*/
public static final String FACTORIES_RESOURCE_LOCATION = "META-INF/spring.factories";
```

再进入最后的 forResourceLocation 方法：

```
public static SpringFactoriesLoader forResourceLocation(@Nullable ClassLoader classLoader,
String resourceLocation) {
    Assert.hasText(resourceLocation, "'resourceLocation' must not be empty");
    ClassLoader resourceClassLoader = (classLoader != null) ? classLoader
            : SpringFactoriesLoader.class.getClassLoader();
    Map<String, SpringFactoriesLoader> loaders = SpringFactoriesLoader.cache.get
(resourceClassLoader);
    if (loaders == null) {
        loaders = new ConcurrentReferenceHashMap<>();
        SpringFactoriesLoader.cache.put(resourceClassLoader, loaders);
    }
    SpringFactoriesLoader loader = loaders.get(resourceLocation);
    if (loader == null) {
        Map<String, List<String>> factories = loadFactoriesResource(resourceClassLoader,
resourceLocation);
        loader = new SpringFactoriesLoader(classLoader, factories);
        loaders.put(resourceLocation, loader);
    }
    return loader;
}

static Map<String, List<String>> loadFactoriesResource(ClassLoader classLoader, String
resourceLocation) {
    Map<String, List<String>> result = new LinkedHashMap<>();
```

```
    try {
        Enumeration<URL> urls = classLoader.getResources(resourceLocation);
        while (urls.hasMoreElements()) {
            UrlResource resource = new UrlResource(urls.nextElement());
            Properties properties = PropertiesLoaderUtils.loadProperties(resource);
            properties.forEach((name, value) -> {
                List<String> implementations = result.computeIfAbsent(((String) name).
trim(), key -> new ArrayList<>());
                Arrays.stream(StringUtils.commaDelimitedListToStringArray((String) value))
                    .map(String::trim).forEach(implementations::add);
            });
        }
        result.replaceAll(SpringFactoriesLoader::toDistinctUnmodifiableList);
    }
    catch (IOException ex) {
        throw new IllegalArgumentException("Unable to load factories from location ["
+ resourceLocation + "]", ex);
    }
    return Collections.unmodifiableMap(result);
}
```

即加载 META-INF/spring.factories 自动配置文件中的资源，然后遍历自动配置文件中的所有行并放到一个 Map<String, List<String>> 中返回。

Spring Boot 自动配置包 spring-boot-autoconfigure-3.0.0.jar 中 META-INF/spring.factories 自动配置文件的所有内容如下：

```
# Initializers
org.springframework.context.ApplicationContextInitializer=\
org.springframework.boot.autoconfigure.SharedMetadataReaderFactoryContextInitializer,\
org.springframework.boot.autoconfigure.logging.ConditionEvaluationReportLoggingListener

# Application Listeners
org.springframework.context.ApplicationListener=\
org.springframework.boot.autoconfigure.BackgroundPreinitializer

# Environment Post Processors
org.springframework.boot.env.EnvironmentPostProcessor=\
org.springframework.boot.autoconfigure.integration.IntegrationPropertiesEnvironmentPostProcessor

# Auto Configuration Import Listeners
org.springframework.boot.autoconfigure.AutoConfigurationImportListener=\
org.springframework.boot.autoconfigure.condition.ConditionEvaluationReportAutoConfigura-
tionImportListener
```

```
# Auto Configuration Import Filters
org.springframework.boot.autoconfigure.AutoConfigurationImportFilter=\
org.springframework.boot.autoconfigure.condition.OnBeanCondition,\
org.springframework.boot.autoconfigure.condition.OnClassCondition,\
org.springframework.boot.autoconfigure.condition.OnWebApplicationCondition

# Failure analyzers
org.springframework.boot.diagnostics.FailureAnalyzer=\
org.springframework.boot.autoconfigure.data.redis.RedisUrlSyntaxFailureAnalyzer,\
org.springframework.boot.autoconfigure.diagnostics.analyzer.NoSuchBeanDefinitionFailureAnalyzer,\
org.springframework.boot.autoconfigure.flyway.FlywayMigrationScriptMissingFailureAnalyzer,\
org.springframework.boot.autoconfigure.jdbc.DataSourceBeanCreationFailureAnalyzer,\
org.springframework.boot.autoconfigure.jdbc.HikariDriverConfigurationFailureAnalyzer,\
org.springframework.boot.autoconfigure.jooq.NoDslContextBeanFailureAnalyzer,\
org.springframework.boot.autoconfigure.r2dbc.ConnectionFactoryBeanCreationFailureAnalyzer,\
org.springframework.boot.autoconfigure.r2dbc.MissingR2dbcPoolDependencyFailureAnalyzer,\
org.springframework.boot.autoconfigure.r2dbc.MultipleConnectionPoolConfigurationsFailureAnalzyer,\
org.springframework.boot.autoconfigure.r2dbc.NoConnectionFactoryBeanFailureAnalyzer

# Template availability providers
org.springframework.boot.autoconfigure.template.TemplateAvailabilityProvider=\
org.springframework.boot.autoconfigure.freemarker.FreeMarkerTemplateAvailabilityProvider,\
org.springframework.boot.autoconfigure.mustache.MustacheTemplateAvailabilityProvider,\
org.springframework.boot.autoconfigure.groovy.template.GroovyTemplateAvailabilityProvider,\
org.springframework.boot.autoconfigure.thymeleaf.ThymeleafTemplateAvailabilityProvider,\
org.springframework.boot.autoconfigure.web.servlet.JspTemplateAvailabilityProvider

# DataSource initializer detectors
org.springframework.boot.sql.init.dependency.DatabaseInitializerDetector=\
org.springframework.boot.autoconfigure.flyway.FlywayMigrationInitializerDatabaseInitializerDetector

# Depends on database initialization detectors
org.springframework.boot.sql.init.dependency.DependsOnDatabaseInitializationDetector=\
org.springframework.boot.autoconfigure.batch.JobRepositoryDependsOnDatabaseInitializationDetector,\
org.springframework.boot.autoconfigure.quartz.SchedulerDependsOnDatabaseInitializationDetector,\
org.springframework.boot.autoconfigure.session.JdbcIndexedSessionRepositoryDependsOnDa-
tabaseInitializationDetector
```

其中配置了各种初始化器、监听器等，这些初始化器和监听器都会在 Spring Boot 启动时完成自动配置。

再看一下获取新的自动配置列表的逻辑，进入 getCandidateConfigurations 方法，再进入导入候选自动配置列表的 ImportCandidates#load 方法：

```
private static final String LOCATION = "META-INF/spring/%s.imports";

public static ImportCandidates load(Class<?> annotation, ClassLoader classLoader) {
    Assert.notNull(annotation, "'annotation' must not be null");
    ClassLoader classLoaderToUse = decideClassloader(classLoader);
    String location = String.format(LOCATION, annotation.getName());
    Enumeration<URL> urls = findUrlsInClasspath(classLoaderToUse, location);
    List<String> importCandidates = new ArrayList<>();
    while (urls.hasMoreElements()) {
        URL url = urls.nextElement();
        importCandidates.addAll(readCandidateConfigurations(url));
    }
    return new ImportCandidates(importCandidates);
}
```

如以上源代码所示，最新的自动配置逻辑是加载 META-INF/spring/%s.imports 配置文件中的自动配置，这个 %s 是占位符，指的是 @AutoConfiguration 注解的类的全路径名称，如下图所示。

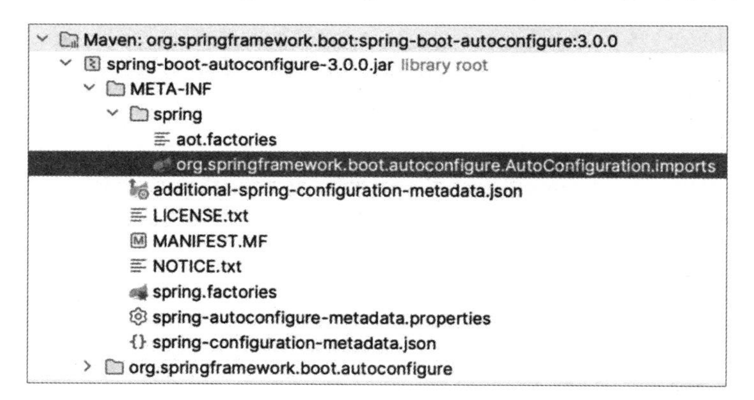

该自动配置文件的内容如下：

```
org.springframework.boot.autoconfigure.admin.SpringApplicationAdminJmxAutoConfiguration
org.springframework.boot.autoconfigure.aop.AopAutoConfiguration
org.springframework.boot.autoconfigure.amqp.RabbitAutoConfiguration
org.springframework.boot.autoconfigure.batch.BatchAutoConfiguration
org.springframework.boot.autoconfigure.cache.CacheAutoConfiguration
org.springframework.boot.autoconfigure.cassandra.CassandraAutoConfiguration
org.springframework.boot.autoconfigure.context.ConfigurationPropertiesAutoConfiguration
org.springframework.boot.autoconfigure.context.LifecycleAutoConfiguration
org.springframework.boot.autoconfigure.context.MessageSourceAutoConfiguration
```

```
org.springframework.boot.autoconfigure.context.PropertyPlaceholderAutoConfiguration
... Spring Boot 3.0.0 共注册了 142 个自动配置类
```

可以看到，编写格式确实简单了，一些应用级别的自动配置类已经全移到这个新配置文件中了，而老的自动配置文件（spring.factories）还保留了一些系统级别的组件。

3.4.4 自动配置原理

自动配置文件被加载后，就会注册里面提供的自动配置类了，比如 3.4.3 节中的下面这条：

```
org.springframework.boot.autoconfigure.jdbc.DataSourceAutoConfiguration
```

这是一个数据源的自动配置类，其核心源码如下：

```
@AutoConfiguration(before = SqlInitializationAutoConfiguration.class)
@ConditionalOnClass({ DataSource.class, EmbeddedDatabaseType.class })
@ConditionalOnMissingBean(type = "io.r2dbc.spi.ConnectionFactory")
@EnableConfigurationProperties(DataSourceProperties.class)
@Import(DataSourcePoolMetadataProvidersConfiguration.class)
public class DataSourceAutoConfiguration {

    @Configuration(proxyBeanMethods = false)
    @Conditional(EmbeddedDatabaseCondition.class)
    @ConditionalOnMissingBean({ DataSource.class, XADataSource.class })
    @Import(EmbeddedDataSourceConfiguration.class)
    protected static class EmbeddedDatabaseConfiguration {

    }

    @Configuration(proxyBeanMethods = false)
    @Conditional(PooledDataSourceCondition.class)
    @ConditionalOnMissingBean({ DataSource.class, XADataSource.class })
    @Import({ DataSourceConfiguration.Hikari.class, DataSourceConfiguration.Tomcat.class,
        DataSourceConfiguration.Dbcp2.class, DataSourceConfiguration.OracleUcp.class,
        DataSourceConfiguration.Generic.class, DataSourceJmxConfiguration.class })
    protected static class PooledDataSourceConfiguration {

    }
    ...
}
```

可以看到，自动配置类是由一系列注解组成的。

1. @AutoConfiguration 注解

自动配置注解 @AutoConfiguration 是 Spring Boot 2.7 中新增的，该注解的源码如下：

```
@Target(ElementType.TYPE)
@Retention(RetentionPolicy.RUNTIME)
@Documented
@Configuration(proxyBeanMethods = false)
@AutoConfigureBefore
@AutoConfigureAfter
public @interface AutoConfiguration {

	/**
	 * Explicitly specify the name of the Spring bean definition associated with the
	 * {@code @AutoConfiguration} class. If left unspecified (the common case), a bean
	 * name will be automatically generated.
	 * <p>
	 * The custom name applies only if the {@code @AutoConfiguration} class is picked up
	 * via component scanning or supplied directly to an
	 * {@link AnnotationConfigApplicationContext}. If the {@code @AutoConfiguration} class
	 * is registered as a traditional XML bean definition, the name/id of the bean element
	 * will take precedence.
	 * @return the explicit component name, if any (or empty String otherwise)
	 * @see AnnotationBeanNameGenerator
	 */
	@AliasFor(annotation = Configuration.class)
	String value() default "";

	/**
	 * The auto-configure classes that should have not yet been applied.
	 * @return the classes
	 */
	@AliasFor(annotation = AutoConfigureBefore.class, attribute = "value")
	Class<?>[] before() default {};

	/**
	 * The names of the auto-configure classes that should have not yet been applied.
	 * @return the class names
	 */
	@AliasFor(annotation = AutoConfigureBefore.class, attribute = "name")
	String[] beforeName() default {};

	/**
	 * The auto-configure classes that should have already been applied.
```

```
     * @return the classes
     */
    @AliasFor(annotation = AutoConfigureAfter.class, attribute = "value")
    Class<?>[] after() default {};

    /**
     * The names of the auto-configure classes that should have already been applied.
     * @return the class names
     */
    @AliasFor(annotation = AutoConfigureAfter.class, attribute = "name")
    String[] afterName() default {};

}
```

自动配置注解 @AutoConfiguration 组合了以下三个注解：

- @Configuration(proxyBeanMethods = false)：配置类注解，并不代理 @Bean 方法。
- @AutoConfigureBefore：自动配置在 XX 配置之前。
- @AutoConfigureAfter：自动配置在 XX 配置之后。

这个注解是专门为自动配置定制的专用注解，其实就是用来代替之前的 @Configuration、@AutoConfigureAfter 和 @AutoConfigureBefore 三个注解，通过这三个注解共同完成了 @Configuration 配置类，以及自动配置类需要在指定的自动配置类完成配置之前或者之后再自动配置。

之前三个注解用起来确实比较烦琐，现在统一使用一个注解，自动配置类有了它自己的注解，这样也可以用来区分用 @Configuration 标识的普通配置类。

所以，DataSourceAutoConfiguration 自动配置类上的 @AutoConfiguration(before=SqlInitialization- AutoConfiguration.class) 就表示这是一个自动配置类，需要在 SqlInitializationAuto Configuration 自动配置类配置之前进行配置。

2. @ConditionalOn*

在自动配置类中还有各种 @ConditionalOn* 的注解，这是一种条件注解，表示在满足指定条件时才会进行自动配置，这也是 Spring Boot 能实现自动配置的核心注解。

以第一个注解 @ConditionalOnClass({ DataSource.class, EmbeddedDatabaseType.class }) 为例，它表示如果类路径下有那两个类才开启这个类的自动配置，否则不开启自动配置。

这个注解的源码如下：

```
@Target({ ElementType.TYPE, ElementType.METHOD })
```

```
@Retention(RetentionPolicy.RUNTIME)
@Documented
@Conditional(OnClassCondition.class)
public @interface ConditionalOnClass {

	/**
	 * The classes that must be present. Since this annotation is parsed by loading class
	 * bytecode, it is safe to specify classes here that may ultimately not be on the
	 * classpath, only if this annotation is directly on the affected component and
	 * <b>not</b> if this annotation is used as a composed, meta-annotation. In order to
	 * use this annotation as a meta-annotation, only use the {@link #name} attribute.
	 * @return the classes that must be present
	 */
	Class<?>[] value() default {};

	/**
	 * The classes names that must be present.
	 * @return the class names that must be present.
	 */
	String[] name() default {};

}
```

它组合了一个最基础的 @Conditional(OnClassCondition.class) 条件注解，并指定了这个 @ConditionalOnClass 条件注解的条件类为 OnClassCondition 类，对应的条件类需要实现 Condition 接口。

基础条件注解 @Conditional 最初是在 Spring 4 中引进的，然后 Spring Boot 基于它组装了各种子条件注解，了解了上面介绍的 @ConditionalOnClass 条件注解，其他条件注解也容易理解，原理是一样的。

更多的条件注解如下表所示。

注解	描述	条件类
@Conditional	基础条件注解	自由指定
@ConditionalOnClass	类路径下有指定的类才开启配置	OnClassCondition
@ConditionalOnMissingClass	类路径下没有指定的类才开启配置	OnClassCondition
@ConditionalOnBean	Spring 容器中有指定的 Bean 才开启配置	OnBeanCondition
@ConditionalOnMissingBean	Spring 容器中没有指定的 Bean 才开启配置	OnBeanCondition
@ConditionalOnProperty	Spring 环境中有指定的参数才开启配置	OnPropertyCondition

续表

注解	描述	条件类
@ConditionalOnExpression	指定的 SpEL 表达式结果为 true 时才开启配置	OnExpressionCondition
@ConditionalOnResource	类路径下有指定的资源才开启配置	OnResourceCondition
@ConditionalOnWebApplication	是指定的 Web 应用类型才开启自动配置	OnWebApplicationCondition
@ConditionalOnJava	是指定的 JDK 版本才开启配置	OnJavaCondition
@ConditionalOnNotWebApplication	不是指定的 Web 应用类型才开启自动配置	OnWebApplicationCondition
@ConditionalOnCloudPlatform	是指定的云平台才开启配置	OnCloudPlatformCondition

以上总结的条件注解覆盖了大部分使用场景，不同的版本可能会有一些变更，更多、更新的条件注解可以查看 Spring Boot 自动配置包下的 condition 目录，如下图所示。

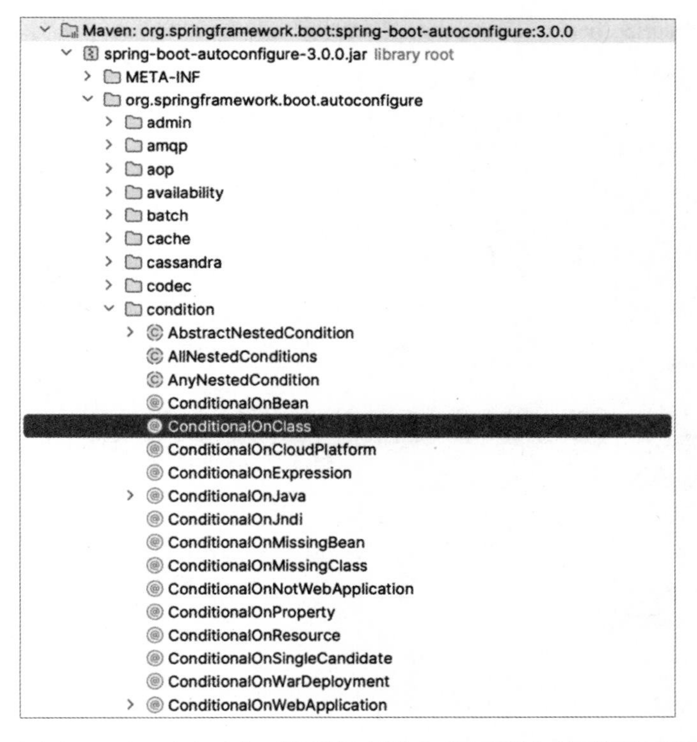

了解了以上各种注解，再回过来看上面的数据源的自动配置源码就能理解了，无非就是根据各种条件开启自动配置而已，比如数据源需要自动配置，类路径下必须得有数据源的相关类才能进行配置，而这些相关的类都是通过 Starter 引入的。

3.4.5 自动配置报告

Spring Boot 提供了查看自动配置报告的方法，包括激活的和未激活的自动配置，如果不确定自动配置类是否被加载了，则可以通过查看自动配置报告的方式来定位问题。

自动配置报告可以在应用的 debug 模式下查看：

```
$ java -jar xx.jar –debug
```

或者：

```
$ java -jar xx.jar –Ddebug
```

如果在 IDE 中启动应用，则可以直接在应用配置文件中指定：

```
debug: true
```

或者在启动命令行参数中指定：

```
spring-boot:run -Dspring-boot.run.arguments="--debug"
```

以及在环境变量中指定，如下图所示。

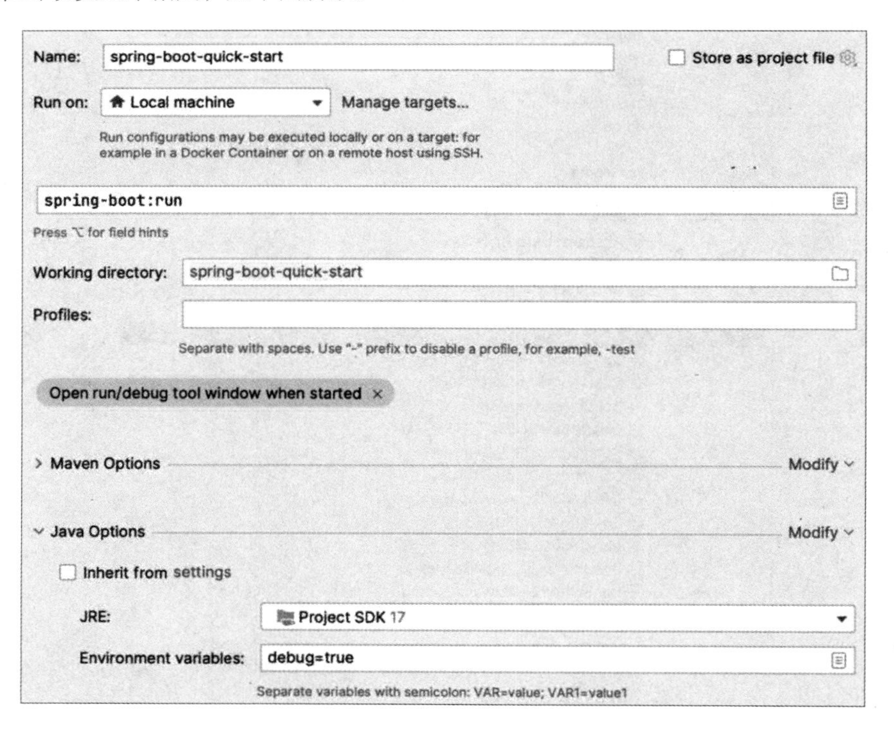

> 方法很多，只要能设置到 Spring 环境参数中就行。另外，如果集成了 spring-boot-starter-actuator 依赖，那么通过 conditions 端点也可以查看，这个 actuator 会在后面的章节介绍。

启动应用后会在控制台看到以下自动配置报告信息：

```
============================
CONDITIONS EVALUATION REPORT
============================

Positive matches:
-----------------
 AopAutoConfiguration matched:
    - @ConditionalOnProperty (spring.aop.auto=true) matched (OnPropertyCondition)

 AopAutoConfiguration.ClassProxyingConfiguration matched:
    - @ConditionalOnMissingClass did not find unwanted class 'org.aspectj.weaver.Advice'
(OnClassCondition)
    - @ConditionalOnProperty (spring.aop.proxy-target-class=true) matched
(OnPropertyCondition)

 DispatcherServletAutoConfiguration matched:
    - @ConditionalOnClass found required class 'org.springframework.web.servlet.
DispatcherServlet' (OnClassCondition)
    - found 'session' scope (OnWebApplicationCondition)

Negative matches:
-----------------

 AopAutoConfiguration.AspectJAutoProxyingConfiguration:
    Did not match:
       - @ConditionalOnClass did not find required class 'org.aspectj.weaver.Advice'
(OnClassCondition)

 ArtemisAutoConfiguration:
    Did not match:
       - @ConditionalOnClass did not find required class 'jakarta.jms.ConnectionFactory'
(OnClassCondition)
 ...

Exclusions:
----------

  None
```

```
Unconditional classes:
---------------------

  org.springframework.boot.autoconfigure.context.ConfigurationPropertiesAutoConfiguration

  org.springframework.boot.autoconfigure.context.LifecycleAutoConfiguration

  org.springframework.boot.autoconfigure.context.PropertyPlaceholderAutoConfiguration

  ...
```

下面是报告中几个项目的说明：

- Positive matches：已经启用的自动配置。
- Negative matches：未启用的自动配置。
- Exclusions：被排除的自动配置。
- Unconditional classes：没有条件的配置类。

知道了这个自动配置报告，在后续的章节中就可以派上用场了。

3.4.6 排除自动配置

Spring Boot 提供的自动配置的功能非常强大，在某些情况下，自动配置的功能可能不符合我们的需求，需要我们自定义配置。比如数据源，不想使用 Spring Boot 默认提供的 DataSourceAutoConfiguration 自动配置，完全想自己创建自定义配置，这时就需要先排除 / 禁用默认的数据源自动配置类。

不同的使用情况，排除自动配置的方法也不一样，但也可以有统一的排除方案。

1. 使用 @SpringBootApplication 注解

使用 @SpringBootApplication 注解的时候，可以使用 exclude 属性排除指定的类：

```
@SpringBootApplication(exclude = { DataSourceAutoConfiguration.class })
public class Application {
    // ...
}
```

也可以使用 excludeName 属性排除指定的类名全路径：

```
@SpringBootApplication(excludeName = { "org.springframework.boot.autoconfigure.jdbc.
DataSourceAutoConfiguration" })
public class Application {
    // ...
}
```

2. 使用 @EnableAutoConfiguration 注解

独立使用 @EnableAutoConfiguration 注解的时候，用法和 @SpringBootApplication 注解是一样的，因为前者就是组合使用了后者的功能。

```
@...
@EnableAutoConfiguration(exclude = { DataSourceAutoConfiguration.class })
public class Application {
    // ...
}
```

也可以使用 excludeName 属性排除指定的类名全路径：

```
@...
@EnableAutoConfiguration(excludeName = { "org.springframework.boot.autoconfigure.jdbc.
DataSourceAutoConfiguration" })
public class Application {
    // ...
}
```

3. 使用 @SpringCloudApplication 注解

使用 Spring Cloud 和 @SpringCloudApplication 注解的时候，Spring Cloud 必须建立在 Spring Boot 应用之上，所以直接使用 @EnableAutoConfiguration 注解排除即可：

```
@...
@EnableAutoConfiguration(exclude = { DataSourceAutoConfiguration.class })
@SpringCloudApplication
public class Application {
    // ...
}
```

但是，这个 @SpringCloudApplication 注解已经在 3.0.1 版本中废除了，该源码如下：

```
/**
* @author Spencer Gibb
* @deprecated This annotation has been deprecated as of the 3.0.1 release.
```

```
* <code>@EnableDiscoveryClient</code> is no longer needed, discovery client
* implementations are enabled as long as an implementation is on the classpath.
* <code>@EnableCircuitBreaker</code> is no longer used now that Hystrix has been removed
* from Spring Cloud.
*/
@Deprecated
@Target(ElementType.TYPE)
@Retention(RetentionPolicy.RUNTIME)
@Documented
@Inherited
@SpringBootApplication
@EnableDiscoveryClient
public @interface SpringCloudApplication {

}
```

因为它组合了 @SpringBootApplication 和 @EnableDiscoveryClient 注解，现在的逻辑是只要客户端实现类在类路径上就会自动启用服务发现，所以组合的 @EnableDiscoveryClient 注解就不再需要了，只组合一个 @SpringBootApplication 注解就没意义了。

4. 使用统一的排除方案

在应用配置文件中指定参数 spring.autoconfigure.exclude 统一排除：

```
spring:
  autoconfigure:
    exclude:
      - org.springframework.boot.autoconfigure.jdbc.DataSourceAutoConfiguration
```

不管是 Spring Boot 应用还是 Spring Cloud 应用，又或者使用的哪个注解，这种配置方案都能解决问题。

最后，具体使用哪种方案可以根据实际应用场景判断，被排除后的自动配置类会在应用运行时的自动配置报告中显示，比如：

```
...

Exclusions:
-----------

  org.springframework.boot.autoconfigure.jdbc.DataSourceAutoConfiguration

...
```

3.4.7 替换自动配置

自动配置是非侵入式的，我们完全可以自定义配置来替换 Spring Boot 默认自动配置中的部分组件配置，我们再来看一下之前的数据源的自动配置类源码，再进入导入的 Hikari 连接池 DataSourceConfiguration.Hikari.class 源码：

```
@Configuration(proxyBeanMethods = false)
@ConditionalOnClass(HikariDataSource.class)
@ConditionalOnMissingBean(DataSource.class)
@ConditionalOnProperty(name = "spring.datasource.type", havingValue = "com.zaxxer.
hikari.HikariDataSource",
        matchIfMissing = true)
static class Hikari {

    @Bean
    @ConfigurationProperties(prefix = "spring.datasource.hikari")
    HikariDataSource dataSource(DataSourceProperties properties) {
        HikariDataSource dataSource = createDataSource(properties, HikariDataSource.class);
        if (StringUtils.hasText(properties.getName())) {
            dataSource.setPoolName(properties.getName());
        }
        return dataSource;
    }

}
```

每个自动配置类、方法上都有一个 @ConditionalOnMissingBean 注解，Spring Boot 应用自己创建的 @Bean 会优先注册，如果应用有自己的 @Bean，则默认自动配置类中的 @Bean 就不会再重复注册。

所以，如果 Spring Boot 提供的默认配置不满足要求，那么完全可以在应用中提供自定义的配置来替换默认的配置。比如，如果使用了 Hikari 连接池，应用注册了一个自定义的 DataSource，那么连接中的 HikariDataSource 就不会重复注册，这就是 @ConditionalOnMissingBean 注解的意义。

3.5 邮件 Starter

3.5.1 概述

前面了解了 Starter 的基本概念、命名规范、分类，以及自动配置原理，也介绍了 Spring Boot

内置的各种各样的 Starter，本节介绍 spring-boot-starter-mail 邮件启动器的集成与应用。

Spring 框架提供了一个发送邮件的抽象和实现：

```
org.springframework.mail.javamail.JavaMailSender
org.springframework.mail.javamail.JavaMailSenderImpl
```

没有 Spring Boot 之前，开发者需要手动构建 JavaMailSenderImpl 实例类，而有了 Spring Boot，就没有必要手工构建了，Spring Boot 提供了一个邮件自动配置类：

```
org.springframework.boot.autoconfigure.mail.MailSenderAutoConfiguration
```

它同样被注册在新的 org.springframework.boot.autoconfigure.AutoConfiguration.imports 自动配置文件中，该自动配置类的源码如下：

```
@AutoConfiguration
@ConditionalOnClass({ MimeMessage.class, MimeType.class, MailSender.class })
@ConditionalOnMissingBean(MailSender.class)
@Conditional(MailSenderCondition.class)
@EnableConfigurationProperties(MailProperties.class)
@Import({ MailSenderJndiConfiguration.class, MailSenderPropertiesConfiguration.class })
public class MailSenderAutoConfiguration {

    /**
     * Condition to trigger the creation of a {@link MailSender}. This kicks in if either
     * the host or jndi name property is set.
     */
    static class MailSenderCondition extends AnyNestedCondition {

        MailSenderCondition() {
            super(ConfigurationPhase.PARSE_CONFIGURATION);
        }

        @ConditionalOnProperty(prefix = "spring.mail", name = "host")
        static class HostProperty {

        }

        @ConditionalOnProperty(prefix = "spring.mail", name = "jndi-name")
        static class JndiNameProperty {

        }

    }

}
```

自动配置类上使用了各种 @Conditional* 注解，在保证默认自动配置机制的情况下，也会在有自定义配置的情况下忽略默认配置。邮件自动配置类使用 @Import 注解引入了 JNDI 和 Properties 两种配置注册方式，在引入的其中一个 MailSenderPropertiesConfiguration 配置类中可以看到自动构建的 JavaMailSenderImpl 实例：

```
@Configuration(proxyBeanMethods = false)
@ConditionalOnProperty(prefix = "spring.mail", name = "host")
class MailSenderPropertiesConfiguration {

    @Bean
    @ConditionalOnMissingBean(JavaMailSender.class)
    JavaMailSenderImpl mailSender(MailProperties properties) {
        JavaMailSenderImpl sender = new JavaMailSenderImpl();
        applyProperties(properties, sender);
        return sender;
    }

    private void applyProperties(MailProperties properties, JavaMailSenderImpl sender) {

      ...

   }

}
```

对应的参数绑定类为 MailProperties，需要 spring.mail.host 配置参数及 spring-boot-starter-mail 依赖，略微配置 spring.mail.* 自定义参数就能自动注册这个 Bean，然后就可以在应用中直接注入 JavaMailSender 并使用了，没有必要重复造轮子。

3.5.2 发邮件实践

首先在 Maven 的 pom.xml 配置文件中加入 spring-boot-starter-mail 邮件启动器依赖：

```
<dependency>
    <groupId>org.springframework.boot</groupId>
    <artifactId>spring-boot-starter-mail</artifactId>
</dependency>
```

然后在应用的 application.properties 配置文件中加入以下邮件自动配置参数：

```
spring:
```

```
  mail:
    host: smtp.exmail.qq.com
    username: xxx@xxx.com
    password: xxx
    properties:
      "[mail.smtp.socketFactory.class]": javax.net.ssl.SSLSocketFactory
      "[mail.smtp.socketFactory.fallback]": false
      "[mail.smtp.socketFactory.port]": 465
      "[mail.smtp.connectiontimeout]": 5000
      "[mail.smtp.timeout]": 3000
      "[mail.smtp.writetimeout]": 5000

mail:
  from: xxx@xxx.com
  personal: 栈长
  bcc: xxx@xxx.com
  subject: Spring Boot 发邮件测试主题
```

> 其中 xxx 为敏感处理内容，需要替换为真实的邮件信息。

笔者写了一个简单的发送邮件的小例子，邮件发送成功后返回 true，若发送失败则返回 false。

```
@Slf4j
@RequiredArgsConstructor
@RestController
public class EmailController {

    private final JavaMailSender javaMailSender;

    private final MailProperties mailProperties;

    @RequestMapping("/sendEmail")
    @ResponseBody
    public boolean sendEmail(@RequestParam("email") String email,
                             @RequestParam("text") String text) {
        try {
            MimeMessage msg = createMimeMsg(email, text, "java.png");
            javaMailSender.send(msg);
        } catch (Exception ex) {
            log.error(" 邮件发送失败: ", ex);
            return false;
        }
        return true;
    }
```

```
    /**
     * 创建复杂邮件
     * @param email
     * @param text
     * @param attachmentClassPath
     * @return
     * @throws MessagingException
     * @throws UnsupportedEncodingException
     */
    private MimeMessage createMimeMsg(String email, String text, String attachmentClassPath)
throws MessagingException, UnsupportedEncodingException {
        MimeMessage msg = javaMailSender.createMimeMessage();
        MimeMessageHelper mimeMessageHelper = new MimeMessageHelper(msg, true);
        mimeMessageHelper.setFrom(mailProperties.getFrom(), mailProperties.getPersonal());
        mimeMessageHelper.setTo(email);
        mimeMessageHelper.setBcc(mailProperties.getBcc());
        mimeMessageHelper.setSubject(mailProperties.getSubject());
        mimeMessageHelper.setText(text);
        mimeMessageHelper.addAttachment("附件",
                new ClassPathResource(attachmentClassPath));
        return msg;
    }

    /**
     * 创建简单邮件
     * @param email
     * @param text
     * @return
     */
    private SimpleMailMessage createSimpleMsg(String email, String text) {
        SimpleMailMessage msg = new SimpleMailMessage();
        msg.setFrom(mailProperties.getFrom());
        msg.setTo(email);
        msg.setBcc(mailProperties.getBcc());
        msg.setSubject(mailProperties.getSubject());
        msg.setText(text);
        return msg;
    }

}
```

可以看到，JavaMailSender是直接注入就能使用了，另外这里提供了两个创建邮件消息的示例方法：

- SimpleMailMessage：用于构建简单的纯文本邮件。
- MimeMessage：用于构建复杂邮件，比如要支持 HTML 内容邮件，以及发送附件等。

MailProperties 参数类用于绑定一些邮件及发送人的固定信息，如发件人、昵称等，可以根据自身情况提供参数配置化。

启动应用后，调用发送邮件接口：

```
http://localhost:8080/sendEmail?email=xxx@xxx.com&text=hello
```

复杂邮件发送正常了，附件也能正常下载和显示，如下图所示。

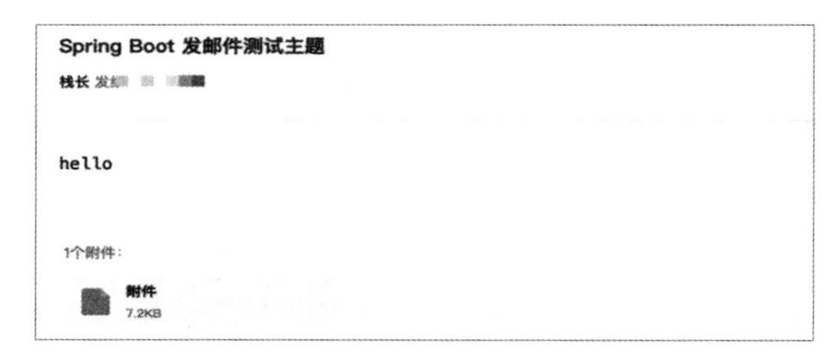

当然，JavaMailSender 也是可以自定义的，比如在需要多个 JavaMailSender 的情况下，这时默认注册的一个 JavaMailSender 实例便不满足要求。

3.6 自定义 Starter

前面介绍了 Starter 的实现机制，以及 Spring Boot 封装的发邮件自动配置原理，可以看到，自定义一个 Starter 并不难，本节实现一个简单的自定义 Spring Boot Starter。

3.6.1 创建 Starter 工程

根据 Starter 定义的规范，一般一个第三方的应用应该以 *-spring-boot-starter 的形式命名，所以需要先创建一个 javastack-spring-boot-starter 的工程。除了命名要注意，其他不再赘述，就是一个普通的 Spring Boot 应用，创建过程略。

3.6.2 创建自动配置类

下面创建了一个简单的自动配置类：

```
@AutoConfiguration
```

```
@ConditionalOnProperty(prefix = "javastack.starter", name = "enabled", havingValue = "true")
public class TestServiceAutoConfiguration {

    @Bean
    public TestService testService() {
        return new TestService();
    }

}
```

这个自动配置类很简单，就是判断 Spring 环境配置中是否有 javastack.starter.enabled=true 这个参数的值，如果有就配置一个 TestService 的 Bean。

TestService 类的示例代码如下：

```
public class TestService {

    public String getServiceName() {
        return "Java 技术栈 ";
    }

}
```

这个类只有一个方法 getServiceName，仅返回一个字符串：Java 技术栈。

3.6.3　注册自动配置类（spring.factories）

老的自动配置类注册方式（spring.factories）已经退出历史舞台了，但主流应用并不会很快升级 Spring Boot 3.0.0，市面上大多的应用还是以 Spring Boot 2.x 为主，所以笔者这里也提供了老的自动配置类注册规范的实现方式。

先在 Spring Boot 应用的 resource 资源目录下创建 spring.factories 自动配置文件：

```
META-INF/spring.factories
```

工程结构如下图所示。

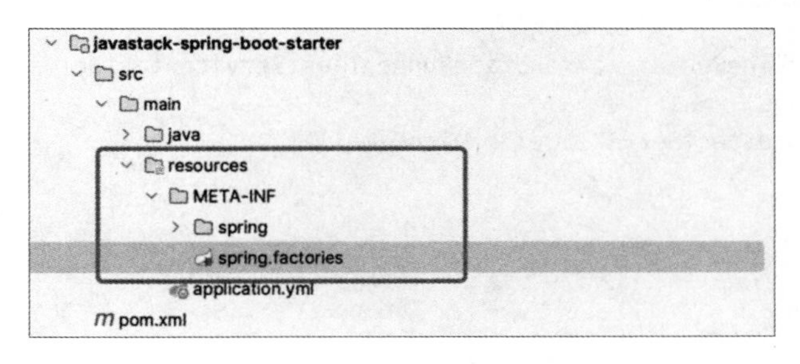

然后添加要注册的自动配置类：

```
org.springframework.boot.autoconfigure.EnableAutoConfiguration=\
cn.javastack.springboot.starter.config.TestServiceAutoConfiguration
```

> 这种老的自动配置类的注册方式适用于 Spring Boot 2.7 及以下版本，Spring Boot 3.0+ 的自动配置类注册方式见 3.6.5 节。

3.6.4 使用 Starter

自定义的 Spring Boot Starter 应用搭建好之后，可以测试一下它是否生效了，一般是把它打成 jar 包并上传到 Maven 仓库供其他同事调用，这里笔者为了测试，先在本地打包，再新建另外一个应用并引用它进行测试。所以，再新建一个 javastack-spring-boot-starter-sample 应用，并添加这个自定义 Starter 的依赖：

```
<dependencies>
  <dependency>
    <groupId>cn.javastack</groupId>
    <artifactId>javastack-spring-boot-starter</artifactId>
    <version>1.0</version>
    <scope>compile</scope>
  </dependency>
</dependencies>
```

在启动类中注入该 Starter 自动配置类中的 TestService 并输出其方法的值：

```
@Slf4j
@SpringBootApplication
public class Application {

    public static void main(String[] args) {
        SpringApplication.run(Application.class);
    }

    @Bean
    public CommandLineRunner commandLineRunner(TestService testService) {
        return (args) -> {
            log.info(testService.getServiceName());
        };
    }

}
```

自定义 Starter 的时候定义了需要有 javastack.starter.enabled=true 这个参数的值才会自动配置，所以在测试应用 application.yml 配置文件中添加这个配置参数：

```
javastack:
  starter:
enabled: true
```

最后启动应用：

```
c.j.s.starter.sample.Application         : Java 技术栈
```

结果正常输出，说明引入的自定义 Starter 自动配置成功了，在应用启动时的自动配置报告中也看到了自定义的自动配置，如下图所示。

```
TaskSchedulingAutoConfiguration matched:
   - @ConditionalOnClass found required class 'org.springframework.scheduling.concurrent.Th

TaskSchedulingAutoConfiguration#taskSchedulerBuilder matched:
   - @ConditionalOnMissingBean (types: org.springframework.boot.task.TaskSchedulerBuilder;

TestServiceAutoConfiguration matched:
   - @ConditionalOnProperty (javastack.starter.enabled=true) matched (OnPropertyCondition)
```

再把配置故意改错进行测试：

```
javastack:
  starter:
    enabled: false
```

然后重新启动应用：

```
***************************
APPLICATION FAILED TO START
***************************

Description:

Parameter 0 of method commandLineRunner in cn.javastack.springboot.starter.sample.
Application required a bean of type 'cn.javastack.springboot.starter.service.TestService'
that could not be found.

Action:
```

```
Consider defining a bean of type 'cn.javastack.springboot.starter.service.TestService'
in your configuration.

Process finished with exit code 1
```

此时应用运行报错，因为把自动配置需要的参数改为了 false，自定义的 Starter 就不能完成自动配置，Spring 容器中没有 TestService 这个实例就抛出异常了。

本节简单演示了如何自定义一个 Spring Boot Starter，根据某个参数的值来决定是否自动配置，还可以根据是否有某个类、某个 Bean 等系列条件注解完成更加复杂的 Starter 自动配置。

3.6.5 注册自动配置类（新规范）

前面的章节使用的是老规范自动配置文件，Spring Boot 2.7 开始约定了自动配置类注册文件新规范，并且在 Spring Boot 3.0.0 中正式废除了老的自动配置类注册方式，所以建议读者在 Spring Boot 2.7+ 中使用新的自动配置文件注册自动配置类，避免后续升级产生兼容问题。

先在 javastack-spring-boot-starter 工程中创建新规范自动配置文件：

```
META-INF/spring/org.springframework.boot.autoconfigure.AutoConfiguration.imports
```

工程结构如下图所示。

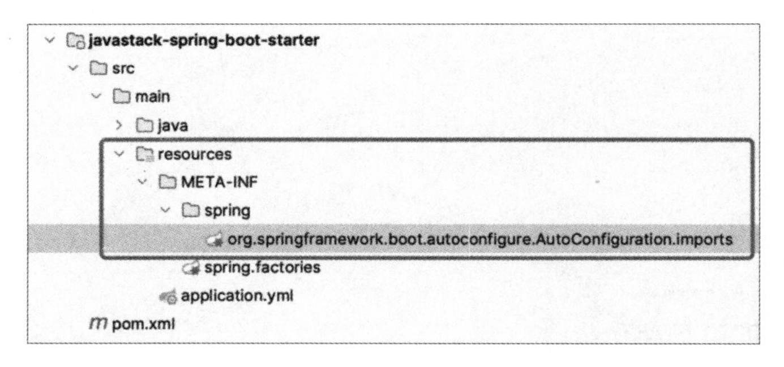

创建完直接在新的自动配置类注册文件中输入自动配置类：

```
cn.javastack.springboot.starter.config.TestServiceAutoConfiguration
```

多个自动配置类直接“回车”配置即可，简单方便，然后重启 javastack-spring-boot-starter-sample 应用：

```
c.j.s.starter.sample.Application          : Java 技术栈
```

应用正常启动，结果也正常输出，说明新规范自动配置文件生效了。

第 4 章 Spring Boot 启动过程与扩展应用

Spring Boot 框架的启动过程与扩展应用是学习 Spring Boot 框架必不可少的部分，本章会介绍 Spring Boot 启动过程中需要掌握的内容，包括 Spring Boot 应用的引导类、启动类、启动方法、核心注解、启动失败分析、启动图案、启动事件和监听器、*Runners 运行器，以及启动流程及源码分析等，这些内容可以帮助开发者更好地运用 Spring Boot。

4.1 启动入口

4.1.1 应用启动类

在 Spring Boot 基础入门一章中，我们看到了 Spring 网站自动生成的 Application 启动类，它是 Spring Boot 应用的启动入口类。如果没有必要，一般建议启动入口类要放置于根目录下，以便使用注解扫描并管理所有子目录包的组件。

Spring Boot 启动类必须包含一个标准的 main 方法，这也是启动 Java 应用的入口方法，然后在 main 方法中添加 SpringApplication.run 方法启动 Spring Boot 应用：

```
@SpringBootApplication
public class Application {

    public static void main(String[] args) {
        SpringApplication.run(Application.class, args);
    }

}
```

启动类上面的 @SpringBootApplication 注解用于标识应用启动的入口类，也负责启动类统一筹划配置类、组件扫描、自动配置等工作，它是 Spring Boot 启动类上的核心注解，源码如下：

```
@Target(ElementType.TYPE)
@Retention(RetentionPolicy.RUNTIME)
@Documented
@Inherited
@SpringBootConfiguration
@EnableAutoConfiguration
@ComponentScan(excludeFilters = { @Filter(type = FilterType.CUSTOM, classes =
TypeExcludeFilter.class),
        @Filter(type = FilterType.CUSTOM, classes = AutoConfigurationExcludeFilter.class) })
public @interface SpringBootApplication {
```

它主要组合了以下 3 个注解：

- @SpringBootConfiguration：标识为一个 Spring Boot 配置类。
- @EnableAutoConfiguration：开启应用的自动配置功能。
- @ComponentScan：开启自动扫描 Spring 组件功能。

类关系如下图所示。

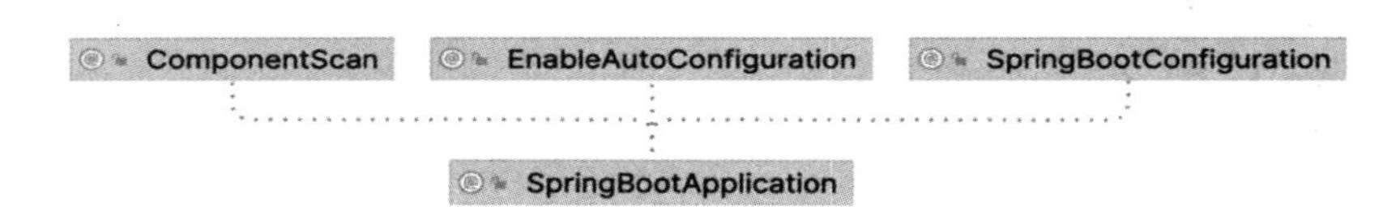

1. @SpringBootConfiguration 注解

@SpringBootConfiguration 注解是 Spring Boot 配置类注解，组合了 @Configuration 注解，该注解的源码如下：

```
@Target(ElementType.TYPE)
@Retention(RetentionPolicy.RUNTIME)
@Documented
@Configuration
@Indexed
public @interface SpringBootConfiguration {
```

@SpringBootConfiguration 注解在 Spring Boot 配置管理一章有详细介绍，它只是用来修饰 Spring Boot 配置类而已，或者可用于 Spring Boot 后续的扩展，也就是说，如果在 Application 启动

类中使用了 @SpringBootApplication 或者 @SpringBootConfiguration 注解，则其可以作为一个配置类使用。

2. @EnableAutoConfiguration 注解

@EnableAutoConfiguration 注解表示开启 Spring Boot 的自动配置功能，一般不会单独使用，如果单独使用，则可以用来关闭某个自动配置的选项，如关闭数据源自动配置功能：

```
import org.springframework.boot.autoconfigure.SpringBootApplication;
import org.springframework.boot.autoconfigure.jdbc.DataSourceAutoConfiguration;

@SpringBootApplication(exclude = { DataSourceAutoConfiguration.class })
public class Application {

}
```

3. @ComponentScan 注解

@ComponentScan 是 Spring 框架扫描 Spring 组件并进行注册的注解，这是 Spring 3.1 添加的一个注解，用来代替配置文件中的 component-scan 配置，默认会扫描当前目录及其所有子目录，这也是为什么推荐把启动类放在根目录下的原因。另外，@ComponentScan 是可重复注解，即可以配置多个，用来扫描并注册不同的子包。

了解了以上三个子注解，我们也就知道了这个 @SpringBootApplication 核心注解并不是必需的，可以分开单独使用。

4.1.2 应用启动方法

1. SpringApplication

Spring Boot 应用的启动是通过一个带有 main 方法的启动类完成的，SpringApplication 默认调用的是 run 静态方法，第一个参数一般为应用启动类，也可以是多个其他的 @Configuration 类或者 @Component 类，第二个参数为传递给应用的 main 方法参数。

这个方法的源码如下：

```
public static ConfigurableApplicationContext run(Class<?>[] primarySources, String[] args) {
    return new SpringApplication(primarySources).run(args);
}
```

它最终也构建了一个默认的 SpringApplication 实例，并调用实例的 run 普通方法，所以，不仅可以在 application 配置文件中配置启动环境，还可以自定义创建 SpringApplication 实例的方式：

```
public static void main(String[] args) {
    SpringApplication springApplication = new SpringApplication(Application.class);

    // 自定义设置
    springApplication.xxx();

    springApplication.run(args);
}
```

2. SpringApplicationBuilder

除了使用 SpringApplication 启动，还能使用链式的 API：

```
public static void main(String[] args) {
    new SpringApplicationBuilder()
        .sources(Parent.class)
        .child(Application.class)
        .bannerMode(Banner.Mode.OFF)
        .run(args);
}
```

SpringApplicationBuilder 的构建器的源码如下：

```
public SpringApplicationBuilder(Class<?>... sources) {
    this(null, sources);
}

public SpringApplicationBuilder(ResourceLoader resourceLoader, Class<?>... sources) {
    this.application = createSpringApplication(resourceLoader, sources);
}

protected SpringApplication createSpringApplication(ResourceLoader resourceLoader,
Class<?>... sources) {
    return new SpringApplication(resourceLoader, sources);
}
```

可以看到 SpringApplicationBuilder 其实也是对 SpringApplication 类的包装，其中的 sources 和 child 方法用于构建多应用上下文，两者共享同一个 Environment 环境，但 Web 组件必须包含在 child 上下文环境中。

了解了本节的内容，就知道如何通过 Java 类在 Spring Boot 中定制启动环境，对了解后续章节的内容也有帮助。

4.1.3　启动引导类

Spring Boot 应用虽然会有一个带有 main 方法的 Application 启动类，但它一般只会在开发工具中执行启动，也就是在 Spring Boot 应用未打包的情况下运行，如果打包成可执行的 jar 包运行，它就不是启动 Spring Boot 应用最前置的入口了。

> Spring Boot 应用打包可以参考第 11 章 Spring Boot 打包和部署。

应用打包成可执行的 jar 包后，可以直接通过 java 命令运行，如下所示。

```
$ java -jar spring-boot-quick-start-1.0.jar
```

这个 jar 包的目录结构如下：

```
.
├── BOOT-INF
│   ├── classes
│   │   └── cn
│   │       └── javastack
│   │           └── springboot
│   │               └── quckstart
│   └── lib
├── META-INF
│   └── maven
│       └── cn.javastack
│           └── spring-boot-quick-start
└── org
    └── springframework
        └── boot
            └── loader
                ├── archive
                ├── data
                ├── jar
                ├── jarmode
                └── util
```

主要包含以下三个目录：

- BOOT-INF：包含应用启动所需要的类库。
- META-INF：包含应用打包的相关描述文件。
- org：包含 Spring Boot 启动所需要的引导类。

在 META-INF 目录下可以找到 MANIFEST.MF 文件：

```
Manifest-Version: 1.0
Created-By: Maven JAR Plugin 3.3.0
Build-Jdk-Spec: 17
Implementation-Title: spring-boot-quick-start
Implementation-Version: 1.0
Main-Class: org.springframework.boot.loader.JarLauncher
Start-Class: cn.javastack.springboot.quckstart.Application
Spring-Boot-Version: 3.0.0
Spring-Boot-Classes: BOOT-INF/classes/
Spring-Boot-Lib: BOOT-INF/lib/
Spring-Boot-Classpath-Index: BOOT-INF/classpath.idx
Spring-Boot-Layers-Index: BOOT-INF/layers.idx
```

它包括一个 jar 包的相关描述信息，除了基本的项目构建信息，其中还有两个最重要的应用启动信息：

- **Main-Class**：应用的启动引导类，即打包后最前置的启动入口类，启动 jar 包的引导类为 org.springframework.boot.loader.JarLauncher，启动 war 包的引导类为 org.springframework.boot.loader.WarLauncher。
- **Start-Class**：即前面章节描述的 Application 启动类。

Main-Class 参数指定了 Spring Boot 应用具体要使用的引导类，Spring Boot 应用的引导机制是由 spring-boot-loader 模块实现的，引导实现类顶层的是 Launcher 抽象类，Launcher 引导类的结构如下图所示，

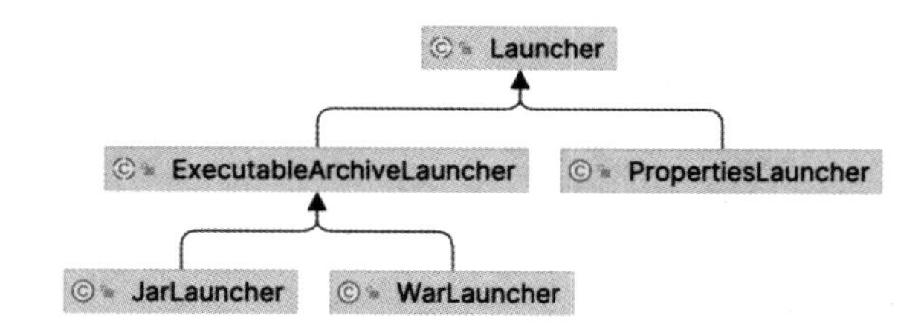

Launcher 引导类主要分为以下两种引导类型：

- **根据可执行文件引导**：ExecutableArchiveLauncher 抽象类，按打包类型又分为 JarLauncher 和 WarLauncher 引导实现类。
- **根据参数文件引导**：PropertiesLauncher 引导实现类。

可执行文件 JarLauncher 引导类相关的源码如下：

```
public class JarLauncher extends ExecutableArchiveLauncher {

    ...

    public static void main(String[] args) throws Exception {
        new JarLauncher().launch(args);
    }

}
```

JarLauncher 引导类调用的 launch 方法是 Launcher 抽象父类中的方法，如下面的 Launcher 抽象类的源码所示。

```
public abstract class Launcher {

    private static final String JAR_MODE_LAUNCHER = "org.springframework.boot.loader.
jarmode.JarModeLauncher";

    /**
     * Launch the application. This method is the initial entry point that should be
     * called by a subclass {@code public static void main(String[] args)} method.
     * @param args the incoming arguments
     * @throws Exception if the application fails to launch
     */
    protected void launch(String[] args) throws Exception {
        if (!isExploded()) {
            JarFile.registerUrlProtocolHandler();
        }
        ClassLoader classLoader = createClassLoader(getClassPathArchivesIterator());
        String jarMode = System.getProperty("jarmode");
        String launchClass = (jarMode != null && !jarMode.isEmpty()) ? JAR_MODE_LAUNCHER:
 getMainClass();
        launch(args, launchClass, classLoader);
    }

    ...

    /**
     * Returns the main class that should be launched.
     * @return the name of the main class
     * @throws Exception if the main class cannot be obtained
```

```
     */
    protected abstract String getMainClass() throws Exception;

    ...

}
```

Launcher#launch 引导方法提供了一个 getMainClass 的抽象方法，用于获取实际的应用启动类，这个便是在 ExecutableArchiveLauncher 可执行文件引导抽象类中实现的，实现的逻辑则是获取应用描述文件 META-INF/MANIFEST.MF 中的 Start-Class 参数的值，如下面的 ExecutableArchiveLauncher 类的源码所示。

```
public abstract class ExecutableArchiveLauncher extends Launcher {

    private static final String START_CLASS_ATTRIBUTE = "Start-Class";

    ...

    @Override
    protected String getMainClass() throws Exception {
        Manifest manifest = this.archive.getManifest();
        String mainClass = null;
        if (manifest != null) {
            mainClass = manifest.getMainAttributes().getValue(START_CLASS_ATTRIBUTE);
        }
        if (mainClass == null) {
            throw new IllegalStateException("No 'Start-Class' manifest entry specified in"
+ this);
        }
        return mainClass;
    }

    ...

}
```

获取到应用启动类之后，进入 Launcher#launch 方法：

```
public abstract class Launcher {

    ...

    protected void launch(String[] args, String launchClass, ClassLoader classLoader)
throws Exception {
```

```
        Thread.currentThread().setContextClassLoader(classLoader);
        createMainMethodRunner(launchClass, args, classLoader).run();
    }

    protected MainMethodRunner createMainMethodRunner(String mainClass, String[] args,
ClassLoader classLoader) {
        return new MainMethodRunner(mainClass, args);
    }

    ...

}
```

首先通过上一个 launch 方法创建的一个类加载器传递并绑定到当前线程，即上下文类加载器，然后创建一个 MainMethodRunner 实例并调用其 run 方法，如下面 MainMethodRunner 类的源码所示。

```
public class MainMethodRunner {

    private final String mainClassName;

    private final String[] args;

    /**
     * Create a new {@link MainMethodRunner} instance.
     * @param mainClass the main class
     * @param args incoming arguments
     */
    public MainMethodRunner(String mainClass, String[] args) {
        this.mainClassName = mainClass;
        this.args = (args != null) ? args.clone() : null;
    }

    public void run() throws Exception {
        Class<?> mainClass = Class.forName(this.mainClassName, false, Thread.currentThread()
.getContextClassLoader());
        Method mainMethod = mainClass.getDeclaredMethod("main", String[].class);
        mainMethod.setAccessible(true);
        mainMethod.invoke(null, new Object[] { this.args });
    }

}
```

即先把启动类和入口参数传过来，然后在运行 run 方法时，通过 Java 反射的方式调用启动类中的 main 方法并运行。所以，Main-Class 和 Start-Class 两者是先后调用的关系，先由 Main-Class 参数指定引导类，引导类执行时再调用 Start-Class 参数指定的应用启动类。这样处理的好处是，应用只需要一个启动类，不同类型的应用由不同的引导类引导启动。

4.2 关闭启动日志

Spring Boot 启动过程中，系统默认会输出一些详细的启动日志：

```
  .   ____          _            __ _ _
 /\\ / ___'_ __ _ _(_)_ __  __ _ \ \ \ \
( ( )\___ | '_ | '_| | '_ \/ _` | \ \ \ \
 \\/  ___)| |_)| | | | | || (_| |  ) ) ) )
  '  |____| .__|_| |_|_| |_\__, | / / / /
 =========|_|==============|___/=/_/_/_/
 :: Spring Boot ::                (v3.0.0)

..... : .....
..... : No active profile set, falling back to 1 default profile: "default"
..... : Tomcat initialized with port(s): 8081 (http)
..... : Starting service [Tomcat]
..... : Starting Servlet engine: [Apache Tomcat/10.0.21]
..... : Initializing Spring embedded WebApplicationContext
..... : Root WebApplicationContext: initialization completed in 830 ms
..... : Tomcat started on port(s): 8081 (http) with context path '/javastack'
..... : Started Application in 1.6 seconds (process running for 1.884)
```

> **Spring Boot 3.0 新变化：**
> 在应用的启动日志中不再记录主机名，这样可以避免不必要的网络查找，有助于减少应用的启动时间。

如果不想输出详细的启动日志，则可以通过以下参数关闭：

```
spring:
  main:
    log-startup-info: false
```

或者在启动方法上禁用：

```
public static void main(String[] args) {
    SpringApplication springApplication = new SpringApplication(Application.class);
```

```
    springApplication.setLogStartupInfo(false);
    springApplication.run(args);
}
```

4.3 启动失败分析

4.3.1 失败分析器

Spring Boot 应用经常启动失败就会显示一大堆错误信息，比如我们重复启动同一个端口，应用启动时就会输出端口已经在使用的异常：

```
***************************
APPLICATION FAILED TO START
***************************

Description:

Embedded servlet container failed to start. Port 8080 was already in use.

Action:

Identify and stop the process that's listening on port 8080 or configure this
application to listen on another port.
```

Spring Boot 应用启动失败的场景是由 FailureAnalyzer 失败分析器拦截并处理的，Spring Boot 注册了许多 FailureAnalyzer，它能分析启动失败异常并给用户显示有用的诊断信息。

FailureAnalyzer 失败分析器对应的是 FailureAnalyzer 接口：

```
/**
 * A {@code FailureAnalyzer} is used to analyze a failure and provide diagnostic
 * information that can be displayed to the user.
 *
 * @author Andy Wilkinson
 * @since 1.4.0
 */
@FunctionalInterface
public interface FailureAnalyzer {

    /**
```

```
     * Returns an analysis of the given {@code failure}, or {@code null} if no analysis
     * was possible.
     * @param failure the failure
     * @return the analysis or {@code null}
     */
    FailureAnalysis analyze(Throwable failure);

}
```

Spring Boot 内置的失败分析器目前还注册在老的自动配置类注册文件中，比如 Spring Boot 基础包中的 spring.factories 文件：

```
spring-boot-3.0.0.jar!/META-INF/spring.factories
```

其中注册的所有系统级的失败分析器列表如下：

```
# Failure Analyzers
org.springframework.boot.diagnostics.FailureAnalyzer=\
org.springframework.boot.context.config.ConfigDataNotFoundFailureAnalyzer,\
org.springframework.boot.context.properties.IncompatibleConfigurationFailureAnalyzer,\
org.springframework.boot.context.properties.NotConstructorBoundInjectionFailureAnalyzer,\
org.springframework.boot.diagnostics.analyzer.BeanCurrentlyInCreationFailureAnalyzer,\
org.springframework.boot.diagnostics.analyzer.BeanDefinitionOverrideFailureAnalyzer,\
org.springframework.boot.diagnostics.analyzer.BeanNotOfRequiredTypeFailureAnalyzer,\
org.springframework.boot.diagnostics.analyzer.BindFailureAnalyzer,\
org.springframework.boot.diagnostics.analyzer.BindValidationFailureAnalyzer,\
org.springframework.boot.diagnostics.analyzer.UnboundConfigurationPropertyFailureAnalyzer,\
org.springframework.boot.diagnostics.analyzer.MutuallyExclusiveConfigurationProperties-
FailureAnalyzer,\
org.springframework.boot.diagnostics.analyzer.NoSuchMethodFailureAnalyzer,\
org.springframework.boot.diagnostics.analyzer.NoUniqueBeanDefinitionFailureAnalyzer,\
org.springframework.boot.diagnostics.analyzer.PortInUseFailureAnalyzer,\
org.springframework.boot.diagnostics.analyzer.ValidationExceptionFailureAnalyzer,\
org.springframework.boot.diagnostics.analyzer.InvalidConfigurationPropertyNameFailureAn-
alyzer,\
org.springframework.boot.diagnostics.analyzer.InvalidConfigurationPropertyValueFailure-
Analyzer,\
org.springframework.boot.diagnostics.analyzer.PatternParseFailureAnalyzer,\
org.springframework.boot.liquibase.LiquibaseChangelogMissingFailureAnalyzer,\
org.springframework.boot.web.context.MissingWebServerFactoryBeanFailureAnalyzer,\
org.springframework.boot.web.embedded.tomcat.ConnectorStartFailureAnalyzer
```

再回到上面的端口重复启动失败异常，它其实就是注册了 PortInUseFailureAnalyzer 失败分析器，可以看到 PortInUseFailureAnalyzer 失败分析器就在注册列表中。

PortInUseFailureAnalyzer 的源码如下：

```
/**
 * A {@code FailureAnalyzer} that performs analysis of failures caused by a
 * {@code PortInUseException}.
 *
 * @author Andy Wilkinson
 */
class PortInUseFailureAnalyzer extends AbstractFailureAnalyzer<PortInUseException> {

    @Override
    protected FailureAnalysis analyze(Throwable rootFailure, PortInUseException cause) {
        return new FailureAnalysis("Web server failed to start. Port " + cause.getPort() +
 " was already in use.",
                "Identify and stop the process that's listening on port " + cause.getPort()+
"or configure this"
                        + "application to listen on another port.",
                cause);
    }

}
```

只要应用启动过程中抛出了 PortInUseException 异常就会被这个失败分析器拦截并输出可读性的错误信息。

4.3.2　自定义失败分析器

默认的错误信息对于开发人员来说是很明显的，但不是所有的错误信息都很好理解，或者对于其他需要查看日志的非开发人员就不是很友好，所以，可以通过扩展 FailureAnalyzer 失败分析器接口来达到自定义错误信息的目的。

但从上面内置的端口重复使用失败分析器这一点可以发现，所有的分析器都继承了这个 AbstractFailureAnalyzer 抽象类，然后由它实现 FailureAnalyzer 接口，所以一般基于这个抽象基类就可以实现自定义失败分析器的扩展。

比如说上面的 PortInUseFailureAnalyzer 失败分析器输出的信息是英文，不是很直观，接下来笔者自定义实现一个类似的中文失败分析器。

首先创建一个 PortInUseFailureAnalyzer 失败分析器继承 AbstractFailureAnalyzer 抽象类：

```
public class PortInUseFailureAnalyzer extends AbstractFailureAnalyzer<PortInUseException> {
```

```
    @Override
    protected FailureAnalysis analyze(Throwable rootFailure, PortInUseException cause) {
        return new FailureAnalysis("你启动的端口 " + cause.getPort() + " 被占用了.",
                                   "快检查下端口 " + cause.getPort() + " 被哪个程序占用了,
                                   或者强制杀掉进程.",
                                   cause);
    }

}
```

重写 analyze 方法，并返回一个 FailureAnalysis 对象，FailureAnalysis 类的三个主要信息如下面的源码所示。

```
public FailureAnalysis(String description, String action, Throwable cause) {
    this.description = description;
    this.action = action;
    this.cause = cause;
}
```

它们分别表示要展示的三个信息如下：

- 可读性的错误描述。
- 建议的检查修复动作。
- 原始异常。

然后在应用的 META-INF/spring.factories 配置文件中进行注册：

```
org.springframework.boot.diagnostics.FailureAnalyzer=\
cn.javastack.springboot.features.analyzer.PortInUseFailureAnalyzer
```

> **需要注意的是：**
> 失败分析器还需要在 spring.factories 文件中注册，这个注册逻辑没有任何变化。
> 如果把失败分析器放在新的自动配置类注册文件中，则不会生效。虽然自动配置类的注册文件已经从 spring.factories 移到另外一个新文件中了，但它仅用于注册自动配置类，不用能注册其他系统组件。

然后重复启动多个相同端口的应用，输出日志如下：

```
***************************
APPLICATION FAILED TO START
***************************
```

```
Description:
你启动的端口 8080 被占用了.
Action:
快检查下端口 8080 被哪个程序占用了，或者强制杀掉进程.
```

下面再自定义一个全新的失败分析器，这样可以更清楚地认识失败分析器。我们在创建 Bean 的过程中手动抛出一个自定义的异常：

```
@Bean
public CommandLineRunner commandLineRunner() {
    return (args) -> {
        throw new JavastackException("Java 技术栈异常");
    };
}
```

如果不注册该失败分析器，则这个自定义的异常就不会被内置的失败分析器拦截，应用启动后会输出大堆的异常信息：

```
java.lang.IllegalStateException: Failed to execute CommandLineRunner
    at org.springframework.boot.SpringApplication.callRunner (SpringApplication.
java:780) ~[spring-boot-3.0.0-M3.jar:3.0.0-M3]
    at org.springframework.boot.SpringApplication.callRunners (SpringApplication.
java:761) ~[spring-boot-3.0.0-M3.jar:3.0.0-M3]
    at org.springframework.boot.SpringApplication.run(SpringApplication.java:317)
~[spring-boot-3.0.0-M3.jar:3.0.0-M3]
    at cn.javastack.springboot.features.Application.main(Application.java:30)
~[classes/:na]
Caused by: cn.javastack.springboot.features.analyzer.JavastackException: Java 技术栈异常
    at cn.javastack.springboot.features.Application.lambda$commandLineRunner$0
(Application.java:41) ~[classes/:na]
    at org.springframework.boot.SpringApplication.callRunner(SpringApplication.
java:777) ~[spring-boot-3.0.0-M3.jar:3.0.0-M3]
    ... 3 common frames omitted
```

如果不想看到大篇幅的启动错误，则可以添加一个失败分析器拦截该异常：

```
public class JavastackFailureAnalyzer extends AbstractFailureAnalyzer<JavastackException> {

    @Override
    protected FailureAnalysis analyze(Throwable rootFailure, JavastackException cause) {
        return new FailureAnalysis("Java 技术栈发生异常了……",
                                   "赶快去检查一下吧！",
                                   cause);
```

```
    }
}
```

添加失败分析器并注册：

```
org.springframework.boot.diagnostics.FailureAnalyzer=\
cn.javastack.springboot.features.analyzer.PortInUseFailureAnalyzer,\
cn.javastack.springboot.features.analyzer.JavastackFailureAnalyzer
```

然后重新启动应用：

```
***************************
APPLICATION FAILED TO START
***************************
Description:
Java 技术栈发生异常了……
Action:
赶快去检查一下吧！
```

Spring Boot 提供的失败分析器以友好的错误信息和修复建议代替了大堆的错误异常信息，使用失败分析器能很直观地看出是什么错误及如何修复这个错误，可以帮助我们更直观地定位应用启动故障。

4.4 全局懒加载

Spring Boot 可以开启全局懒加载，懒加载的意思是 Bean 不会在应用启动时全部创建，只会在真正使用时才创建，Spring Boot 默认不开启懒加载，如需开启，则可以在应用配置文件中配置：

```
spring:
  main:
lazy-initialization: true
```

也可以在 SpringApplication 或者 SpringApplicationBuilder 启动类上设置，这里不再赘述。

Spring Boot 懒加载可以大大减少应用的启动时间，还能节省系统资源，但也会造成一些在应用启动时就能发现的问题不能被及时发现，而要等到懒加载时才被发现，比如可能会引发以下几个问题：

- JVM 内存配置的问题，因为不是全部加载，所以不能在启动时判断所需内存的大小，在应用真正运行时可能引发内存不足等问题。

- 运行时找不到指定的类等运行时异常，或者是配置错误引发的系列问题，如果能在启动时及时发现问题，就不要等到运行时再去处理。
- 如果是 HTTP 服务，则可能因为临时实例化而造成第一个请求变慢，响应延迟，体验不是很好，对负载均衡和自动伸缩也可能造成影响。

因为懒加载是真正使用的时候才去实例化，所以是否开启全局懒加载完全看应用类型，如果是 HTTP 服务为主的应用，要求即时响应，则不建议开启全局懒加载。

4.5　启动图案

4.5.1　默认图案

Spring Boot 应用在启动的时候会显示一个默认的 Spring 图案：

```
  .   ____          _            __ _ _
 /\\ / ___'_ __ _ _(_)_ __  __ _ \ \ \ \
( ( )\___ | '_ | '_| | '_ \/ _` | \ \ \ \
 \\/  ___)| |_)| | | | | || (_| |  ) ) ) )
  '  |____| .__|_| |_|_| |_\__, | / / / /
 =========|_|==============|___/=/_/_/_/
 :: Spring Boot ::                (v3.0.0)
```

4.5.2　输出模式

Spring Boot 是通过 Banner 接口在日志中输出启动图案的：

```
@FunctionalInterface
public interface Banner {

    /**
    * Print the banner to the specified print stream.
    * @param environment the spring environment
    * @param sourceClass the source class for the application
    * @param out the output print stream
    */
    void printBanner(Environment environment, Class<?> sourceClass, PrintStream out);
```

```
    /**
    * An enumeration of possible values for configuring the Banner.
    */
    enum Mode {

        /**
        * Disable printing of the banner.
        */
        OFF,

        /**
        * Print the banner to System.out.
        */
        CONSOLE,

        /**
        * Print the banner to the log file.
        */
        LOG

    }

}
```

从 Banner.Mode 接口枚举可以看到支持的输出模式：

- OFF：关闭图案。
- CONSOLE（**默认**）：输出到 System.out。
- LOG：输出到日志文件。

所以，如果想关闭启动图案，或者将启动图案输出到日志文件，那么只需要在启动类上设置即可，比如关闭启动图案：

```
public static void main(String[] args) {
    SpringApplication springApplication = new SpringApplication(Application.class);

    // 关闭图案
    springApplication.setBannerMode(Banner.Mode.OFF);

    springApplication.run(args);
}
```

也可以通过以下参数设置：

```
spring:
  main:
    banner-mode: OFF
```

4.5.3　图案实现类

Spring Boot 2.7 中默认有三个实现类，用于不同的实现，类结构如下图所示。

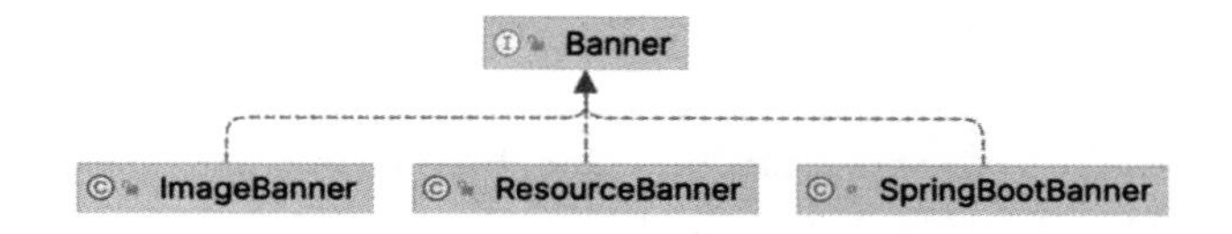

Spring Boot 3.0.0 新变化：

Spring Boot 3.0.0 已经删除了 ImageBanner 图案类，即不再支持 spring.banner.image.* 系列参数配置 Image 图像来作为启动图案，Spring Boot 3.0.0 只支持纯文本形式的启动图案了，类结构如下图所示。

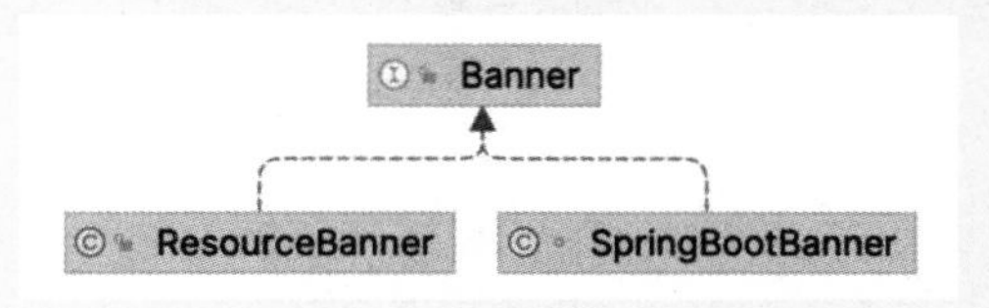

默认的是 SpringBootBanner 实现类，源码如下：

```
class SpringBootBanner implements Banner {

    private static final String[] BANNER = { "", "  .   ____          _            __ _ _",
            " /\\\\ / ___'_ __ _ _(_)_ __  __ _ \\ \\ \\ \\", "( ( )\\___ | '_ | '_| |
'_ \\/ _` | \\ \\ \\ \\",
            " \\\\/  ___)| |_)| | | | | || (_| |  ) ) ) )", "  '  |____| .__|_| |_|_|
|_\\__, | / / / /",
            " =========|_|==============|___/=/_/_/_/" };

    private static final String SPRING_BOOT = " :: Spring Boot :: ";

    private static final int STRAP_LINE_SIZE = 42;

    @Override
```

```
    public void printBanner(Environment environment, Class<?> sourceClass, PrintStream
printStream) {
        for (String line : BANNER) {
            printStream.println(line);
        }
        String version = SpringBootVersion.getVersion();
        version = (version != null) ? " (v" + version + ")" : "";
        StringBuilder padding = new StringBuilder();
        while (padding.length() < STRAP_LINE_SIZE - (version.length() + SPRING_BOOT.
length())) {
            padding.append(" ");
        }

        printStream.println(AnsiOutput.toString(AnsiColor.GREEN, SPRING_BOOT,
AnsiColor.DEFAULT, padding.toString(),
                AnsiStyle.FAINT, version));
        printStream.println();
    }

}
```

从源码可以看到，默认的 Spring 启动图案来源于这个实现类。

4.5.4 自定义图案

如果需要改变应用默认图案，则可以直接在应用资源根目录下创建一个 banner.txt 图案文件：

```
     ____.                             __                 __
    |    |____ ___  _______    _______/  |______    ____ |  | __
    |    \__  \\  \/ /\__  \  /  ___/\   __\__  \_/  ___\|  |/ /
/\__|    |/ __ \\   /  / __ \_\___ \  |  |  / __ \\  \___|    <
\________(____  /\_/  (____  /____  > |__| (____  /\___  >__|_ \
              \/           \/     \/            \/     \/     \/
```

也可以通过 spring.banner.location 参数指定图案文件的位置，如果不是 UTF-8 图案方案，则需要使用 spring.banner.charset 指定具体编码。

另外在 banner.txt 图案文件中，也可以使用环境占位符，比如：

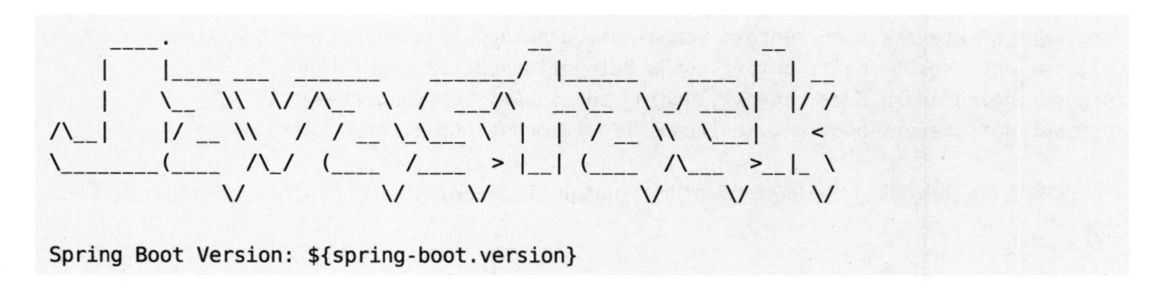
Spring Boot Version: ${spring-boot.version}

启动应用：

```
     ____.                         __                 __
    |    |_____ ___  _______    _______/  |______    ____ |  | __
    |    \__  \\  \/ /\__  \  /  ___/\   __\__  \ _/ ___\|  |/ /
/\__|    |/ __ \\   /  / __ \_\___ \  |  |  / __ \\  \___|    <
\________(____  /\_/  (____  /____  > |__| (____  /\___  >__|_ \
              \/           \/     \/            \/     \/     \/

Spring Boot Version: 3.0.0
```

自定义图案在实际工作中很少使用，不过在启动图案上输出一些应用的特定版本信息还是略有参考意义的。

4.6　启动事件和监听器

Spring Framework 框架中的事件和监听器无处不在，一般来说，我们很少会使用应用的事件和监听器，但我们也不要忘了它们的存在，Spring 框架内部使用了各种不同的事件来处理不同的任务，比如 ContextRefreshedEvent。

除了常用的 Spring 框架中的事件，Spring Boot 在启动过程中还会发送一系列其他的事件，比如 Spring Boot 基础包中的 spring.factories 文件：

```
spring-boot-3.0.0.jar!/META-INF/spring.factories
```

其中注册的所有系统级的监听器列表如下：

```
# Application Listeners
org.springframework.context.ApplicationListener=\
org.springframework.boot.ClearCachesApplicationListener,\
org.springframework.boot.builder.ParentContextCloserApplicationListener,\
org.springframework.boot.context.FileEncodingApplicationListener,\
```

```
org.springframework.boot.context.config.AnsiOutputApplicationListener,\
org.springframework.boot.context.config.DelegatingApplicationListener,\
org.springframework.boot.context.logging.LoggingApplicationListener,\
org.springframework.boot.env.EnvironmentPostProcessorApplicationListener
```

这些监听器都实现了 Spring 框架中的 ApplicationListener 应用监听器接口，监听和处理不同的事件。

4.6.1 启动事件的顺序

当 Spring Boot 应用启动后，事件会按以下顺序发送：

- ApplicationStartingEvent

这个事件在 Spring Boot 应用运行开始时，且进行任何处理之前发送（除了监听器和初始化器注册）。

- ApplicationEnvironmentPreparedEvent

这个事件在当已知要在上下文中使用 Spring 环境 Environment 时，在 Spring 上下文 context 创建之前发送。

- ApplicationContextInitializedEvent

这个事件在 Spring 应用上下文 ApplicationContext 准备好了，并且应用初始化器 ApplicationContextInitializers 已经被调用，在 bean definitions 被加载之前发送。

- ApplicationPreparedEvent

这个事件是在 Spring 上下文 context 刷新之前，且在 bean definitions 被加载之后发送。

- ApplicationStartedEvent

这个事件是在 Spring 上下文 context 刷新之后，且在任何 application/ command-line runners 被调用之前发送。

- AvailabilityChangeEvent

这个事件紧随上个事件之后发送，激活状态 ReadinessState.CORRECT，表示应用已处于活动状态。

- ApplicationReadyEvent

这个事件在任何 application/ command-line runners 调用之后发送。

- AvailabilityChangeEvent

这个事件紧随上个事件之后发送，激活状态 ReadinessState.ACCEPTING_TRAFFIC，表示应用可以开始准备接收请求了。

- ApplicationFailedEvent

这个事件在应用启动异常时发送。

上面介绍的这些事件列表仅包括 SpringApplication 启动类发出的 SpringApplicationEvents 事件，除了这些事件，以下事件也会在 ApplicationPreparedEvent 之后和 ApplicationStartedEvent 之前发送：

- WebServerInitializedEvent

这个 Web 服务器初始化事件在 WebServer 启动之后发送，实现事件类包括 ServletWebServerInitializedEvent（Servlet Web 服务器初始化事件）、ReactiveWebServerInitializedEvent（响应式 Web 服务器初始化事件）。

- ContextRefreshedEvent

这个上下文刷新事件在 Spring 应用上下文 ApplicationContext 刷新之后发送。

> 关于事件的具体发送流程，可见 4.8.2 节中的启动流程源码分析。

4.6.2　自定义事件监听器

前面介绍了 Spring Boot 在启动过程中的各个事件，如果想在每个环节处理一些我们想做的事情，那么只需要自定义一个监听器来监听某个事件就可以了。

比如我们想监听上面的第 8 个启动事件（AvailabilityChangeEvent），即应用启动完成可以接收请求了，我们定义一个该事件的监听器并简单输出一个成功标识。

首先新建一个自定义监听器：

```
@Slf4j
@Component
public class JavastackListener implements ApplicationListener<AvailabilityChangeEvent> {

    @Override
    public void onApplicationEvent(AvailabilityChangeEvent event) {
        log.info(" 监听到事件：" + event);
        if (ReadinessState.ACCEPTING_TRAFFIC == event.getState()){
            log.info(" 应用启动完成，可以接收请求了……");
        }
    }

}
```

监听器需要实现 ApplicationListener 接口，泛型 AvailabilityChangeEvent 表示仅监听 AvailabilityChangeEvent 事件。

方法中的逻辑是，因启动事件 AvailabilityChangeEvent 发送了两次，每次状态不同，所以监听该事件时可以根据事件的状态来区分到底是哪一个环节的事件。

因为是 Spring 应用上下文 ApplicationContext 创建之后的事件，所以可以直接在监听器上使用 @Component 注解注册。如果 ApplicationContext 并未创建，那么这时的 Bean 是不能被加载的，所以这时不能使用 @Component 注解注册，需要在 META-INF/spring.factories 配置文件中注册监听器：

```
org.springframework.context.ApplicationListener=\
cn.javastack.springboot.features.listener.JavastackListener
```

或者可以使用 SpringApplication.addListeners(…) 方法的自动注册。

启动应用来看一下启动日志：

```
... JavastackListener        : 监听到事件：org.springframework.boot.availability.
AvailabilityChangeEvent...
... JavastackListener        : 监听到事件：org.springframework.boot.availability.
AvailabilityChangeEvent...
... JavastackListener        : 应用启动完成，可以接收请求了……
```

可以看到同时输出了两次监听日志，但只输出最后一步启动完成的日志，自定义事件监听器实现成功。

了解了 Spring Boot 启动过程中的各个事件及监听机制，读者可以“依葫芦画瓢”实现 Spring Boot 启动过程中的各个自定义操作，比如在启动过程上实现动态注册、移除 Bean 等。

一般不建议使用事件和监听器实现比较耗时和繁重的任务，这样会影响应用的正常启动，Spring Boot 推荐使用 Runners 进行实现，即在应用启动后再处理，具体见下一节的介绍。

4.7 启动运行器

4.7.1 概述

前面很多章节中都介绍了 Runner 运行器的使用，它允许 Spring Boot 应用启动完成之后，在接收请求之前运行一些特定的代码逻辑。

Spring Boot Runner 运行器可以实现以下两个接口：

- ApplicationRunner。
- CommandLineRunner。

下面看一下两个接口的源码。

```
ApplicationRunner：

@FunctionalInterface
public interface ApplicationRunner {

    /**
    * Callback used to run the bean.
    * @param args incoming application arguments
    * @throws Exception on error
    */
    void run(ApplicationArguments args) throws Exception;

}

CommandLineRunner：

@FunctionalInterface
public interface CommandLineRunner {

    /**
    * Callback used to run the bean.
    * @param args incoming main method arguments
    * @throws Exception on error
    */
    void run(String... args) throws Exception;

}
```

这两个接口的工作原理是一样的，它们都只提供了一个 run 运行方法，只是接收的方法参数不一样而已，ApplicationRunner 用来接收 ApplicationArguments 参数，即 SpringApplication 启动时的应用参数，CommandLineRunner 则用来接收 String 可变参数，所以 CommandLineRunner 的使用要更灵活一些。

4.7.2 使用方式

之前的章节中也都有示例，既可以是传统的实现接口定义方式：

```
@Component
public class JavastackRunner implements CommandLineRunner {

    public void run(String... args) {
        // Do something...
    }

}
```

也可以是结合 Lambda 表达式的 @Bean 方法的定义方式：

```
@Bean
public CommandLineRunner javastackRunner() {
    return (args) -> {
        // Do something...
    };
}
```

一般推荐使用第二种定义方式，简单、清晰，也节约代码和文件量。

另外，如果应用启动时定义了多个 ApplicationRunner 和 CommandLineRunner，则默认是按它们被加载的顺序启动的，如果想控制它们的启动顺序，则可以实现 org.springframework. core.Ordered 接口或者使用 org.springframework.core.annotation.Order 注解，这里不再赘述。

4.8 应用启动流程

前面的章节介绍了一些 Spring Boot 启动相关的扩展应用，本节通过分析启动源码的方式介绍 Spring Boot 应用的具体启动流程，看一下 Spring Boot 启动过程中是如何处理以上扩展应用的。

4.8.1 实例化流程

通过之前的章节我们知道，Spring Boot 应用的启动入口就在 SpringApplication # run 方法中，跟踪 run 方法进入 SpringApplication 最核心的构造方法：

```
public class SpringApplication {
```

```
    ...

    public static ConfigurableApplicationContext run(Class<?> primarySource, String... args) {
        return run(new Class<?>[] { primarySource }, args);
    }

    public static ConfigurableApplicationContext run(Class<?>[] primarySources, String[] args) {
        return new SpringApplication(primarySources).run(args);
    }

    public SpringApplication(Class<?>... primarySources) {
        this(null, primarySources);
    }

    public SpringApplication(ResourceLoader resourceLoader, Class<?>... primarySources) {
        // 1
        this.resourceLoader = resourceLoader;

        // 2
        Assert.notNull(primarySources, "PrimarySources must not be null");

        // 3
        this.primarySources = new LinkedHashSet<>(Arrays.asList(primarySources));

        // 4
        this.webApplicationType = WebApplicationType.deduceFromClasspath();

        // 5
        this.bootstrapRegistryInitializers = new ArrayList<>(
                getSpringFactoriesInstances(BootstrapRegistryInitializer.class));

        // 6
        setInitializers((Collection) getSpringFactoriesInstances(ApplicationContextIni-
tializer.class));

        // 7
        setListeners((Collection) getSpringFactoriesInstances(ApplicationListener.class));

        // 8
        this.mainApplicationClass = deduceMainApplicationClass();
    }

      ...

}
```

1. 设置资源加载器

```
this.resourceLoader = resourceLoader;
```

ResourceLoader 是加载资源的策略接口，默认为 null，不需要指定，如果需要自定义资源加载策略，则可以通过手动 new SpringApplication(...) 的方式实例化 SpringApplication。

2. 判断主资源类是否存在

```
Assert.notNull(primarySources, "PrimarySources must not be null");
```

这个主资源类一般是指应用启动类，如 SpringApplication#run 方法所示。

```
@SpringBootApplication
public class Application {

    public static void main(String[] args) {
        SpringApplication.run(Application.class, args);
    }

}
```

这里的 Application.class 便是主资源类，主资源类可以指定多个，但不能为 null，否则启动报错。

3. 设置主资源类集合

```
this.primarySources = new LinkedHashSet<>(Arrays.asList(primarySources));
```

如果第二步中主资源类不为 null，则使用 LinkedHashSet(List) 构造方式初始化主资源类并去重。

4. 推断并设置 Web 应用类型

```
this.webApplicationType = WebApplicationType.deduceFromClasspath();

deduceFromClasspath 方法的相关源码如下：

static WebApplicationType deduceFromClasspath() {
    if (ClassUtils.isPresent(WEBFLUX_INDICATOR_CLASS, null) && !ClassUtils.isPresent
(WEBMVC_INDICATOR_CLASS, null)
            && !ClassUtils.isPresent(JERSEY_INDICATOR_CLASS, null)) {
        return WebApplicationType.REACTIVE;
    }
```

```
    for (String className : SERVLET_INDICATOR_CLASSES) {
        if (!ClassUtils.isPresent(className, null)) {
            return WebApplicationType.NONE;
        }
    }
    return WebApplicationType.SERVLET;
}
```

根据类路径下是否有对应应用类型的 Class 类推断出对应的 Web 应用类型，Spring Boot 目前支持的 Web 应用类型枚举如下面的 WebApplicationType 类源码所示。

```
public enum WebApplicationType {

    /**
     * The application should not run as a web application and should not start an
     * embedded web server.
     */
    NONE,

    /**
     * The application should run as a servlet-based web application and should start an
     * embedded servlet web server.
     */
    SERVLET,

    /**
     * The application should run as a reactive web application and should start an
     * embedded reactive web server.
     */
    REACTIVE;

    ...

}
```

Spring Boot 应用主要包括 Servlet 和 Reactive 两种 Web 应用类型，现在的应用主要还是以传统的 Servlet 类型为主，比如，加入 spring-boot-starter-web 依赖，应用就能判断出是 Servlet 应用，加入 spring-boot-starter-webflux 依赖，应用就能判断出是 Reactive 应用。

5. 设置引导注册初始化器

```
this.bootstrapRegistryInitializers = new ArrayList<>(
                getSpringFactoriesInstances(BootstrapRegistryInitializer.class));
```

从 spring.factories 文件中获取注册的 BootstrapRegistryInitializer 接口的实现类，该接口的源码如下所示。

```
/**
 * Callback interface that can be used to initialize a {@link BootstrapRegistry} before it
 * is used.
 *
 * @author Phillip Webb
 * @since 2.4.5
 * @see SpringApplication#addBootstrapRegistryInitializer(BootstrapRegistryInitializer)
 * @see BootstrapRegistry
 */
@FunctionalInterface
public interface BootstrapRegistryInitializer {

	/**
	 * Initialize the given {@link BootstrapRegistry} with any required registrations.
	 * @param registry the registry to initialize
	 */
	void initialize(BootstrapRegistry registry);

}
```

BootstrapRegistryInitializer 是一个回调接口，主要用于在后面创建 BootstrapRegistry 实例时对它进行初始化回调操作，如下面的 DefaultBootstrapContext 实例的源码所示。

```
private DefaultBootstrapContext createBootstrapContext() {
    DefaultBootstrapContext bootstrapContext = new DefaultBootstrapContext();
    this.bootstrapRegistryInitializers.forEach((initializer) -> initializer.initialize
(bootstrapContext));
    return bootstrapContext;
}
```

DefaultBootstrapContext 是 BootstrapContext 和 BootstrapRegistry 接口默认的实例，类结构所下图所示。

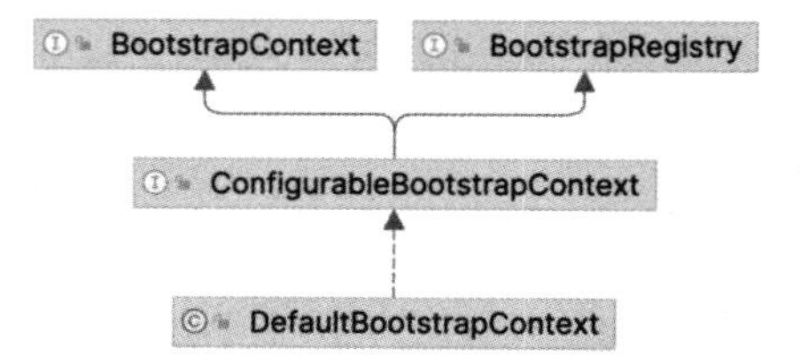

DefaultBootstrapContext 实现的两个重要接口：

- BootstrapContext：它是一个简单的引导上下文接口，在应用启动期间和环境后处理（Environment post-processing）期间可用，直到准备好 ApplicationContext 为止，对创建成本很高或者需要在 ApplicationContext 可用之前共享的单例对象提供延迟访问。
- BootstrapRegistry：它是一个简单的对象注册表接口，在应用启动期间和环境后处理（Environment post-processing）期间可用，直到准备好 ApplicationContext 为止，用于注册创建成本较高的实例，或者在 ApplicationContext 可用之前需要共享的实例。

虽然 createBootstrapContext 方法有初始化 BootstrapRegistry 的回调逻辑，但据笔者研究，从 Spring Boot 2.7 到 3.0.0，这个 BootstrapRegistryInitializer 接口目前还没有任何实现类，也没有在 spring.factories 文件中注册任何实现类，断点调试所下图所示。

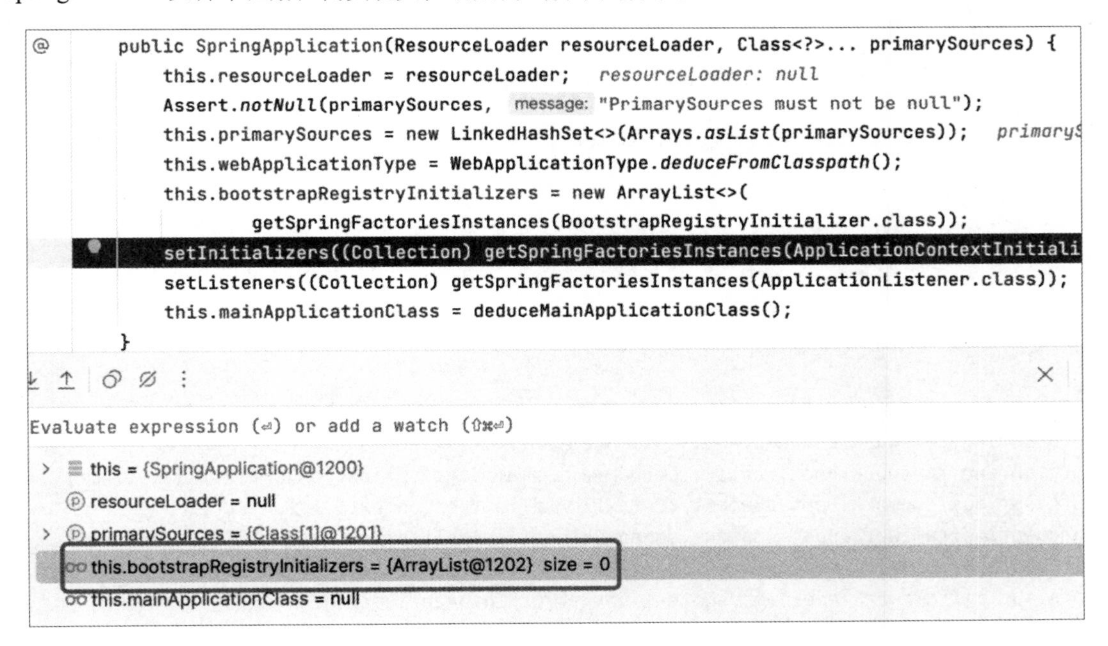

这个可能是保留逻辑，有待后续的版本更新。

6. 设置应用上下文初始化器

```
setInitializers((Collection) getSpringFactoriesInstances(ApplicationContextInitializer.class));
```

应用上下文初始化器 ApplicationContextInitializer 接口的源码如下所示。

```
/**
 * Callback interface for initializing a Spring {@link ConfigurableApplicationContext}
 * prior to being {@linkplain ConfigurableApplicationContext#refresh() refreshed}.
 *
 * <p>Typically used within web applications that require some programmatic initialization
 * of the application context. For example, registering property sources or activating
 * profiles against the {@linkplain ConfigurableApplicationContext#getEnvironment()
 * context's environment}. See {@code ContextLoader} and {@code FrameworkServlet} support
 * for declaring a "contextInitializerClasses" context-param and init-param, respectively.
 * ...
 */
@FunctionalInterface
public interface ApplicationContextInitializer<C extends ConfigurableApplicationContext> {

	/**
	 * Initialize the given application context.
	 * @param applicationContext the application to configure
	 */
	void initialize(C applicationContext);

}
```

这是一个回调接口，主要用来在 Spring 应用上下文 ConfigurableApplicationContext 刷新之前进行一些初始化操作，如注册属性资源、激活 Profile 等，它在 Spring Boot 各个包中都注册了多个不同的实现，如 Spring Boot 基础包中的 spring.factories 文件：

```
# Application Context Initializers
org.springframework.context.ApplicationContextInitializer=\
org.springframework.boot.context.ConfigurationWarningsApplicationContextInitializer,\
org.springframework.boot.context.ContextIdApplicationContextInitializer,\
org.springframework.boot.context.config.DelegatingApplicationContextInitializer,\
org.springframework.boot.rsocket.context.RSocketPortInfoApplicationContextInitializer,\
org.springframework.boot.web.context.ServerPortInfoApplicationContextInitializer
```

如 Spring Boot 自动配置包中的 spring.factories 文件：

```
# Initializers
org.springframework.context.ApplicationContextInitializer=\
org.springframework.boot.autoconfigure.SharedMetadataReaderFactoryContextInitializer,\
org.springframework.boot.autoconfigure.logging.ConditionEvaluationReportLoggingListener
```

一共注册了 8个应用上下文初始化器，这一点从断点调试也可以看出来，如下图所示。

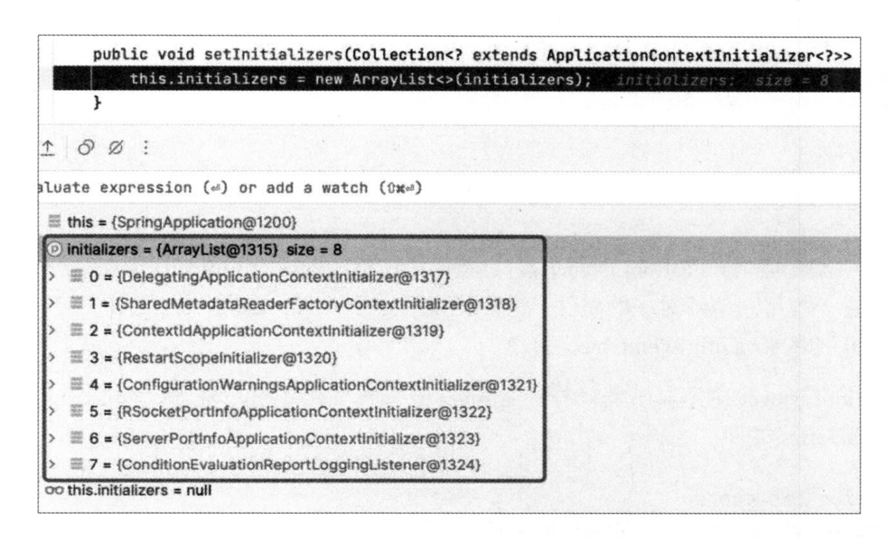

获取 spring.factories 配置文件的原理在 Spring Boot 自动配置章节中有详细讲解，这里不再赘述。

7. 设置应用事件监听器

```
setListeners((Collection) getSpringFactoriesInstances(ApplicationListener.class));
```

Spring 中的事件监听器是 ApplicationListener 接口，是所有 Spring 事件的抽象，如下面的源码所示。

```
@FunctionalInterface
public interface ApplicationListener<E extends ApplicationEvent> extends EventListener {

    /**
    * Handle an application event.
    * @param event the event to respond to
    */
    void onApplicationEvent(E event);


    /**
    * Create a new {@code ApplicationListener} for the given payload consumer.
    * @param consumer the event payload consumer
    * @param <T> the type of the event payload
    * @return a corresponding {@code ApplicationListener} instance
    * @since 5.3
    * @see PayloadApplicationEvent
    */
```

```
    static <T> ApplicationListener<PayloadApplicationEvent<T>> forPayload(Consumer<T> consumer) {
        return event -> consumer.accept(event.getPayload());
    }

}
```

这个事件监听器ApplicationListener接口继承了JDK的java.util.EventListener接口，实现了观察者模式，它一般用来监听对应的事件，事件类型限定于ApplicationEvent及其子接口，这个接口同样继承了JDK的java.util.EventObject接口。

ApplicationListener应用监听器同样在spring.factories文件中进行注册，如Spring Boot基础包中的spring.factories文件：

```
# Application Listeners
org.springframework.context.ApplicationListener=\
org.springframework.boot.ClearCachesApplicationListener,\
org.springframework.boot.builder.ParentContextCloserApplicationListener,\
org.springframework.boot.context.FileEncodingApplicationListener,\
org.springframework.boot.context.config.AnsiOutputApplicationListener,\
org.springframework.boot.context.config.DelegatingApplicationListener,\
org.springframework.boot.context.logging.LoggingApplicationListener,\
org.springframework.boot.env.EnvironmentPostProcessorApplicationListener
```

除此之外，在Spring Boot所有核心包中一共注册了11个事件监听器，这一点从断点调试也可以看出来，如下图所示。

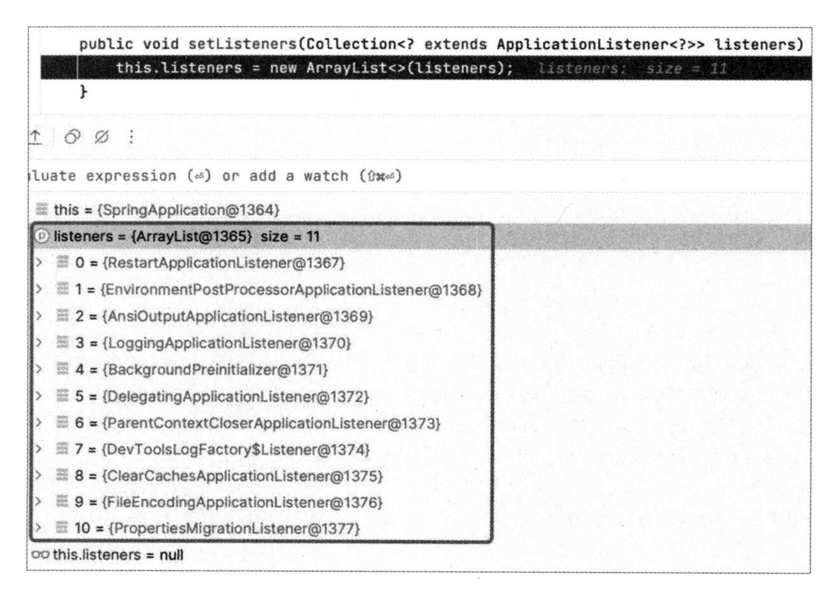

每个监听器监听的事件和启动阶段可能不一样，这一点在 4.6 节中介绍过了，这里不再赘述，但是在后续的启动过程分析中可以看到监听的事件是从哪个启动阶段发出来的。

8. 推断并设置应用启动类

```
this.mainApplicationClass = deduceMainApplicationClass();

deduceMainApplicationClass 方法的源码如下:

private Class<?> deduceMainApplicationClass() {
    return StackWalker.getInstance(StackWalker.Option.RETAIN_CLASS_REFERENCE).walk
(this::findMainClass)
            .orElse(null);
}

private Optional<Class<?>> findMainClass(Stream<StackFrame> stack) {
    return stack.filter((frame) -> Objects.equals(frame.getMethodName(), "main"))
.findFirst()
            .map(StackWalker.StackFrame::getDeclaringClass);
}
```

推断应用启动类的方式有点特别，也很新颖，和 Spring Boot 2.7 中的实现方式完全不同，Spring Boot 3.0.0 使用了最新的 Java 9+ 中的 StackWalker 栈帧跟踪技术，先创建一个包含 Class 对象的 StackWalker 实例，然后筛选出第一个包含 main 方法的 Class 类。

最后总结一下 SpringApplication 实例化的大概流程，如下图所示。

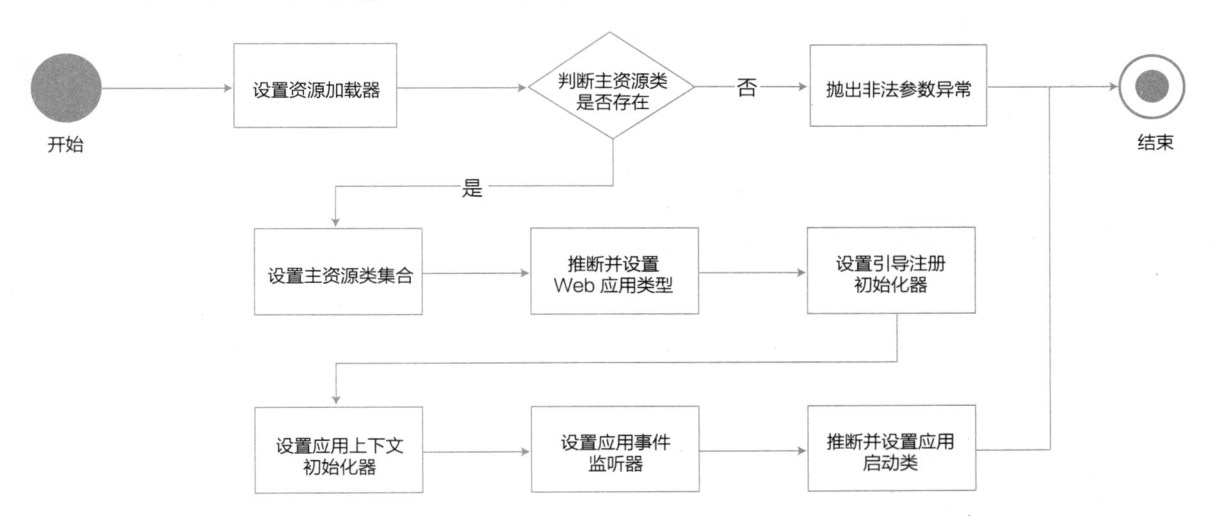

4.8.2 启动流程

前面介绍了 SpringApplication 构造器的实例化流程，它只是初始化一些应用的必要信息，下面进入真正的应用启动流程，Spring Boot 3.0.0 和 Spring Boot 2.7 又略有不同，如下面最新的 Spring Boot 3.0.0 的 SpringApplication#run 方法的源码所示。

```
public ConfigurableApplicationContext run(String... args) {
    // 1
    long startTime = System.nanoTime();

    // 2
    DefaultBootstrapContext bootstrapContext = createBootstrapContext();
    ConfigurableApplicationContext context = null;

    // 3
    configureHeadlessProperty();

    // 4
    SpringApplicationRunListeners listeners = getRunListeners(args);
    listeners.starting(bootstrapContext, this.mainApplicationClass);

    try {
        // 5
        ApplicationArguments applicationArguments = new DefaultApplicationArguments(args);

        // 6
        ConfigurableEnvironment environment = prepareEnvironment(listeners,
bootstrapContext, applicationArguments);

        // 7
        Banner printedBanner = printBanner(environment);

        // 8
        context = createApplicationContext();
        context.setApplicationStartup(this.applicationStartup);

        // 9
        prepareContext(bootstrapContext, context, environment, listeners,
applicationArguments, printedBanner);

        // 10
        refreshContext(context);
```

```
        // 11
        afterRefresh(context, applicationArguments);

        // 12
        Duration timeTakenToStartup = Duration.ofNanos(System.nanoTime() - startTime);
        if (this.logStartupInfo) {
            new StartupInfoLogger(this.mainApplicationClass).logStarted
(getApplicationLog(), timeTakenToStartup);
        }
        listeners.started(context, timeTakenToStartup);

        // 13
        callRunners(context, applicationArguments);
    }
    catch (Throwable ex) {
        if (ex instanceof AbandonedRunException) {
            throw ex;
        }
        handleRunFailure(context, ex, listeners);
        throw new IllegalStateException(ex);
    }
    try {
        // 14
        if (context.isRunning()) {
            Duration timeTakenToReady = Duration.ofNanos(System.nanoTime() - startTime);
            listeners.ready(context, timeTakenToReady);
        }
    }
    catch (Throwable ex) {
        if (ex instanceof AbandonedRunException) {
            throw ex;
        }
        handleRunFailure(context, ex, null);
        throw new IllegalStateException(ex);
    }
    return context;
}
```

1. 初始化启动开始时间

```
long startTime = System.nanoTime();
```

初始化一个开始时间，通过 System 类获取，单位为纳秒。

2. 初始化默认引导上下文

```
DefaultBootstrapContext bootstrapContext = createBootstrapContext();
ConfigurableApplicationContext context = null;
```

创建一个默认的引导上下文 DefaultBootstrapContext，源码如下：

```
private DefaultBootstrapContext createBootstrapContext() {
    DefaultBootstrapContext bootstrapContext = new DefaultBootstrapContext();
    this.bootstrapRegistryInitializers.forEach((initializer) -> initializer.initialize
(bootstrapContext));
    return bootstrapContext;
}
```

具体内容在前面实例化流程一节中介绍过了，这里不再赘述。

3. 设置 Headless 模式

```
configureHeadlessProperty();
```

设置 Headless 模式，该方法的源码如下：

```
private static final String SYSTEM_PROPERTY_JAVA_AWT_HEADLESS = "java.awt.headless";

private boolean headless = true;

private void configureHeadlessProperty() {
    System.setProperty(SYSTEM_PROPERTY_JAVA_AWT_HEADLESS,
            System.getProperty(SYSTEM_PROPERTY_JAVA_AWT_HEADLESS, Boolean.toString
(this.headless)));
}
```

这里的逻辑是，如果没有设置 java.awt.headless 参数的值，则将它默认设置为 true。

> **为什么要设置 java.awt.headless = true？**
> AWT 是 Java 生态中用于构建 GUI 应用的标准 API 接口，一般用于 Java 标准模式下。而 Headless 是 Java 中另一套配置模式，指的是没有显示屏、鼠标或者键盘时的系统模式，当设置 java.awt.headless = true 时，可以强制应用使用 Headless 版本的 AWT 实现类，从而避免缺失相应的设备而导出应用出错，后端服务器一般都没有这些设备，所以后端应用一般都要将这个参数设置为 true。

4. 初始化启动方法监听器

```
SpringApplicationRunListeners listeners = getRunListeners(args);
listeners.starting(bootstrapContext, this.mainApplicationClass);
```

从 spring.factories 文件中获取所有 SpringApplicationRunListener 类型的监听器，并封装成 SpringApplicationRunListeners 监听器包装类，SpringApplicationRunListener 监听器目前只有一个实现类，类结构如下图所示。

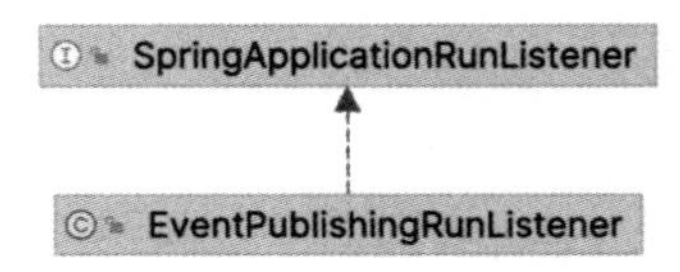

SpringApplicationRunListeners 包装类主要组合了以下信息：

- Log：日志对象。
- List<SpringApplicationRunListener>：SpringApplicationRunListener 监听器集合。
- ApplicationStartup：应用启动管理类，它能步骤化监视应用的启动阶段。

接下来调用 SpringApplicationRunListeners#starting 方法，如下面的源码所示。

```
class SpringApplicationRunListeners {

    private final Log log;

      private final List<SpringApplicationRunListener> listeners;

      private final ApplicationStartup applicationStartup;

    ...

    void starting(ConfigurableBootstrapContext bootstrapContext, Class<?>
mainApplicationClass) {
        doWithListeners("spring.boot.application.starting", (listener) -> listener.
starting (bootstrapContext),
                (step) -> {
                    if (mainApplicationClass != null) {
                        step.tag("mainApplicationClass", mainApplicationClass.getName());
                    }
```

```
                });
    }

    private void doWithListeners(String stepName, Consumer<SpringApplicationRunListener>
listenerAction,
            Consumer<StartupStep> stepAction) {
        StartupStep step = this.applicationStartup.start(stepName);
        this.listeners.forEach(listenerAction);
        if (stepAction != null) {
            stepAction.accept(step);
        }
        step.end();
    }

    ...

}
```

先标记启动步骤，然后执行 EventPublishingRunListener 监听器的 starting 方法，如下面的源码所示。

```
class EventPublishingRunListener implements SpringApplicationRunListener, Ordered {

    private final SpringApplication application;

    private final String[] args;

    private final SimpleApplicationEventMulticaster initialMulticaster;

    EventPublishingRunListener(SpringApplication application, String[] args) {
        this.application = application;
        this.args = args;
        this.initialMulticaster = new SimpleApplicationEventMulticaster();
    }

    ...

    @Override
    public void starting(ConfigurableBootstrapContext bootstrapContext) {
        multicastInitialEvent(new ApplicationStartingEvent(bootstrapContext,
this.application, this.args));
    }
```

```
    ...
}
```

如上所示，这里发布了 ApplicationStartingEvent 应用启动事件。

5. 初始化应用参数类

```
ApplicationArguments applicationArguments = new DefaultApplicationArguments(args);
```

先通过 main 方法提供的入口参数构建一个默认的 DefaultApplicationArguments 应用参数类，后续可以通过这个类来获取应用参数。

6. 准备 Spring 环境

```
ConfigurableEnvironment environment = prepareEnvironment(listeners, bootstrapContext,
applicationArguments);
```

准备 Spring 环境，prepareEnvironment 方法的源码如下：

```
private ConfigurableEnvironment prepareEnvironment(SpringApplicationRunListeners listeners,
        DefaultBootstrapContext bootstrapContext, ApplicationArguments applicationArgu-
ments) {
    // (1)
    // Create and configure the environment
    ConfigurableEnvironment environment = getOrCreateEnvironment();

    // (2)
    configureEnvironment(environment, applicationArguments.getSourceArgs());
    ConfigurationPropertySources.attach(environment);

    // (3)
    listeners.environmentPrepared(bootstrapContext, environment);
    DefaultPropertiesPropertySource.moveToEnd(environment);
    Assert.state(!environment.containsProperty("spring.main.environment-prefix"),
            "Environment prefix cannot be set via properties.");

    // (4)
    bindToSpringApplication(environment);
```

```
    // (5)
    if (!this.isCustomEnvironment) {
        EnvironmentConverter environmentConverter = new EnvironmentConverter
(getClassLoader());
        environment = environmentConverter.convertEnvironmentIfNecessary(environment,
deduceEnvironmentClass());
    }
    ConfigurationPropertySources.attach(environment);
    return environment;
}
```

这里主要是根据上面获得的 listeners、bootstrapContext、applicationArguments 再去准备 Spring Environment 环境，大概分为以下几个步骤：

（1）创建对应 Web类型的 Spring Environment 环境。

（2）把 ApplicationArguments 参数类及其他应用参数设置到 Spring Environment 环境中。

（3）调用监听器的 environmentPrepared 方法，发布 ApplicationEnvironmentPreparedEvent 应用环境已准备好事件。

（4）把 Spring Environment 环境绑定到 SpringApplication 类。

（5）非自定义的 Spring Environment 环境，如果有需要还会进行环境转换。

7. 打印启动图案

```
private Banner printBanner(ConfigurableEnvironment environment) {
    if (this.bannerMode == Banner.Mode.OFF) {
        return null;
    }
    ResourceLoader resourceLoader = (this.resourceLoader != null) ? this.resourceLoader
            : new DefaultResourceLoader(null);
    SpringApplicationBannerPrinter bannerPrinter = new SpringApplicationBannerPrinter
(resourceLoader, this.banner);
    if (this.bannerMode == Mode.LOG) {
        return bannerPrinter.print(environment, this.mainApplicationClass, logger);
    }
    return bannerPrinter.print(environment, this.mainApplicationClass, System.out);
}
```

如果应用开启了打印启动图案的功能，则创建一个默认的 DefaultResourceLoader 资源加载器，用于加载类路径下的 Banner 图案文件，然后封装成一个 SpringApplicationBannerPrinter 图案打印类，判断不同的图案输出模式，最后打印 Spring Boot 启动图案。如果没有自定义启动图案，则使用的是 SpringBootBanner 实现类中的默认图案，如下所示。

```
  .   ____          _            __ _ _
 /\\ / ___'_ __ _ _(_)_ __  __ _ \ \ \ \
( ( )\___ | '_ | '_| | '_ \/ _` | \ \ \ \
 \\/  ___)| |_)| | | | | || (_| |  ) ) ) )
  '  |____| .__|_| |_|_| |_\__, | / / / /
 =========|_|==============|___/=/_/_/_/
 :: Spring Boot ::                (v3.0.0)
```

8. 创建 Spring 上下文

```
context = createApplicationContext();
context.setApplicationStartup(this.applicationStartup);
```

先创建一个对应 Web类型的 Spring 上下文，然后创建一个默认的 ApplicationStartup 应用启动管理类并设置到该上下文对象中。Spring 上下文是通过 ApplicationContextFactory 工厂创建的，如下面的 createApplicationContext 方法的源码所示。

```
protected ConfigurableApplicationContext createApplicationContext() {
    return this.applicationContextFactory.create(this.webApplicationType);
}
```

ApplicationContextFactory 上下文工厂有三种实现类：

- DefaultApplicationContextFactory：默认的应用上下文工厂类。
- ServletWebServerApplicationContextFactory：Servlet 的应用上下文工厂类。
- ReactiveWebServerApplicationContextFactory：Reactive 的应用上下文工厂类。

类结构如下图所示。

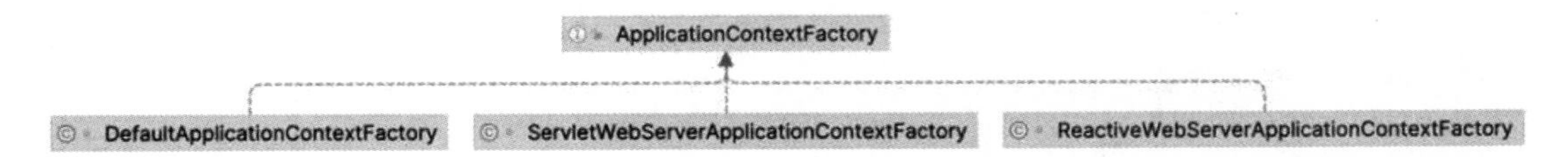

Spring Boot默认为 DefaultApplicationContextFactory 工厂类，由它来驱动其他两个不同类型的上下文工厂类，具体见下面的源码分析：

```
public class SpringApplication {

    ...

    private ApplicationContextFactory applicationContextFactory = ApplicationContextFactory.DEFAULT;

    ...

}
```

DefaultApplicationContextFactory 创建上下文对象的核心源代码如下所示。

```
class DefaultApplicationContextFactory implements ApplicationContextFactory {

    ...

    @Override
    public ConfigurableApplicationContext create(WebApplicationType webApplicationType) {
        try {
            return getFromSpringFactories(webApplicationType, ApplicationContextFactory::
                    create, this::createDefaultApplicationContext);
        }
        catch (Exception ex) {
            throw new IllegalStateException("Unable create a default ApplicationContext
                    instance, "+"you may need a custom ApplicationContextFactory", ex);
        }
    }

    ...

    private <T> T getFromSpringFactories(WebApplicationType webApplicationType,
            BiFunction<ApplicationContextFactory, WebApplicationType, T> action,
Supplier<T> defaultResult) {
        for (ApplicationContextFactory candidate : SpringFactoriesLoader.loadFactories
(ApplicationContextFactory.class,
                getClass().getClassLoader())) {
            T result = action.apply(candidate, webApplicationType);
            if (result != null) {
                return result;
            }
```

```
        }
        return (defaultResult != null) ? defaultResult.get() : null;
    }

}
```

核心逻辑就是，DefaultApplicationContextFactory 工厂在创建上下文时会加载 Spring Boot 基础包的 spring.factories 文件中的 ApplicationContextFactory 注册的工厂实现类，具体如下所示。

```
# Application Context Factories
org.springframework.boot.ApplicationContextFactory=\
org.springframework.boot.web.reactive.context.ReactiveWebServerApplicationContextFactory,\
org.springframework.boot.web.servlet.context.ServletWebServerApplicationContextFactory
```

然后循环所有注册的 ApplicationContextFactory 工厂实现类，再通过 ApplicationContextFactory::create 方法引用一一判断当前 Web 应用类型是否成功创建上下文对象，如果创建成功则直接返回，否则返回默认的上下文对象。因为这里是 Servlet 应用，所以它会成功匹配 ServletWebServerApplicationContextFactory 上下文工厂，如该工厂的源码所示。

```
class ServletWebServerApplicationContextFactory implements ApplicationContextFactory {

    ...

    @Override
    public ConfigurableApplicationContext create(WebApplicationType webApplicationType) {
        return (webApplicationType != WebApplicationType.SERVLET) ? null : createContext();
    }

    private ConfigurableApplicationContext createContext() {
        if (!AotDetector.useGeneratedArtifacts()) {
            return new AnnotationConfigServletWebServerApplicationContext();
        }
        return new ServletWebServerApplicationContext();
    }

}
```

因为这里不涉及 AOT 技术，所以它最终会创建并返回一个 AnnotationConfigServletWebServerApplicationContext 上下文对象。

9. 准备 Spring上下文

```
prepareContext(bootstrapContext, context, environment, listeners, applicationArguments,
printedBanner);
```

到了准备 Spring 上下文阶段了，prepareContext 方法的源码如下所示。

```
private void prepareContext(DefaultBootstrapContext bootstrapContext,
ConfigurableApplicationContext context,
        ConfigurableEnvironment environment, SpringApplicationRunListeners listeners,
        ApplicationArguments applicationArguments, Banner printedBanner) {
    // (1)
    context.setEnvironment(environment);
    postProcessApplicationContext(context);
    addAotGeneratedInitializerIfNecessary(this.initializers);

    // (2)
    applyInitializers(context);

    // (3)
    listeners.contextPrepared(context);

    // (4)
    bootstrapContext.close(context);
    if (this.logStartupInfo) {
        logStartupInfo(context.getParent() == null);
        logStartupProfileInfo(context);
    }

    // (5)
    // Add boot specific singleton beans
    ConfigurableListableBeanFactory beanFactory = context.getBeanFactory();
    beanFactory.registerSingleton("springApplicationArguments", applicationArguments);
    if (printedBanner != null) {
        beanFactory.registerSingleton("springBootBanner", printedBanner);
    }
    if (beanFactory instanceof AbstractAutowireCapableBeanFactory autowireCapableBean-
Factory) {
        autowireCapableBeanFactory.setAllowCircularReferences(this.allowCircularReferences);
        if (beanFactory instanceof DefaultListableBeanFactory listableBeanFactory) {
```

```
            listableBeanFactory.setAllowBeanDefinitionOverriding (this.allowBeanDefinition
Overriding);
        }
    }

    // (6)
    if (this.lazyInitialization) {
        context.addBeanFactoryPostProcessor(new LazyInitializationBeanFactoryPostProcessor());
    }
    context.addBeanFactoryPostProcessor(new PropertySourceOrderingBeanFactoryPostProcessor
(context));

    // (7)
    if (!AotDetector.useGeneratedArtifacts()) {
        // Load the sources
        Set<Object> sources = getAllSources();
        Assert.notEmpty(sources, "Sources must not be empty");
        load(context, sources.toArray(new Object[0]));
    }

    // (8)
    listeners.contextLoaded(context);
}
```

这里的逻辑比较复杂，笔者简单概括一下。

（1）绑定 Spring环境到 Spring 应用上下文，以及相关后置处理。

（2）遍历所有 ApplicationContextInitializer 上下文初始化器，逐个初始化 Spring 上下文。

（3）调用 SpringApplicationRunListeners#contextPrepared 方法，发布 ApplicationContext-InitializedEvent 应用上下文初始化事件。

（4）调用 DefaultBootstrapContext#close 方法，发布 BootstrapContextClosedEvent 引导上下文关闭事件。

（5）获取 BeanFactory，先注册 springApplicationArguments、springBootBanner 注入单例对象，然后设置是否允许循环引用、是否允许覆盖 Bean 参数。

（6）为 Spring 上下文添加 BeanFactoryPostProcessor 相关实现类，此接口一般可用来扩展或者修改 Bean 的定义，比如这里添加了 Bean 懒加载处理器及 PropertySource 重排序处理器。

（7）如果是非 AOT 模式，则获取所有资源类并把 Beans 加载到上下文环境中。

（8）调用 SpringApplicationRunListeners#contextLoaded 方法，发布 ApplicationPreparedEvent 应用已准备好事件。

10. 刷新 Spring 上下文

```
refreshContext(context);
```

Spring 上下文准备好之后就进入上下文刷新阶段了，refreshContext 方法的源码如下所示。

```
public class SpringApplication {

    ...

    private boolean registerShutdownHook = true;

    static final SpringApplicationShutdownHook shutdownHook = new SpringApplicationShutdownHook();

    private void refreshContext(ConfigurableApplicationContext context) {
        if (this.registerShutdownHook) {
            shutdownHook.registerApplicationContext(context);
        }
        refresh(context);
    }

    protected void refresh(ConfigurableApplicationContext applicationContext) {
        applicationContext.refresh();
    }

    ...

}
```

SpringApplicationShutdownHook 是一个实现了 Runnable 接口的线程类，它是优雅关闭 Spring Boot 应用时的一个关闭钩子，这个关闭钩子在这里将 Spring 上下文进行了注册，默认为允许注册。

然后 refresh 方法再调用 ApplicationContext#refresh 方法对应用上下文进行刷新，Servlet 应用则进入 AnnotationConfigServletWebServerApplicationContext 父类 ServletWebServerApplication-Context#refresh 方法，如下面的源码所示。

```
public class ServletWebServerApplicationContext extends GenericWebApplicationContext
        implements ConfigurableWebServerApplicationContext {

    ...

    @Override
    public final void refresh() throws BeansException, IllegalStateException {
        try {
            super.refresh();
        }
        catch (RuntimeException ex) {
            WebServer webServer = this.webServer;
            if (webServer != null) {
                webServer.stop();
            }
            throw ex;
        }
    }

    @Override
    protected void onRefresh() {
        super.onRefresh();
        try {
            createWebServer();
        }
        catch (Throwable ex) {
            throw new ApplicationContextException("Unable to start web server", ex);
        }
    }

    ...

}
```

执行 Spring 上下文的刷新操作，这里会通过重写 Spring 中的 AbstractApplicationContext 抽象类的 onRefresh 方法来创建一个 Web Server 服务器，并发布 ContextRefreshedEvent、WebServerInitializedEvent 事件，如果创建成功后发生异常则立即停止运行 Web Server 服务器。

11. 刷新 Spring 上下文之后

```
afterRefresh(context, applicationArguments);
```

这是刷新 Spring 上下文之后的动作，该 afterRefresh 方法目前还没有任何实现逻辑，该方法的源码如下所示。

```
protected void afterRefresh(ConfigurableApplicationContext context, ApplicationArguments args) {
}
```

可以看到 afterRefresh 方法是 protected 类型，说明可以继承 SpringApplication 类来自定义实现这个逻辑。

12. 应用启动成功

```
Duration timeTakenToStartup = Duration.ofNanos(System.nanoTime() - startTime);
if (this.logStartupInfo) {
    new StartupInfoLogger(this.mainApplicationClass).logStarted(getApplicationLog(),
timeTakenToStartup);
}
listeners.started(context, timeTakenToStartup);
```

到此，应用已经启动成功，先计算出应用的启动总耗时，再输出应用启动日志，最后调用 SpringApplicationRunListeners#started 方法，发布 ApplicationStartedEvent 应用已成功启动事件。

13. 运行 Spring Boot Runner

```
callRunners(context, applicationArguments);
```

应用启动成功之后，再执行所有的 Spring Boot Runner 运行器，如该方法的源码所示。

```
private void callRunners(ApplicationContext context, ApplicationArguments args) {
    List<Object> runners = new ArrayList<>();
    runners.addAll(context.getBeansOfType(ApplicationRunner.class).values());
    runners.addAll(context.getBeansOfType(CommandLineRunner.class).values());
    AnnotationAwareOrderComparator.sort(runners);
    for (Object runner : new LinkedHashSet<>(runners)) {
        if (runner instanceof ApplicationRunner applicationRunner) {
            callRunner(applicationRunner, args);
        }
```

```
        if (runner instanceof CommandLineRunner commandLineRunner) {
            callRunner(commandLineRunner, args);
        }
    }
}

private void callRunner(ApplicationRunner runner, ApplicationArguments args) {
    try {
        (runner).run(args);
    }
    catch (Exception ex) {
        throw new IllegalStateException("Failed to execute ApplicationRunner", ex);
    }
}

private void callRunner(CommandLineRunner runner, ApplicationArguments args) {
    try {
        (runner).run(args.getSourceArgs());
    }
    catch (Exception ex) {
        throw new IllegalStateException("Failed to execute CommandLineRunner", ex);
    }
}
```

先将所有的 ApplicationRunner 和 CommandLineRunner 实例添加到一个集合中，再排序并去重，最后遍历该集合并依次调用其 run 方法。

14. 应用已就绪

```
if (context.isRunning()) {
    Duration timeTakenToReady = Duration.ofNanos(System.nanoTime() - startTime);
    listeners.ready(context, timeTakenToReady);
}
```

检查 Spring 上下文是否正在运行中，如果是，则调用 SpringApplicationRunListeners#ready 方法，并发布以下两个事件：

- ApplicationReadyEvent：应用已就绪事件。
- AvailabilityChangeEvent：应用可用性变更事件，可用状态变更为 ReadinessState.ACCEPTING_TRAFFIC，即应用可以接收请求了。

15. 启动流程异常处理

```
catch (Throwable ex) {
    if (ex instanceof AbandonedRunException) {
        throw ex;
    }
    handleRunFailure(context, ex, listeners);
    throw new IllegalStateException(ex);
}
```

在执行上述启动流程中，会捕捉 Throwable 总异常，如下面的 handleRunFailure 方法的源码所示。

```
private void handleRunFailure(ConfigurableApplicationContext context, Throwable exception,
        SpringApplicationRunListeners listeners) {
    try {
        try {
            // (1)
            handleExitCode(context, exception);

            // (2)
            if (listeners != null) {
                listeners.failed(context, exception);
            }
        }
        // (3)
        finally {
            reportFailure(getExceptionReporters(context), exception);
            if (context != null) {
                context.close();
                shutdownHook.deregisterFailedApplicationContext(context);
            }
        }
    }
    catch (Exception ex) {
        logger.warn("Unable to close ApplicationContext", ex);
    }
    // (4)
    ReflectionUtils.rethrowRuntimeException(exception);
}
```

大概的失败处理逻辑如下：

（1）处理 ExitCode 退出错误码，发布 ExitCodeEvent 错误码事件。

（2）调用 SpringApplicationRunListeners#failed 方法，发布 ApplicationFailedEvent 应用启动失败事件。

（3）finally 部分先记录失败异常，再关闭 Spring 上下文并取消与之相关的优雅关闭的钩子。

（4）将运行时异常抛出。

最后总结一下 Spring Boot 应用启动的大概流程，如下图所示。

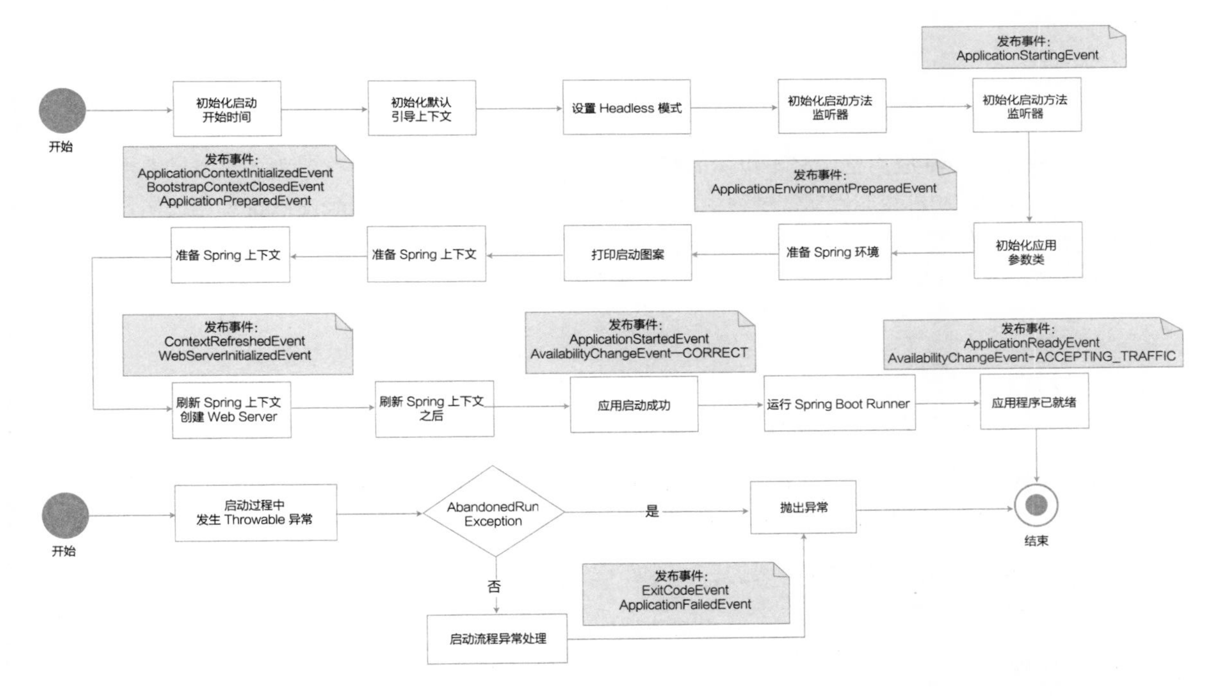

Spring Boot 启动过程中主要创建了内嵌式的 Web Server 服务器，然后结合 Spring 环境集成了丰富的启动特性，底层还运用了大量的启动事件和监听器，以完成一些特定启动环节的逻辑。

第5章 Spring Boot日志管理

日志是每个Java应用框架必不可少的基本组件，也代表着应用的运行轨迹，输出日志可以帮助开发者有效追踪和排查应用运行过程中出现的疑难杂症。本章会介绍Spring Boot如何使用日志，包括支持的日志框架、日志格式、日志级别、日志分组、日志归档、自定义日志配置文件，以及切换其他日志框架及版本、输入彩色日志、日志关闭钩子等。

5.1 概述

Spring Boot框架内部所有日志的记录使用的是Apache开源的Commons Logging日志框架，它只是一种日志门面，提供了一套日志规范接口，底层的日志框架实现是可以自由切换的。

Spring Boot提供的几种日志框架的启动器如下表所示。

日志框架	启动器
Java Util Logging	spring-boot-starter-logging
Logback	spring-boot-starter-logging
Log4j2	spring-boot-starter-log4j2

其中spring-boot-starter-logging日志启动器是Spring Boot Starter启动器默认集成的，无须自行导入，这个可以从任何一个包括spring-boot-starter的启动器包中找到，比如spring-boot-starter-web依赖了spring-boot-starter，而spring-boot-starter就会依赖spring-boot-starter-logging：

```
<dependency>
  <groupId>org.springframework.boot</groupId>
  <artifactId>spring-boot-starter-logging</artifactId>
```

```
  <version>3.0.0</version>
  <scope>compile</scope>
</dependency>
```

所以，任何一个 spring-boot-starter-* 启动器都会默认导入一个 spring-boot-starter-logging 启动器依赖，如果引入了其他官方的 Spring Boot Starter，则无须自行导入 spring-boot-starter-logging 启动器依赖。

Spring Boot 默认集成的是 Logback 日志框架，因为默认的 spring-boot-starter-logging 日志启动器使用的是 Logback 日志框架，并为 Logback 提供了支持 Java Util Logging、Commons Logging、Log4J、SLF4J 的桥接器以便能从这些日志门面中自由切换，因此项目中不管使用哪个日志门面，Logback 都能正常工作。

如下图所示，可以从 spring-boot-starter-web 依赖树中看到它包含了默认日志框架 Logback 及其他桥接器。

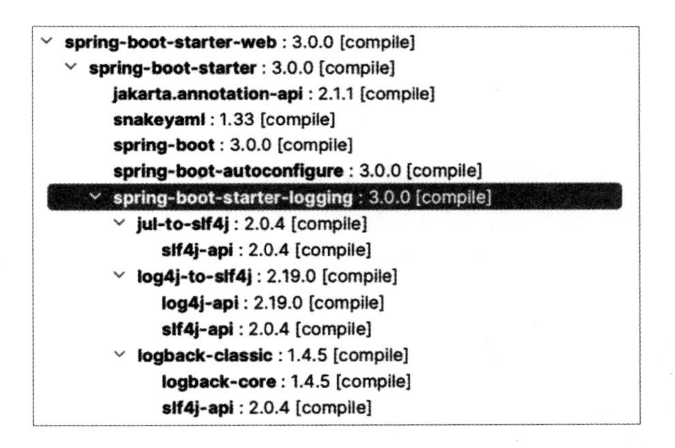

5.2 日志格式

Spring Boot 日志的格式如下：

```
2022-7-17T09:38:15.072+08:00  INFO 66672 --- [  restartedMain] o.s.b.w.embedded.tomcat.
TomcatWebServer : Tomcat started on port(s): 8443 (https) 8080 (http) with context path ''
```

日志主要包含以下信息：

- 日期时间：精确到毫秒，易于排序。
- 日志级别：具体见后面的章节。

- 应用进程 ID。
- 分隔符（---）：用来区分实际开始的日志。
- 线程名称。
- 日志名称：一般是指日志输出的当前类的缩写。
- 日志具体内容。

5.3 控制台日志

Spring Boot 默认从控制台输出日志，默认情况下，会记录 ERROR、WARN、INFO 级别的日志，如果要让应用输出更多日志，则可以在应用启动时使用 --debug 标志来启用调试模式，比如以下示例：

```
$ java –jar javastack.jar ––debug
```

其实还有更多开启应用调试模式的方法，已在第 3 章 Spring Boot Starter 与自动配置的自动配置报告一节中有全面的介绍。

需要注意的是：

启用应用的 Debug 调试模式，并不是让应用启用并输出所有的 DEBUG 级别的日志，而是让应用输出更多的框架核心日志，框架核心日志包括嵌入式容器、Spring Boot 等组件的更多的日志。

另外，还可以开启应用的 trace 追踪模式，日志要比 Debug 调试模式更详细，使用方法、注意事项和 Debug 调试模式是一样的，比如以下示例：

```
$ java -jar javastack.jar –trace
```

5.4 日志文件

Spring Boot 默认从控制台输出日志，而不会将日志写到日志文件中，这在实际项目中肯定是不可行的，在实际项目中不可能一直开着控制台，控制台日志随时会挂失，日志输出到日志文件可有利于归档并排查 N 天之前的日志。

将日志输出到日志文件，可以设置以下两个属性，如下表所示。

配置参数	说明	示例
logging.file.name	指定需要输出的日志文件名称	javastack.log

续表

配置参数	说明	示例
logging.file.path	指定需要输出的日志文件路径	/var/log 此时默认日志文件为 spring.log

Spring Boot 根据以上两个配置参数来输出日志文件，可以任用一个，也可以两个结合使用，可以是指定的路径，也可以是当前目录的相对路径，比如以下日志文件配置：

```
logging:
  file:
name: ./logs/javastack.log
```

应用启动之后就会在当前目录下自动创建 logs/javastack.log 目录及日志文件，如下图所示。

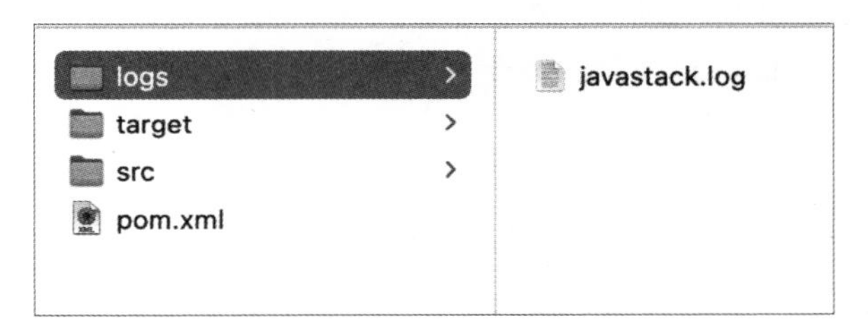

具体原理就是在Spring Boot主包的自动配置文件中注册一个LoggingApplicationListener监听器，然后监听 ApplicationEnvironmentPreparedEvent 事件：

```
private void onApplicationEnvironmentPreparedEvent(ApplicationEnvironmentPreparedEvent event) {
    SpringApplication springApplication = event.getSpringApplication();
    if (this.loggingSystem == null) {
        this.loggingSystem = LoggingSystem.get(springApplication.getClassLoader());
    }
    initialize(event.getEnvironment(), springApplication.getClassLoader());
}

protected void initialize(ConfigurableEnvironment environment, ClassLoader classLoader) {
    getLoggingSystemProperties(environment).apply();
    this.logFile = LogFile.get(environment);
    if (this.logFile != null) {
        this.logFile.applyToSystemProperties();
    }
    this.loggerGroups = new LoggerGroups(DEFAULT_GROUP_LOGGERS);
    initializeEarlyLoggingLevel(environment);
    initializeSystem(environment, this.loggingSystem, this.logFile);
    initializeFinalLoggingLevels(environment, this.loggingSystem);
```

```
    registerShutdownHookIfNecessary(environment, this.loggingSystem);
}
```

然后在 initialize 初始化方法中通过 LogFile#get 方法获取具体的日志文件：

```
public static LogFile get(PropertyResolver propertyResolver) {
    String file = propertyResolver.getProperty(FILE_NAME_PROPERTY);
    String path = propertyResolver.getProperty(FILE_PATH_PROPERTY);
    if (StringUtils.hasLength(file) || StringUtils.hasLength(path)) {
        return new LogFile(file, path);
    }
    return null;
}
```

根据前面的 Spring Boot 启动过程与扩展应用一章可以知道，这个 ApplicationEnvironment-PreparedEvent 事件是在 prepareEnvironment 方法准备环境阶段时发布的。

Spring Boot 在获取日志配置的时候为什么不是用 @ConfigurationProperties 注解绑定配置，而是手动获取配置呢？这是因为日志是在 ApplicationContext 创建之前初始化的，所以此时还无法获取环境中的配置。

5.5 日志级别

Spring Boot 支持的所有日志级别如下表所示。

日志级别	说明
OFF	关闭日志
TRACE	追踪
DEBUG	调试
INFO	信息
WARN	警告
ERROR	错误
FATAL	致命

Spring Boot 支持的最新的所有日志级别可以查看 LogLevel 这个类的源码：

```
public enum LogLevel {

    TRACE, DEBUG, INFO, WARN, ERROR, FATAL, OFF

}
```

Spring Boot 默认会输出 INFO 级别及以上的日志，另外，Logback 不支持 FATAL 日志级别，它会映射到 ERROR 级别上面。

日志级别的配置格式为：

```
logging.level.<logger-name>=<level>
```

根包的日志配置格式为：

```
logging.level.root=<level>
```

比如以下日志级别的配置：

```
logging:
  level:
    root: INFO
org.springframework: WARN
```

这里配置了根包的日志级别为 INFO，org.springframework 包的日志级别为 WARN。

除了在应用配置文件中配置，还可以通过环境变量配置日志级别，比如添加一个环境变量 LOGGING_LEVEL_ORG_SPRINGFRAMEWORK=DEBUG，它就会将 org.springframework 包日志级别调整为 DEBUG。

日志级别是在 LoggingApplicationListener 监听器的 initialize > ... > setLogLevels 初始化方法中设置的，如该方法的源码所示。

```
protected void setLogLevels(LoggingSystem system, ConfigurableEnvironment environment) {
    BiConsumer<String, LogLevel> customizer = getLogLevelConfigurer(system);
    Binder binder = Binder.get(environment);
    Map<String, LogLevel> levels = binder.bind(LOGGING_LEVEL, STRING_LOGLEVEL_MAP).
orElseGet(Collections::emptyMap);
    levels.forEach((name, level) -> configureLogLevel(name, level, customizer));
}
```

5.6　日志分组

上一节在配置对应包的日志级别的时候，需要一个个包地指定日志级别，如果要指定的包比较多，再为它们批量更换日志级别就比较麻烦，所以，Spring Boot 提供了日志分组功能，可以把相关的包配置成一个日志组，再针对日志组统一设置日志级别。

配置示例如下：

```
logging:
  group:
    tomcat: org.apache.catalina,org.apache.coyote,org.apache.tomcat
  level:
    tomcat: ERROR
```

这里配置了一个 Tomcat 组，组下面包含了三个包，然后把该日志分组设置为 ERROR 日志级别，即组下面的所有包都统一设置了，如果现在要更换这些包的日志级别，则只需要更换该组的日志级别，简单方便。

Spring Boot 也提供了几个可“开箱即用”的日志组，如下表所示。

日志组名称	包含的包
web	org.springframework.core.codec org.springframework.http org.springframework.web org.springframework.boot.actuate.endpoint.web org.springframework.boot.web.servlet.ServletContextInitializerBeans
sql	org.springframework.jdbc.core org.hibernate.SQL org.jooq.tools.LoggerListener

这些日志组可以直接配置日志级别，不需要再额外配置日志组，比如以下配置示例：

```
logging:
  level:
    web: INFO
```

这样配置后，Web 日志组下面的所有包都是 INFO 日志级别。

5.7 日志归档

Spring Boot 的日志在达到默认的 10MB 之后就会自动归档，如果使用的是默认的 Logback 日志框架，则可以直接在 application 配置文件中配置日志滚动规则，其他日志框架需要在其日志配置文件中配置（如 Log4j2）。

Logback 日志归档的配置参数对应的参数类为 LogbackLoggingSystemProperties 类，具体参数如下表所示。

日志滚动配置参数	说明
logging.logback.rollingpolicy.file-name-pattern	创建日志归档的文件名模式
logging.logback.rollingpolicy.clean-history-on-start	应用启动时是否进行日志归档清理
logging.logback.rollingpolicy.max-file-size	日志归档文件的最大文件容量，超过就进行归档（默认为 10MB）
logging.logback.rollingpolicy.total-size-cap	日志归档文件可占用的最大容量，超过会删除旧的归档日志文件
logging.logback.rollingpolicy.max-history	日志归档文件要保留的最长天数（默认为 7 天）

参考配置如下：

```
logging:
  file:
    name: ./logs/javastack.log
  logback:
    rollingpolicy:
      max-file-size: 1KB
```

这里我们指定文件日志超到 1KB 就进行归档，实际项目中请根据自己的需求进行配置。

应用重启后再来看一下日志目录，如下图所示。

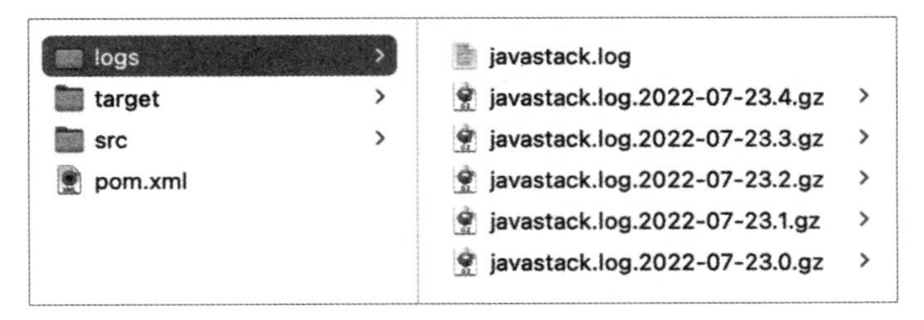

可以看到应用生成了好几个日志归档文件，日志归档文件默认是按日期进行归档的，同一天的多次归档文件又会按索引进行归档，索引从 0 开始。

5.8　日志配置文件

5.8.1　概述

虽然 Spring Boot 可以通过参数配置的方式来定义日志的行为，但也仅限于一些常用的使用场景，如果涉及复杂的日志配置，则需要通过日志配置文件的方式来处理。

Spring Boot 默认加载的日志配置文件如下表所示。

日志框架	配置文件
Logback	logback-spring.xml logback-spring.groovy logback.xml logback.groovy
Log4j2	log4j2-spring.xml log4j2.xml
Java Util Logging	logging.properties

Spring Boot 建议使用 *-spring 格式的日志配置文件，比如 logback-spring.xml，而不是使用 logback.xml，因为标准的 logback.xml 日志配置文件的初始化太早了，会导致 Spring 无法完全控制日志的初始化，使用 *-spring 格式的日志配置文件的好处是可以使用 Spring 中的日志扩展功能。

前面的章节也提到了，因为日志的初始化较早，所以无法读取 Spring 环境中的配置参数，但是可以通过 System Property 系统参数的方式来配置，Spring Boot 提供了以下对应的配置参数，如下表所示。

Spring Environment	System Property
logging.exception-conversion-word	LOG_EXCEPTION_CONVERSION_WORD
logging.file.name	LOG_FILE
logging.file.path	LOG_PATH
logging.pattern.console	CONSOLE_LOG_PATTERN
logging.pattern.dateformat	LOG_DATEFORMAT_PATTERN
logging.charset.console	CONSOLE_LOG_CHARSET
logging.pattern.file	FILE_LOG_PATTERN
logging.charset.file	FILE_LOG_CHARSET
logging.pattern.level	LOG_LEVEL_PATTERN
PID	PID

另外，日志框架自身的配置也有对应的系统参数，比如 Logback 的 logging.logback.rollingpolicy.file-name-pattern 配置参数对应 LOGBACK_ROLLINGPOLICY_FILE_NAME_PATTERN 参数，等等。

5.8.2 日志配置模板

Spring Boot 提供了以下几个 Logback 默认的内置配置模板：

- defaults.xml。
- console-appender.xml。
- file-appender.xml。

此外，为了与早期的 Spring Boot 版本兼容，Spring Boot 在 Logback 包中还提供了一个遗留的 base.xml 模板文件，这些模板都定义在 org/springframework/boot/logging/logback/ 目录下，如下图所示。

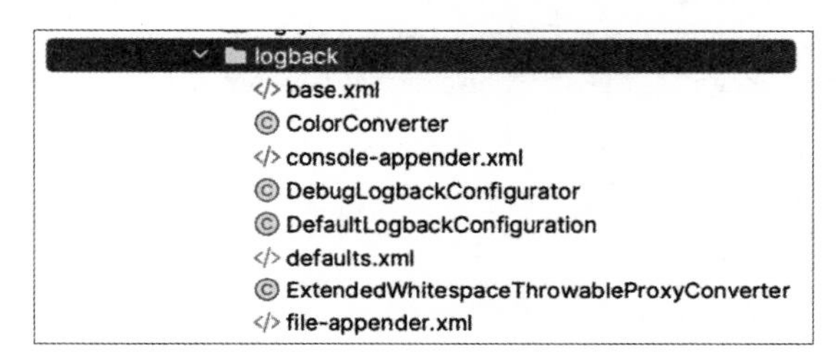

它们都可以包含在自定义日志配置文件中，这些模板中的内容如下所示。

```
defaults.xml:

<?xml version="1.0" encoding="UTF-8"?>

<!--
Default logback configuration provided for import
-->

<included>
    <conversionRule conversionWord="clr" converterClass="org.springframework.boot.
logging.logback.ColorConverter" />
    <conversionRule conversionWord="wex" converterClass="org.springframework.boot.
logging.logback.WhitespaceThrowableProxyConverter" />
    <conversionRule conversionWord="wEx" converterClass="org.springframework.boot.
logging.logback.ExtendedWhitespaceThrowableProxyConverter" />

    <property name="CONSOLE_LOG_PATTERN" value="${CONSOLE_LOG_PATTERN:-%clr
(%d{${LOG_DATEFORMAT_PATTERN:-yyyy-MM-dd'T'HH:mm:ss.SSSXXX}}){faint} %clr
(${LOG_LEVEL_PATTERN:-%5p}) %clr(${PID:- }){magenta} %clr(---){faint} %clr([%15.15t])
{faint} %clr(%-40.40logger{39}){cyan} %clr(:){faint} %m%n${LOG_EXCEPTION_CONVERSION_
WORD:-%wEx}}"/>
    <property name="CONSOLE_LOG_CHARSET" value="${CONSOLE_LOG_CHARSET:-${file.encoding:
-UTF-8}}"/>
    <property name="FILE_LOG_PATTERN" value="${FILE_LOG_PATTERN:-%d{${LOG_DATEFORMAT_
PATTERN:-yyyy-MM-dd'T'HH:mm:ss.SSSXXX}} ${LOG_LEVEL_PATTERN:-%5p} ${PID:- } --- [%t]
%-40.40logger{39} : %m%n${LOG_EXCEPTION_CONVERSION_WORD:-%wEx}}"/>
```

```
    <property name="FILE_LOG_CHARSET" value="${FILE_LOG_CHARSET:-${file.encoding:-UTF-8}}"/>

    <logger name="org.apache.catalina.startup.DigesterFactory" level="ERROR"/>
    <logger name="org.apache.catalina.util.LifecycleBase" level="ERROR"/>
    <logger name="org.apache.coyote.http11.Http11NioProtocol" level="WARN"/>
    <logger name="org.apache.sshd.common.util.SecurityUtils" level="WARN"/>
    <logger name="org.apache.tomcat.util.net.NioSelectorPool" level="WARN"/>
    <logger name="org.eclipse.jetty.util.component.AbstractLifeCycle" level="ERROR"/>
    <logger name="org.hibernate.validator.internal.util.Version" level="WARN"/>
    <logger name="org.springframework.boot.actuate.endpoint.jmx" level="WARN"/>
</included>
```

可以看到，defaults.xml 配置模板中提供了一些转换规则、模式参数和常用的包的日志级别配置。

```
console-appender.xml:

<?xml version="1.0" encoding="UTF-8"?>

<!--
Console appender logback configuration provided for import, equivalent to the programmatic
initialization performed by Boot
-->

<included>
    <appender name="CONSOLE" class="ch.qos.logback.core.ConsoleAppender">
        <encoder>
            <pattern>${CONSOLE_LOG_PATTERN}</pattern>
            <charset>${CONSOLE_LOG_CHARSET}</charset>
        </encoder>
    </appender>
</included>
```

可以看到，console-appender.xml 配置模板使用 CONSOLE_LOG_PATTERN 参数添加了一个 ConsoleAppender 配置，这意味着如果要在控制台输出日志，那么只需要在自己的日志配置文件中引入该配置模板即可，反之如果要禁用控制台输出日志，则不包含此配置模板即可。

```
file-appender.xml:

<?xml version="1.0" encoding="UTF-8"?>

<!--
File appender logback configuration provided for import, equivalent to the programmatic
initialization performed by Boot
```

```
-->

<included>
    <appender name="FILE" class="ch.qos.logback.core.rolling.RollingFileAppender">
        <encoder>
            <pattern>${FILE_LOG_PATTERN}</pattern>
            <charset>${FILE_LOG_CHARSET}</charset>
        </encoder>
        <file>${LOG_FILE}</file>
        <rollingPolicy class="ch.qos.logback.core.rolling.SizeAndTimeBasedRollingPolicy">
            <fileNamePattern>${LOGBACK_ROLLINGPOLICY_FILE_NAME_PATTERN: -${LOG_FILE}.%d
{yyyy-MM-dd}.%i.gz}</fileNamePattern>
            <cleanHistoryOnStart>${LOGBACK_ROLLINGPOLICY_CLEAN_HISTORY_ON_START: -false}
</cleanHistoryOnStart>
            <maxFileSize>${LOGBACK_ROLLINGPOLICY_MAX_FILE_SIZE:-10MB}</maxFileSize>
            <totalSizeCap>${LOGBACK_ROLLINGPOLICY_TOTAL_SIZE_CAP:-0}</totalSizeCap>
            <maxHistory>${LOGBACK_ROLLINGPOLICY_MAX_HISTORY:-7}</maxHistory>
        </rollingPolicy>
    </appender>
</included>
```

可以看到，file-appender.xml 配置模板使用 FILE_LOG_PATTERN 和 ROLLING_FILE_ NAME_PATTERN 参数添加了一个 RollingFileAppender 配置，这意味着如果要在日志文件中输出日志并进行归档，则只需要在自己的日志配置文件中引入该配置模板即可，反之亦然。

以上三个内置的日志配置模板能满足大部分的应用需求，直接拿来用即可，如果不合适，就不需要引入，再参考自定义的配置即可，这也是 Spring Boot 为了简化配置所做出的努力。

5.8.3　自定义日志配置文件

了解了上面几节的内容，本节就可以自定义一个 Logback 日志配置文件。首先在应用的 src/main/resources/ 资源目录下创建一个 logback-spring.xml 日志配置文件，参考配置如下：

```
<?xml version="1.0" encoding="UTF-8"?>
<configuration>
    <include resource="org/springframework/boot/logging/logback/defaults.xml"/>
    <include resource="org/springframework/boot/logging/logback/console-appender.xml" />

    <property name="LOG_FILE" value="${LOG_FILE:-${LOG_PATH:-${LOG_TEMP:
-${java.io.tmpdir:-/tmp}}}/javastack.log}"/>
    <include resource="org/springframework/boot/logging/logback/file-appender.xml" />
```

```
    <root level="INFO">
        <appender-ref ref="CONSOLE" />
        <appender-ref ref="FILE" />
    </root>

    <springProfile name="dev | test">
        <logger name="cn.javastack" level="DEBUG"/>
    </springProfile>

    <springProfile name="prod">
        <logger name="cn.javastack" level="INFO"/>
    </springProfile>

    <logger name="org.springframework.web" level="INFO"/>
</configuration>
```

通过 include 标签引用内置的模板配置文件，通过 springProfile 扩展标签可以指定在 profiles 环境下才生效的日志配置。

如果想自定义日志配置参数，那么从模板内容可以看到是有 ${} 参数占位符的，其中“:-”表示如果没有此参数的值就继续再取后面参数的值，以此类推，如果都没有，那么会使用最后配置的临时目录，所以可以通过添加对应的系统参数来使用模板。

以下是相关测试代码：

```
@SpringBootApplication
public class Application {

    public static void main(String[] args) {
        System.setProperty("LOG_PATH", "./logs");
        System.setProperty("LOGBACK_ROLLINGPOLICY_MAX_FILE_SIZE", "1KB");
        SpringApplication.run(Application.class);
    }

    private static final org.apache.commons.logging.Log logger1 = org.apache.commons.logging
            .LogFactory
            .getLog(Application.class);

    private static final org.slf4j.Logger logger2 = org.slf4j.LoggerFactory
            .getLogger(Application.class);

    @Bean
    public CommandLineRunner commandLineRunner() {
```

```
        return (args) -> {
            logger1.error("commons logging error...");
            logger1.info("commons logging info...");
            logger1.debug("commons logging debug...");

            logger2.error("slf4j info...");
            logger2.info("slf4j info...");
            logger2.debug("slf4j info...");
        };
    }

}
```

这里定义了一个日志文件目录，为了测试，这里还将日志文件归档的最大文件值设置为 1KB，并添加了对应的日志门面 DEBUG 日志输出功能。

然后指定当前 profiles 为 dev 环境：

```
spring:
  profiles:
active: dev
```

注释 application 中的日志配置，删除之前的 logs 目录再重启应用，如下图所示。

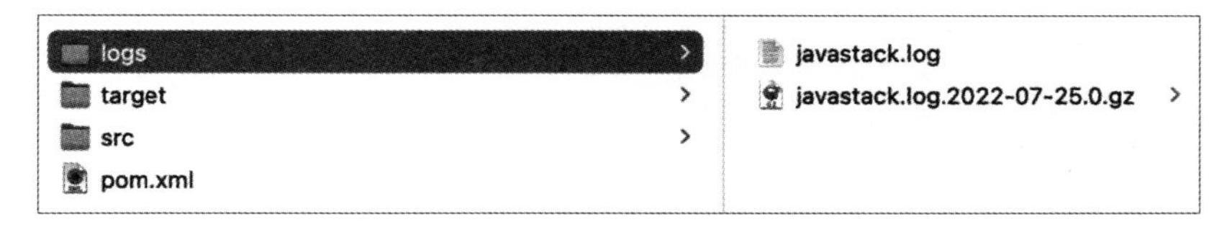

可以看到，目录和日志文件正常生成和归档了，并且当 dev 环境开启时，Debug 日志正常输出了：

```
commons logging error...
commons logging info...
commons logging debug...
slf4j error...
slf4j info...
slf4j debug...
```

如果切换为 prod 环境，则 Debug 日志是不会被输出的，这就是使用 -spring 格式的日志配置文件的好处。

5.9 切换 Log4j2 日志框架

在本章最开始的时候提到了 Spring Boot 支持的日志框架，其中 spring-boot-starter 包中默认使用的是 spring-boot-starter-logging 启动器，对应的是 Logback 日志框架，如果要使用 Log4j2 就必须先排除默认的日志框架，再引入新的 spring-boot-starter-log4j2 日志框架启动器。

以下是推荐的切换方法：

```
<dependency>
    <groupId>org.springframework.boot</groupId>
    <artifactId>spring-boot-starter-web</artifactId>
</dependency>
<dependency>
    <groupId>org.springframework.boot</groupId>
    <artifactId>spring-boot-starter</artifactId>
    <exclusions>
        <exclusion>
            <groupId>org.springframework.boot</groupId>
            <artifactId>spring-boot-starter-logging</artifactId>
        </exclusion>
    </exclusions>
</dependency>
<dependency>
    <groupId>org.springframework.boot</groupId>
    <artifactId>spring-boot-starter-log4j2</artifactId>
</dependency>
```

然后创建对应的 log4j2-spring.xml 日志配置文件即可。

> **Spring Boot 3.0 新变化：**
> Log4j2 的功能增强了，和 Logback 中的扩展功能一样，支持以下几个 Log4j2 的新扩展功能：
> - 支持使用基于 Profile 生效的配置。
> - 支持使用 Spring 环境中的参数。
> - 支持使用系统参数。

5.10 切换日志框架版本

日志框架经常出现漏洞需要升级到最新版本，而 Spring Boot 中的日志框架版本都是固定跟随 Spring Boot 版本升级的，如果默认集成的日志框架版本有问题，那么在 Spring Boot 最新版本发布之前可以自行切换使用更新的日志框架版本。

前面的基础章节中介绍了，如果使用的是继承 spring-boot-starter-parent 的方式，那么直接使用对应的 properties 参数覆盖版本号即可：

```
<log4j2.version>2.18.0</log4j2.version>
```

如果使用导入 spring-boot-dependencies 依赖的集成方式，则需要显式导入依赖并指定版本号：

```
<dependencies>
    <dependency>
        <groupId>org.apache.logging.log4j</groupId>
        <artifactId>log4j-bom</artifactId>
        <version>2.18.0</version>
        <type>pom</type>
        <scope>import</scope>
      </dependency>
 </dependencies>
```

这里只是日志框架依赖的升级示例，更多 Spring Boot 依赖的更新替换都可以参考此种方式。如果不是太大的版本升级，一般情况下可以直接升级具体的依赖，但不排除会与 Spring Boot 当前版本存在兼容性问题，所以也得谨慎升级。

5.11　输出彩色日志

5.11.1　开启彩色日志输出

如果你的终端支持 ANSI 字符标准，则可以设置控制台输出彩色日志来提高可读性。各个日志级别对应的颜色如下表所示。

日志级别	颜色
FATAL	红
ERROR	红
WARN	黄
INFO	绿
DEBUG	绿
TRACE	绿

其实也就是根据日志级别的严重性分为红、黄、绿三种颜色，开启配置的参数为 spring.output.ansi.enabled，可支持的值如下表所示。

选项	说明
ALWAYS	开启 ANSI 颜色输出
DETECT（默认）	尝试检测 ANSI 功能是否可用
NEVER	禁用 ANSI 颜色输出

最新选项可参考 AnsiOutput.Enabled 枚举类，如该枚举类的源码所示。

```
public enum Enabled {

    /**
     * Try to detect whether ANSI coloring capabilities are available. The default
     * value for {@link AnsiOutput}.
     */
    DETECT,

    /**
     * Enable ANSI-colored output.
     */
    ALWAYS,

    /**
     * Disable ANSI-colored output.
     */
    NEVER

}
```

默认值为 DETECT，如 AnsiOutput 类的源码所示。

```
public abstract class AnsiOutput {

    private static final String ENCODE_JOIN = ";";

    private static Enabled enabled = Enabled.DETECT;

    ...

}
```

DETECT 即只进行检测，并不会输出彩色日志，如下图的启动日志所示。

```
416+08:00  INFO 84579 --- [  restartedMain] o.s.b.w.embedded.tomcat.TomcatWebServer  : Tomcat started on port(s): 8080 (http)
425+08:00  INFO 84579 --- [  restartedMain] c.j.springboot.logging.Application       : Started Application in 2.705 seconds (
427+08:00 ERROR 84579 --- [  restartedMain] c.j.springboot.logging.Application       : commons logging error...
427+08:00  WARN 84579 --- [  restartedMain] c.j.springboot.logging.Application       : commons logging warn
427+08:00  INFO 84579 --- [  restartedMain] c.j.springboot.logging.Application       : commons logging info...
428+08:00 DEBUG 84579 --- [  restartedMain] c.j.springboot.logging.Application       : commons logging debug...
428+08:00 ERROR 84579 --- [  restartedMain] c.j.springboot.logging.Application       : slf4j error...
428+08:00  WARN 84579 --- [  restartedMain] c.j.springboot.logging.Application       : commons logging warn
428+08:00  INFO 84579 --- [  restartedMain] c.j.springboot.logging.Application       : slf4j info...
429+08:00 DEBUG 84579 --- [  restartedMain] c.j.springboot.logging.Application       : slf4j debug...
```

笔者这里把日志设置为开启颜色输出，如以下配置所示。

```
spring:
  output:
    ansi:
      enabled: always
```

然后重启应用，查看日志，如下图所示。

```
634+08:00  INFO 83883 --- [  restartedMain] o.s.b.w.embedded.tomcat.TomcatWebServer  : Tomcat started on port(s): 8080 (http)
647+08:00  INFO 83883 --- [  restartedMain] c.j.springboot.logging.Application       : Started Application in 2.734 seconds (
652+08:00 ERROR 83883 --- [  restartedMain] c.j.springboot.logging.Application       : commons logging error...
653+08:00  WARN 83883 --- [  restartedMain] c.j.springboot.logging.Application       : commons logging warn
653+08:00  INFO 83883 --- [  restartedMain] c.j.springboot.logging.Application       : commons logging info...
655+08:00 DEBUG 83883 --- [  restartedMain] c.j.springboot.logging.Application       : commons logging debug...
655+08:00 ERROR 83883 --- [  restartedMain] c.j.springboot.logging.Application       : slf4j error...
655+08:00  WARN 83883 --- [  restartedMain] c.j.springboot.logging.Application       : commons logging warn
655+08:00  INFO 83883 --- [  restartedMain] c.j.springboot.logging.Application       : slf4j info...
655+08:00 DEBUG 83883 --- [  restartedMain] c.j.springboot.logging.Application       : slf4j debug...
```

可以看到日志带颜色输出了（彩色图片下载地址见封底），除了日志级别有颜色，其他关键的信息也有默认的颜色，具体配置可见下一节的介绍。

5.11.2 日志上色原理

Spring Boot 中的日志颜色是使用 %clr 转换词来配置的，配置格式如下：

```
%clr( 日志 ){ 颜色 }
```

可支持配置的颜色如下表所示。

选项	颜色
blue	蓝
cyan	青
faint	模糊
green	绿

续表

选项	颜色
magenta	洋红
red	红
yellow	黄

下面介绍 Spring Boot 默认的控制台配置示例，内置的 defaults.xml 日志模板如下所示。

```
<included>
  <conversionRule conversionWord="clr" converterClass="org.springframework.boot.
logging.logback.ColorConverter" />
  <conversionRule conversionWord="wex" converterClass="org.springframework.boot.
logging.logback.WhitespaceThrowableProxyConverter" />
  <conversionRule conversionWord="wEx" converterClass="org.springframework.boot.
logging.logback.ExtendedWhitespaceThrowableProxyConverter" />

  <property name="CONSOLE_LOG_PATTERN" value="${CONSOLE_LOG_PATTERN:-%clr(%d {${LOG_
DATEFORMAT_PATTERN:-yyyy-MM-dd'T'HH:mm:ss.SSSXXX}}){faint} %clr(${LOG_LEVEL_PATTERN:
-%5p}) %clr(${PID:- }){magenta} %clr(---){faint} %clr([%15.15t]){faint} %clr(%-40.40logger
{39}){cyan} %clr(:){faint} %m%n${LOG_EXCEPTION_CONVERSION_WORD:-%wEx}}"/>

  ...
```

这便是上一节中默认的彩色日志配置，定义了 clr、wex、wEx 三个转换词，关于颜色的转换便是 clr 这个转换词，对应的转换类为 ColorConverter 类，从源码看目前只有 Logback 和 Log4j2 实现了颜色日志，如下图所示。

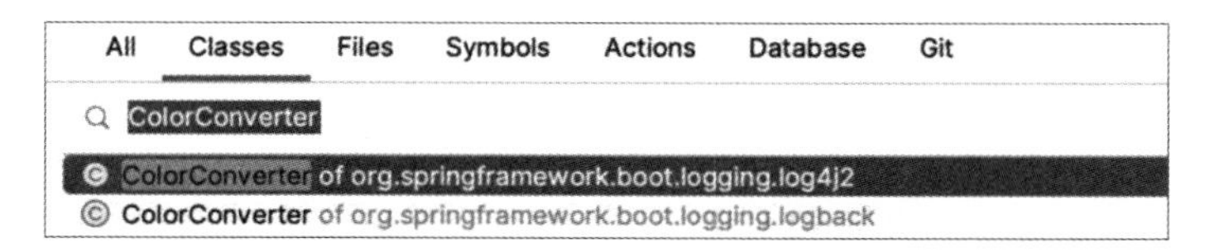

具体上色的过程很简单，以下是 Logback 日志框架对应的 ColorConverter 类的核心源码：

```
public class ColorConverter extends CompositeConverter<ILoggingEvent> {

    ...

    static {
        Map<String, AnsiElement> ansiElements = new HashMap<>();
        ansiElements.put("faint", AnsiStyle.FAINT);
        ansiElements.put("red", AnsiColor.RED);
```

```
        ansiElements.put("green", AnsiColor.GREEN);
        ansiElements.put("yellow", AnsiColor.YELLOW);
        ansiElements.put("blue", AnsiColor.BLUE);
        ansiElements.put("magenta", AnsiColor.MAGENTA);
        ansiElements.put("cyan", AnsiColor.CYAN);
        ELEMENTS = Collections.unmodifiableMap(ansiElements);
    }

    @Override
    protected String transform(ILoggingEvent event, String in) {
        AnsiElement element = ELEMENTS.get(getFirstOption());
        if (element == null) {
            // Assume highlighting
            element = LEVELS.get(event.getLevel().toInteger());
            element = (element != null) ? element : AnsiColor.GREEN;
        }
        return toAnsiString(in, element);
    }

    protected String toAnsiString(String in, AnsiElement element) {
        return AnsiOutput.toString(element, in);
    }

}
```

实现很简单，在 DefaultLogbackConfiguration 配置类中注册了这个颜色转换器。

5.11.3 自定义日志颜色

了解了日志上色的原理和配置方法后，下面介绍一个简单的自定义颜色的示例。在应用的自定义日志配置文件中，修改日志的具体内容为黄色，如下面的配置示例所示。

```
<?xml version="1.0" encoding="UTF-8"?>
<configuration>
    <property name="LOG_FILE" value="${LOG_FILE:-${LOG_PATH:-${LOG_TEMP: -${java.io.tmpdir:
-/tmp}}}/javastack.log}"/>
    <property name="CONSOLE_LOG_PATTERN" value="${CONSOLE_LOG_PATTERN:-%clr (%d{${LOG_
DATEFORMAT_PATTERN:-yyyy-MM-dd'T'HH:mm:ss.SSSXXX}}){faint} %clr(${LOG_LEVEL_PATTERN:
-%5p}) %clr(${PID:- }){magenta} %clr(---){faint} %clr([%15.15t]){faint} %clr
(%-40.40logger{39}){cyan} %clr(:){faint} %clr(%m%n){yellow}${LOG_EXCEPTION_CONVERSION_
WORD:-%wEx}}"/>
```

```
  ...
}
```

笔者新增了一个 CONSOLE_LOG_PATTERN 参数，这样就可以覆盖框架内置的默认的参数值，然后将 %m%n 修改为 %clr(%m%n){yellow}，再重启应用并查看日志，如下图所示。

```
229+08:00  INFO 89504 --- [  restartedMain] o.s.b.w.embedded.tomcat.TomcatWebServer  : Tomcat started on port(s): 8080 (http)
237+08:00  INFO 89504 --- [  restartedMain] c.j.springboot.logging.Application       : Started Application in 1.957 seconds (
239+08:00 ERROR 89504 --- [  restartedMain] c.j.springboot.logging.Application       : commons logging error...
239+08:00  WARN 89504 --- [  restartedMain] c.j.springboot.logging.Application       : commons logging warn
239+08:00  INFO 89504 --- [  restartedMain] c.j.springboot.logging.Application       : commons logging info...
240+08:00 DEBUG 89504 --- [  restartedMain] c.j.springboot.logging.Application       : commons logging debug...
240+08:00 ERROR 89504 --- [  restartedMain] c.j.springboot.logging.Application       : slf4j error...
240+08:00  WARN 89504 --- [  restartedMain] c.j.springboot.logging.Application       : commons logging warn
240+08:00  INFO 89504 --- [  restartedMain] c.j.springboot.logging.Application       : slf4j info...
240+08:00 DEBUG 89504 --- [  restartedMain] c.j.springboot.logging.Application       : slf4j debug...
```

日志具体内容已经成功显示为黄色了，如果想定制自己的控制台日志颜色，那么“依葫芦画瓢”自定义日志配置即可。

5.12 日志关闭钩子

在 Spring Boot 启动过程与扩展应用一章中，我们知道 Spring Boot 提供了一个优雅关闭的钩子，在日志系统中，Spring Boot 也注册了日志相关的关闭钩子，该钩子可以在 JVM 退出时触发日志系统清理，以保证应用终止时及时释放日志使用的资源。

> Spring Boot 应用会自动注册该关闭钩子，以 war 包形式部署的除外。

如果应用具有复杂的上下文层次结构，而 Spring Boot 的日志关闭钩子无法满足需求，则可以禁用这个日志关闭钩子，关闭配置如下所示。

```
logging:
  register-shutdown-hook: false
```

禁用注册日志关闭钩子后，可以考虑直接使用底层日志框架提供的相关资源管理机制，例如，Logback 提供了一种上下文选择器，它可以允许每个 Logger 对象在它们自己的上下文中进行创建，无须 Spring Boot 干预。

第6章 Spring Boot Web 核心应用

在 Java 编程应用中，最重要的一点就是 Web 与后端接口编程，使用 Spring Boot 框架会让这一点变得轻而易举，它可以轻松利用 Web 模块快速搭建 HTTP 接口服务，所以本章会介绍 Spring Boot 与 Web 的方方面面，包括 Web 模块基本介绍、嵌入式容器、自定义配置及注册一些 Web 组件、静态资源处理、模板引擎，以及异常处理、参数校验、国际化、分布式会话、跨域处理、安全性、调用 REST 服务等。

6.1 概述

在之前的 Spring Boot 启动过程分析中，笔者介绍了 Spring Boot 支持的两种 Web 类型：

- SERVLET（即传统的 WebMVC）。
- REACTIVE（响应式的 WebFlux）。

传统的 WebMVC 应用使用的是同步阻塞的 I/O 模型，而 WebFlux 响应式应用则使用异步非阻塞的 I/O 模型，所以 WebFlux 的性能要优于 WebMVC，但 WebFlux 还不是很成熟，生态也不是很完善，现在绝大多数的 Web 应用还是基于 Servlet 规范和容器的，所以，本章绝大多数应用也都是基于 Servlet 的。

通过集成 Spring Boot 提供的 spring-boot-starter-web 启动器可以快速启用一个基于 Servlet 的 HTTP 服务，在 Spring Boot 自动配置包中也提供了对 Spring MVC 框架的自动配置，具体可以查看 WebMvcAutoConfiguration 类的源码，源码较多，笔者在此不粘贴详细的代码，自动配置类的主要内容是：

- 包含对 ContentNegotiatingViewResolver 和 BeanNameViewResolver 的 Bean 注册。
- 支持对静态资源，包括对静态网页 index.html 的支持，以及对 WebJars 的支持。
- 包含对 Converter、GenericConverter、Formatter 的 Bean 自动注册。
- 支持 HttpMessageConverters 消息转换器。
- 支持对 MessageCodesResolver 的自动注册。
- 自动使用 ConfigurableWebBindingInitializer 的 Bean。

既然 Spring Boot 只是提供了对 Spring MVC 的自动配置，所以我们在 Spring Boot 中使用 Spring MVC 和独立使用没有任何区别，比如先加入基于 Servlet 的 Web 依赖启动器 spring-boot-starter-web，然后定义两个接口，分别返回 JSON、XML 格式的数据，如下面的代码所示。

```
@RestController
public class ResponseBodyController {

    @GetMapping(value = "/user/json/{userId}", produces = MediaType.APPLICATION_JSON_VALUE)
    public ResponseEntity getJsonUserInfo(@PathVariable("userId") String userId) {
        User user = new User("Java 技术栈 ", 18);
        user.setId(Long.valueOf(userId));
        return new ResponseEntity(user, HttpStatus.OK);
    }

    @GetMapping(value = "/user/xml/{userId}", produces = MediaType.APPLICATION_XML_VALUE)
    public ResponseEntity getXmlUserInfo(@PathVariable("userId") String userId) {
        UserXml user = new UserXml();
        user.setName(" 栈长 ");
        user.setId(userId);

        List<OrderInfo> orderList = new ArrayList<>();
        OrderInfo orderInfo1 = new OrderInfo("123456001", 999, new Date());
        OrderInfo orderInfo2 = new OrderInfo("123456002", 777, new Date());
        OrderInfo orderInfo3 = new OrderInfo("123456003", 666, new Date());
        orderList.add(orderInfo1);
        orderList.add(orderInfo2);
        orderList.add(orderInfo3);
        user.setOrderList(orderList);

        return new ResponseEntity(user, HttpStatus.OK);
    }

}
```

注解都还是 Spring MVC 那套东西，Spring Boot 只是自动完成了框架上的配置而已，使得开发者更容易上手。

6.2　嵌入式容器

6.2.1　概述

嵌入式容器是 Spring Boot 最重要的特性之一，通过这个特性可以完成 Spring Boot 应用的快速启动和扩容，不需要自己安装和部署 Servlet 容器。

Spring Boot 支持的 Servlet 容器如下表所示。

容器	依赖
Tomcat	spring-boot-starter-tomcat
Jetty	spring-boot-starter-jetty
Undertow	spring-boot-starter-undertow

通过导入 spring-boot-starter-web 启动器依赖，Spring Boot 可以自动配置 Web 应用环境，该启动器导入的依赖如下图所示。

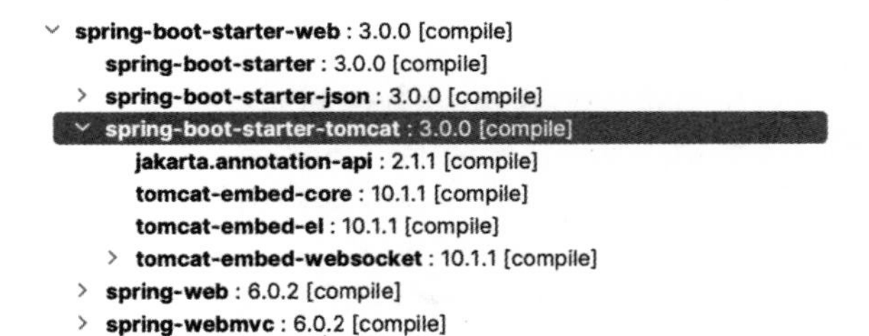

默认的嵌入式容器为 Tomcat，也可以更换为其他两个容器，具体见 6.2.3 节。

6.2.2　容器配置

默认情况下不用做任何参数配置，应用启动绑定的协议与端口默认为 HTTP 与 8080，容器参数绑定类为 ServerProperties 类，可以通过其绑定的以 server.* 开头的参数来配置所有容器的通用配置，比如：

```
server:
```

```
  port: 8081
  servlet:
    context-path: /javastack
```

还有更多容器的详细配置，有需要的读者可以参考 ServerProperties 参数类进行额外配置，比如 Tomcat 容器可以通过 server.tomcat.* 参数配置等。

另外，还能通过 Java 类的方式来自定义 Servlet 容器，只需实现 WebServerFactoryCustomizer< ConfigurableServletWebServerFactory> 接口即可：

```
@Component
public class CustomWebServerFactoryCustomizer implements WebServerFactoryCustomizer
<ConfigurableServletWebServerFactory> {

    @Override
    public void customize(ConfigurableServletWebServerFactory server) {
        server.setPort(8081);
    }

}
```

除了 ConfigurableServletWebServerFactory，还有 TomcatServletWebServerFactory、JettyServlet-WebServerFactory 和 UndertowServletWebServerFactory，ConfigurableServletWebServerFactory 工厂类的结构如下图所示。

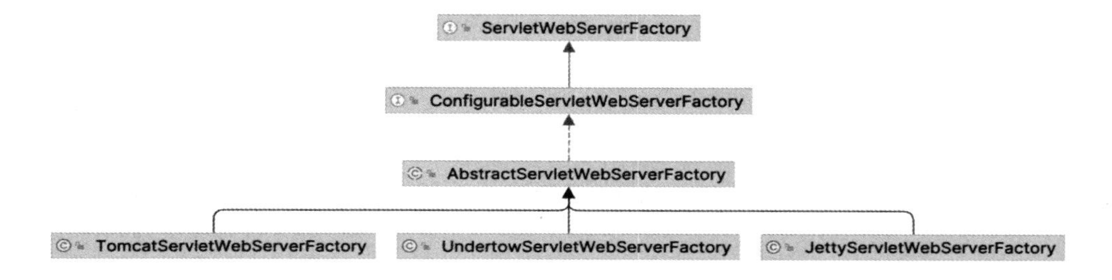

可以看到 ConfigurableServletWebServerFactory 工厂类有三个不同的实现变体，它们分别为 Tomcat、Jetty 和 Undertow 容器提供了额外的自定义设置，比如以下示例提供了对 Tomcat 的特定配置：

```
@Component
public class TomcatWebServerFactoryCustomizer implements WebServerFactoryCustomizer
<TomcatServletWebServerFactory> {

    @Override
```

```
    public void customize(TomcatServletWebServerFactory server) {
        connector.setPort(8088);
        server.addConnectorCustomizers((connector) -> connector.setAsyncTimeout
(Duration.ofSeconds(20).toMillis()));
    }

}
```

6.2.3　切换容器

如果不想使用内置默认的 Tomcat 容器，也可以切换为其他支持的容器，只要排除 spring-boot-starter-web 启动器依赖包中的 Tomcat 依赖，再导入其他容器的依赖即可。

比如想使用 Undertow 作为应用的嵌入式容器，则替换方法如下面的示例所示。

```
<dependency>
    <groupId>org.springframework.boot</groupId>
    <artifactId>spring-boot-starter-web</artifactId>
    <exclusions>
        <!-- Exclude the Tomcat dependency -->
        <exclusion>
            <groupId>org.springframework.boot</groupId>
            <artifactId>spring-boot-starter-tomcat</artifactId>
        </exclusion>
    </exclusions>
</dependency>
<dependency>
    <groupId>org.springframework.boot</groupId>
    <artifactId>spring-boot-starter-undertow</artifactId>
</dependency>
```

6.2.4　随机空闲端口

为了防止端口冲突，Spring Boot 可以配置并使用系统随机空闲的端口，只需要把 server.port 设置为 0 即可：

```
server:
  port: 0
```

Spring Boot 会自动扫描系统空闲的端口，并随机选择一个使用。

6.2.5 SSL

Spring Boot 嵌入式容器也可以通过 SSL 配置以支持 HTTPS 访问，通过一系列 server.ssl.* 参数来配置 SSL。

在本地环境下，可以先用 JDK 的 keytool 工具生成一个自签名的证书：

```
> $ keytool -genkey -alias https -keyalg RSA
输入密钥库口令：
再次输入新口令：
您的名字与姓氏是什么？
  [Unknown]:  test
您的组织单位名称是什么？
  [Unknown]:  test
您的组织名称是什么？
  [Unknown]:  test
您所在的城市或区域名称是什么？
  [Unknown]:  test
您所在的省 / 市 / 自治区名称是什么？
  [Unknown]:  test
该单位的双字母国家 / 地区代码是什么？
  [Unknown]:  test
CN=john, OU=test, O=test, L=test, ST=test, C=test 是否正确？
  [ 否 ]:  y

输入 <https> 的密钥口令
    ( 如果和密钥库口令相同 , 按回车键 ):
再次输入新口令：

Warning:
JKS 密钥库使用专用格式。建议使用 "keytool -importkeystore -srckeystore .keystore -destkeystore
.keystore -deststoretype pkcs12" 迁移到行业标准格式 PKCS12。
> $ keytool -importkeystore -srckeystore .keystore -destkeystore .keystore -deststoretype
pkcs12
输入源密钥库口令：
已成功导入别名 https 的条目。
已完成导入命令： 1 个条目成功导入 , 0 个条目失败或取消
```

然后把证书复制到应用资源根目录下，比如以下参考配置：

```
server:
  port: 8443
  ssl:
```

```
    protocol: TLS
    key-store-type: JKS
    key-store: classpath:.keystore
    key-store-password: javastack
```

上面的例子配置后就能支持 HTTPS 了，默认的 HTTP 就不再支持了，Spring Boot 不能以配置文件的方式同时支持 HTTP 和 HTTPS。

如果需要同时支持这两个协议，就需要把另外一个协议用 Java 配置类以程序化的方式来配置，因为通过程序的方式配置 HTTP 协议更加简单一点，所以，Spring Boot 推荐的做法是把 HTTPS 配置在配置文件中，HTTP 通过程序来配置。

通过程序的方式来额外支持 HTTP 的示例如下：

```
@SpringBootApplication
public class Application {

    @Bean
    public ServletWebServerFactory servletContainer() {
        TomcatServletWebServerFactory tomcat = new TomcatServletWebServerFactory();
        tomcat.addAdditionalTomcatConnectors(createStandardConnector());
        return tomcat;
    }

    private Connector createStandardConnector() {
        Connector connector = new Connector("org.apache.coyote.http11.Http11NioProtocol");
        connector.setPort(8080);
        return connector;
    }

    ...

}
```

启动 Spring Boot 之后就会看到同时支持两个协议的日志：

```
Tomcat started on port(s): 8443 (https) 8080 (http) with context path '/'
```

现在可以同时使用 HTTPS 和 HTTP 来访问应用了，如下图所示。

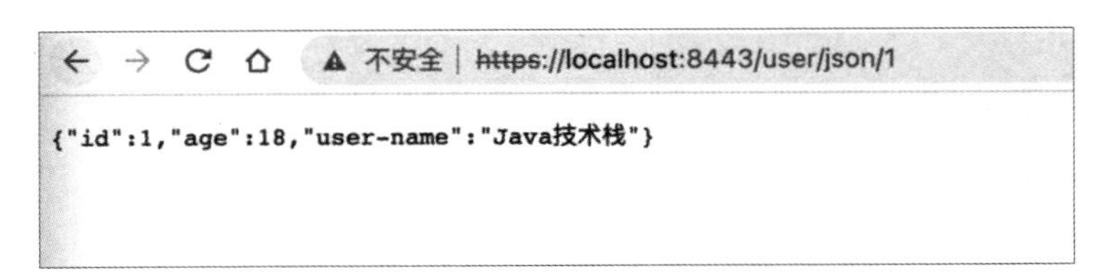

虽然 Spring Boot 支持 HTTPS 是如此简单，但实际应用中却很少使用，因为实际项目肯定要考虑负载均衡、高可用性，不可能由开发人员在每个应用中配置协议，所以 HTTPS 一般由运维人员在 Nginx 负载均衡组件上统一配置，方便运维人员统一管理。

另一方面，Spring Boot 应用一般不会直接对外开放，一般是内网部署的，都需要负载均衡组件或者服务网关进行路由转发，所以 HTTPS 会配置在外网负载均衡的组件上。

6.2.6 持久化

嵌入式容器也是支持会话持久化配置的，比如：

```
server:
  port: 8080
  servlet:
    session:
      # 开启持久化
      persistent: true
      # 持久化目录
      store-dir: /tmp/session-store
      # 追踪模式
      tracking-modes:
        - cookie
        - url
```

更多的配置参数可以查看 org.springframework.boot.web.servlet.server.Session 类。

> **需要注意的是：**
> 持久化需要使用 kill -15 pid 以触发优雅关闭应用，不能使用 kill -9 pid 强制“杀进程”等方式，否则应用无法接收退出信号、无法完成会话持久化。

6.2.7 优雅关闭

Spring Boot 内置的容器都支持优雅关闭，在应用正常退出时可以完成对现有请求的处理并关闭相应的资源，以达到平滑退出的目的。

实现 Spring Boot 的优雅关闭只需要添加以下配置参数：

```
server:
  shutdown: graceful
```

并且还支持一个超时参数设置：

```
spring:
  lifecycle:
timeout-per-shutdown-phase: 20s
```

当优雅关闭 Spring Boot 时，在超时时间限定内将允许完成对现有请求的处理，且停止接收新的请求，具体方式也因使用的 Web 容器而异，比如 Jetty、Reactor Netty、Tomcat 容器将停止接收请求，而 Undertow 会继续接收请求，但会立即响应服务不可用（503）错误。如果超时后请求还没处理完，则 Spring Boot 会强制关闭应用。所以需要注意的是，优雅关闭时一定要考虑当时的业务处理量，所设置的缓冲时间是否能处理完正在处理中的业务。

具体的实现机制可以查看 Spring Boot 启动过程中的 refreshContext 方法的源码：

```
private void refreshContext(ConfigurableApplicationContext context) {
    if (this.registerShutdownHook) {
        shutdownHook.registerApplicationContext(context);
    }
    refresh(context);
}
```

先检查是否注册优雅关闭的钩子，默认是允许的，然后将优雅关闭的钩子注册到 Spring 应用上下文中，最后刷新 Spring 应用上下文。

注册钩子的系列方法的源码：

```
void registerApplicationContext(ConfigurableApplicationContext context) {
    addRuntimeShutdownHookIfNecessary();
    synchronized (SpringApplicationShutdownHook.class) {
        assertNotInProgress();
        context.addApplicationListener(this.contextCloseListener);
        this.contexts.add(context);
    }
}

private void addRuntimeShutdownHookIfNecessary() {
    if (this.shutdownHookAdded.compareAndSet(false, true)) {
```

```
            addRuntimeShutdownHook();
        }
    }

    void addRuntimeShutdownHook() {
        Runtime.getRuntime().addShutdownHook(new Thread(this, "SpringApplicationShutdownHook"));
    }
```

可以看到，Spring Boot 的优雅关闭其实也是调用了 JDK 的 Runtime.getRuntime(). addShutdownHook 方法注册了一个 JVM 关闭的钩子。

> **需要注意的是：**
> 不建议使用 kill -9 pid 强制“杀掉”应用，另外，还可能出现其他异常情况，比如停电、系统崩溃等，一般可以使用 kill -15 pid 的方式以触发应用优雅关闭。

以上介绍了几个常用容器的应用配置，其实还有更多高级的应用，可以根据需要自行配置。

6.3 自定义 Web 配置

如果不想使用 Spring Boot 默认的 Web 配置，那么可以通过实现 Spring MVC 中的 WebMvcConfigurer 接口的方式自定义 Web 配置，比如自定义添加拦截器、格式化器、资源处理器等，WebMvcConfigurer 接口中的所有方法如下图所示。

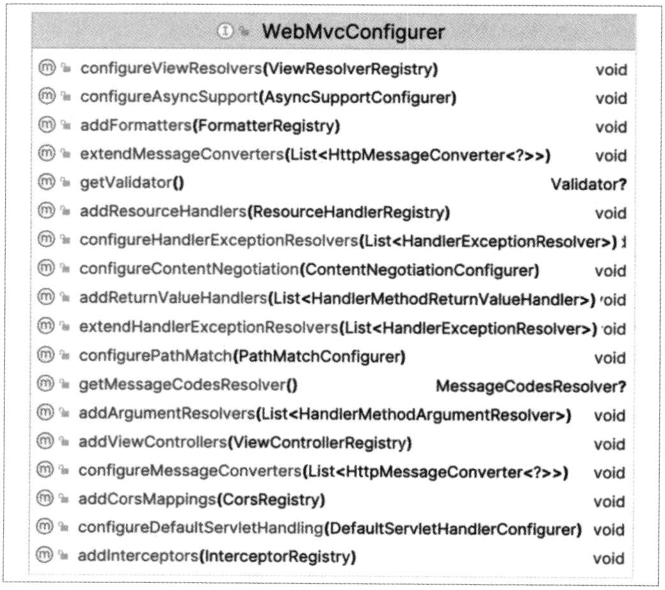

WebMvcConfigurer	
configureViewResolvers(ViewResolverRegistry)	void
configureAsyncSupport(AsyncSupportConfigurer)	void
addFormatters(FormatterRegistry)	void
extendMessageConverters(List<HttpMessageConverter<?>>)	void
getValidator()	Validator?
addResourceHandlers(ResourceHandlerRegistry)	void
configureHandlerExceptionResolvers(List<HandlerExceptionResolver>)	i
configureContentNegotiation(ContentNegotiationConfigurer)	void
addReturnValueHandlers(List<HandlerMethodReturnValueHandler>)	'oid
extendHandlerExceptionResolvers(List<HandlerExceptionResolver>)	'oid
configurePathMatch(PathMatchConfigurer)	void
getMessageCodesResolver()	MessageCodesResolver?
addArgumentResolvers(List<HandlerMethodArgumentResolver>)	void
addViewControllers(ViewControllerRegistry)	void
configureMessageConverters(List<HttpMessageConverter<?>>)	void
addCorsMappings(CorsRegistry)	void
configureDefaultServletHandling(DefaultServletHandlerConfigurer)	void
addInterceptors(InterceptorRegistry)	void

所以，可以创建一个 @Configuration 配置类，并实现 WebMvcConfigurer 接口，然后实现需要扩展的方法即可，如下所示。

```
@Configuration
public class WebConfig implements WebMvcConfigurer {

    ...

}
```

具体内容可见后续的章节。

6.4 注册拦截器

可以通过实现 Spring MVC 中的 HandlerInterceptor 接口创建一个拦截器，笔者这里创建并注入了一个简单的登录拦截器，如下面的示例所示。

```
@Component
public class LoginInterceptor implements HandlerInterceptor {

    @Override
    public boolean preHandle(HttpServletRequest request,
                     HttpServletResponse response, Object handler) throws Exception {
        HttpSession session = request.getSession();
        String userSession = (String) session.getAttribute("userSession");

        if (userSession == null) {
            response.sendRedirect("/login");
            return false;
        } else {
            return true;
        }
    }

}
```

先获取对应的 userSession 会话值，如果该会话值存在则通过，返回 true 再到下一个拦截器，否则跳到登录页并返回 false。

前面的章节中介绍了，自定义的配置可以在自定义 Web 配置类中进行注册，如下面的示例所示。

```
@Configuration
@RequiredArgsConstructor
public class WebConfig implements WebMvcConfigurer {

    private final LoginInterceptor loginInterceptor;

    @Override
    public void addInterceptors(InterceptorRegistry registry) {
        registry.addInterceptor(loginInterceptor)
                .addPathPatterns("/**")
                .excludePathPatterns("/login/**")
                .excludePathPatterns("/static/**");
    }

}
```

先注入该拦截器，再通过 InterceptorRegistry.addInterceptor 方法添加并注册拦截器，还可以对该拦截器配置需要拦截的请求路径，以及要排除的请求路径等。

6.5 注册消息转换器

在 Spring MVC 框架中，使用了一个 HttpMessageConverter 消息转换器接口来转换 HTTP 请求和响应，Spring Boot 默认是"开箱即用"的，例如，JSON 或 XML 数据类型的请求格式可以自动转换为 Java Bean 对象，也可以互转。

Spring Boot 支持以下三种 JSON 库的自动配置：

- Jackson（首选默认）。
- JSON-B。
- Gson。

在 Jackson 的 JacksonHttpMessageConvertersConfiguration 消息转换器的自动配置类中就提供了 JSON、XML 的默认自动配置：

```
@Configuration(proxyBeanMethods = false)
class JacksonHttpMessageConvertersConfiguration {

    @Configuration(proxyBeanMethods = false)
    @ConditionalOnClass(ObjectMapper.class)
```

```
    @ConditionalOnBean(ObjectMapper.class)
    @ConditionalOnProperty(name = HttpMessageConvertersAutoConfiguration.PREFERRED_
            MAPPER_PROPERTY, havingValue = "jackson", matchIfMissing = true)
    static class MappingJackson2HttpMessageConverterConfiguration {

        @Bean
        @ConditionalOnMissingBean(value = MappingJackson2HttpMessageConverter.class,
                ignoredType = {
                        "org.springframework.hateoas.server.mvc.TypeConstrained-
MappingJackson2HttpMessageConverter",
                        "org.springframework.data.rest.webmvc.alps.AlpsJsonHttp-
MessageConverter" })
        MappingJackson2HttpMessageConverter mappingJackson2HttpMessageConverter
(ObjectMapper objectMapper) {
            return new MappingJackson2HttpMessageConverter(objectMapper);
        }

    }

    @Configuration(proxyBeanMethods = false)
    @ConditionalOnClass(XmlMapper.class)
    @ConditionalOnBean(Jackson2ObjectMapperBuilder.class)
    protected static class MappingJackson2XmlHttpMessageConverterConfiguration {

        @Bean
        @ConditionalOnMissingBean
        public MappingJackson2XmlHttpMessageConverter mappingJackson2XmlHttpMessageConverter(
                Jackson2ObjectMapperBuilder builder) {
            return new MappingJackson2XmlHttpMessageConverter(builder.createXmlMapper (true)
.build());
        }

    }

}
```

如果默认的配置不符合要求，则可以在自己的配置文件中进行自定义覆盖，比如下面的示例，覆盖了默认的 JSON 消息转换器：

```
@Configuration
public class WebConfig implements WebMvcConfigurer {

    @Bean
```

```
    public MappingJackson2HttpMessageConverter mappingJackson2HttpMessageConverter() {
        MappingJackson2HttpMessageConverter converter = new MappingJackson2HttpMessage-
Converter();
        ObjectMapper mapper = new ObjectMapper();
        mapper.configure(DeserializationFeature.FAIL_ON_UNKNOWN_PROPERTIES, false);

        SimpleModule module = new SimpleModule();
        module.addDeserializer(String.class, new StringWithoutSpaceDeserializer (String.class));
        mapper.registerModule(module);

        converter.setObjectMapper(mapper);
        return converter;
    }

}

public class StringWithoutSpaceDeserializer extends StdDeserializer<String> {

    private static final long serialVersionUID = -6972065572263950443L;

    public StringWithoutSpaceDeserializer(Class<String> vc) {
        super(vc);
    }

    @Override
    public String deserialize(JsonParser p, DeserializationContext deserializationContext)
throws IOException {
        return StringUtils.trimToEmpty(p.getText());
    }

}
```

这个自定义的 JSON 消息转换器设置了反序列化过程中遇到不明确的参数时不用失败，并且配置了一个字符串反序列化器，用来过滤 HTTP 请求字符串参数首尾的空格。

如果不是覆盖自动配置已有的消息转换器 Bean，而是需要添加额外的自定义转换器，则需要使用 Spring Boot 的 HttpMessageConverters 类，如下面的示例所示。

```
@Configuration
public class WebConfig implements WebMvcConfigurer {

    @Bean
    public HttpMessageConverters customConverters() {
```

```
        HttpMessageConverter<?> additional = new AdditionalHttpMessageConverter();
        HttpMessageConverter<?> another = new AnotherHttpMessageConverter();
        return new HttpMessageConverters(additional, another);
    }

}
```

HttpMessageConverters类可以允许多个自定义HttpMessageConverter消息转换器的添加和注册。

6.6　注册类型转换器

Spring MVC 中的 Converter 接口可以转换参数的类型，比如笔者这里配置了一个 CustomConverter 转换器，如下面的示例所示。

```
@Configuration
public class WebConfig implements WebMvcConfigurer {

    @Override
    public void addFormatters(FormatterRegistry registry) {
        registry.addConverter(new CustomConverter());
    }

}

@Slf4j
public class CustomConverter implements Converter<String, String> {

    @Override
    public String convert(String source) {
        if (StringUtils.isNotEmpty(source)) {
            source = source.trim();
        }
        return source;
    }

}
```

这个转换器的逻辑很简单，就是去掉请求参数中的首尾空格。

6.7 注册 Servlet、Filter、Listener

6.7.1 Spring Boot 的手动注册

Spring Boot 提供了以下三个注册类：

- ServletRegistrationBean。
- FilterRegistrationBean。
- ServletListenerRegistrationBean。

分别用来注册 Servlet、Filter、Listener。

下面是注册 Servlet 的示例代码，首先创建一个 Servlet：

```
public class RegisterServlet extends HttpServlet {

    @Override
    protected void service(HttpServletRequest req, HttpServletResponse resp) throws IOException {
        String name = getServletConfig().getInitParameter("name");
        String sex = getServletConfig().getInitParameter("sex");

        resp.getOutputStream().println("name is " + name);
        resp.getOutputStream().println("sex is " + sex);
    }

}
```

然后在任何一个 @Configuration 配置类中进行注册，如下面的注册示例所示。

```
@Bean
public ServletRegistrationBean registerServlet() {
    ServletRegistrationBean servletRegistrationBean = new ServletRegistrationBean(
        new RegisterServlet(), "/registerServlet");
    servletRegistrationBean.addInitParameter("name", "javastack");
    servletRegistrationBean.addInitParameter("sex", "male");
    return servletRegistrationBean;
}
```

应用启动后访问该 Servlet，结果如下图所示。

```
http://localhost:8080/registerServlet
```

localhost:8080/registerServlet

```
name is registerServlet
sex is male
```

注册 Filter、Listener 也是同样的方法，这里不再赘述。

6.7.2　组件扫描注册

Servlet 3.0 之前，Servlet、Filter、Listener 这些组件都需要在 web.xml 中进行配置，Servlet 3.0 之后不再需要 web.xml 这个配置文件了，所有的组件都可以通过代码配置或者注解来实现相关功能，嵌入式 Tomcat 的核心包如下图所示。

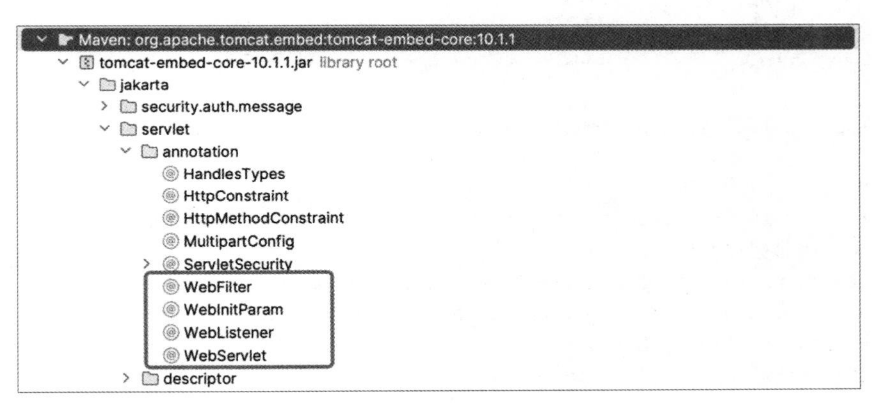

Servlet 3.0 提供了以下 3 个注解来代替相关配置：

- @WebServlet => 代替 servlet 配置。
- @WebFilter => 代替 filter 配置。
- @WebListener => 代替 listener 配置。

以下是配置 Servlet 的示例：

```
@WebServlet(name = "javaServlet", urlPatterns = "/javaServlet", asyncSupported = true,
        initParams = {
                @WebInitParam(name = "name", value = "javaServlet"),
                @WebInitParam(name = "sex", value = "male")})
public class JavaServlet extends HttpServlet {

    @Override
```

```
    protected void service(HttpServletRequest req, HttpServletResponse resp) throws
IOException {
        String name = getServletConfig().getInitParameter("name");
        String sex = getServletConfig().getInitParameter("sex");

        resp.getOutputStream().println("name is " + name);
        resp.getOutputStream().println("sex is " + sex);

    }

}
```

以下是配置 Filter 的示例：

```
@WebFilter(filterName = "javaFilter", urlPatterns = "/*", initParams = {
        @WebInitParam(name = "name", value = "javastack"),
        @WebInitParam(name = "code", value = "123456")})
@Slf4j
public class JavaFilter implements Filter {

    @Override
    public void init(FilterConfig filterConfig) throws ServletException {
        log.info("java filter init.");
        String name = filterConfig.getInitParameter("name");
        String code = filterConfig.getInitParameter("code");
        log.info("name is " + name);
        log.info("code is " + code);
    }

    @Override
    public void doFilter(ServletRequest request, ServletResponse response, FilterChain chain)
            throws IOException, ServletException {
        log.info("java filter processing.");
        chain.doFilter(request, response);
    }

    @Override
    public void destroy() {
        log.info("java filter destroy.");
    }

}
```

Listener 的配置方式与之类似，这里不再赘述。

> **需要注意的是：**
> 为了安全考虑，内嵌服务器不会直接执行 Servlet 3.0 中的 javax.servlet.ServletContainerInitializer 接口，或者 Spring 中的 org.springframework.web.WebApplicationInitializer 接口，否则会终止 Spring Boot 应用。

如果使用的是 Spring Boot 内嵌服务器，则需要在配置类上面添加额外的 @ServletComponentScan 注解来开启 Servlet 组件扫描功能，如果使用的是独立的服务器，则不需要添加，会使用服务器内部的自动发现机制。

现在访问该 Servlet，结果如下图所示。

```
http://localhost:8080/javaServlet
```

localhost:8080/javaServlet

```
name is javaServlet
sex is male
```

过滤器输出以下日志：

```
java filter init.
name is javastack
code is 123456

...

java filter processing.
```

6.7.3 动态注册

如果想在 Spring Boot 中完成 Servlet、Filter、Listener 的初始化操作，则需要在 Spring 中实现 ServletContextInitializer 接口并注册为一个 Bean，然后通过 ServletContext 提供的几个方法动态注册 Web 组件，如下图所示。

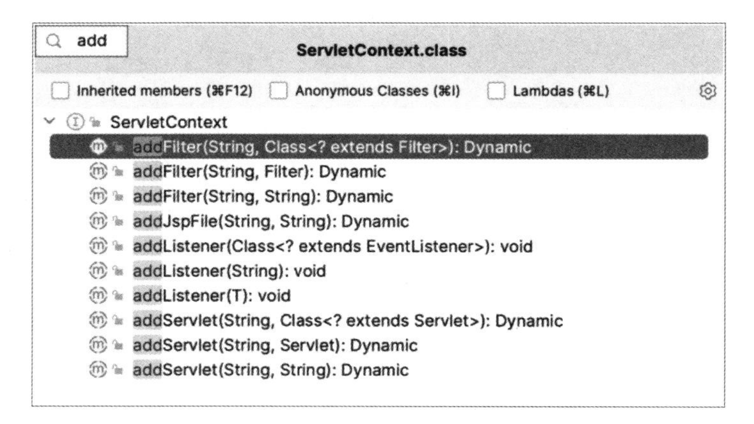

这里创建一个 Servlet：

```
public class InitServlet extends HttpServlet {

    @Override
    protected void service(HttpServletRequest req, HttpServletResponse resp) throws
IOException {
        String name = getServletConfig().getInitParameter("name");
        String sex = getServletConfig().getInitParameter("sex");

        resp.getOutputStream().println("name is " + name);
        resp.getOutputStream().println("sex is " + sex);

    }

}
```

该 Servlet没有在任何地方注册，以下为动态注册该 Servlet 的代码：

```
// 扫描组件的注册方式
@Component
public class ServletConfig implements ServletContextInitializer {

    @Override
    public void onStartup(ServletContext servletContext) {
        ServletRegistration initServlet = servletContext
            .addServlet("initServlet", InitServlet.class);
        initServlet.addMapping("/initServlet");
        initServlet.setInitParameter("name", "initServlet");
        initServlet.setInitParameter("sex", "male");
```

```
    }

}

// Bean 注册
@Bean
public ServletContextInitializer servletContextInitializer() {
    return (servletContext) -> {
        ServletRegistration initServlet = servletContext.addServlet("initServlet",
InitServlet.class);
        initServlet.addMapping("/initServlet");
        initServlet.setInitParameter("name", "initServlet");
        initServlet.setInitParameter("sex", "male");
    };
}
```

以上两种方式是等效的，注册完成后再访问该 Servlet，结果如下图所示。

```
http://localhost:8080/initServlet
```

总结一下，注册 Servlet、Filter、Listener 的方式至少有上述三种，读者可以根据需要自行选择，但一般情况下扫描注册无疑是最简洁的。

6.8 静态资源处理

默认情况下，Spring Boot 加载的 classpath 下的几个静态资源目录如下：

- /static。
- /public。
- /resources。
- /META-INF/resources。

如果需要替换这些默认的资源目录位置，则可以通过 spring.web.resources.static-locations 参数指定自定义的资源目录。

Spring Boot 默认把静态资源映射到 /** 访问路径，但也可以使用 spring.mvc.static-path-pattern 参数对其进行调整，比如以下示例：

```
spring:
  mvc:
static-path-pattern: "/pub/**"
```

这里将所有资源访问路径重定位到 /pub/**，比如以前 static 目录下的静态资源是直接用 /** 访问的，现在需要用 /pub/** 进行访问。

另外，Spring MVC 默认使用 ResourceHttpRequestHandler 处理器类来处理静态资源，也可以实现自己的处理器来改变默认的处理器行为，只需要实现 WebMvcConfigurer 并覆盖 addResourceHandlers 方法即可。

来看下面的示例：

```
@Configuration
public class WebConfig implements WebMvcConfigurer {

    @Override
    public void addResourceHandlers(ResourceHandlerRegistry registry) {
        registry.addResourceHandler("/assets/**").addResourceLocations ("classpath:/
assets/");
        registry.addResourceHandler("/dist/**").addResourceLocations ("classpath:/
dist/");
    }

}
```

这里额外添加了两个 classpath 下静态资源目录的映射。

6.9 模板引擎

Spring Boot 不仅可以作为 HTTP 接口服务使用，还可以用来开发动态网页，Spring Boot 支持以下几个模板引擎的自动配置：

- FreeMarker。
- Thymeleaf。
- Mustache。

- Velocity（Spring Boot 1.5 不再支持）。
- Groovy（Spring Boot 2.7.1 不再支持）。

Spring Boot 早期的版本还支持 Velocity、Groovy，不过后面都废除了。除了以上这些模板引擎的“开箱即用”，Spring Boot 还支持 JSP，但这种古老的技术早已不推荐使用了，如果 JSP 在内嵌容器中运行还会存在很多限制。

现在 Spring Boot 使用比较广泛的是 Thymeleaf 模板引擎，本节以 Thymeleaf 模板引擎为示例进行集成演示。集成模板引擎只需要加入它的 Starter 启动器即可，比如下面添加 Thymeleaf 的 spring-boot-starter-thymeleaf 启动器示例：

```
<dependency>
    <groupId>org.springframework.boot</groupId>
    <artifactId>spring-boot-starter-thymeleaf</artifactId>
</dependency>
```

Spring Boot 会完成 Thymeleaf 模板引擎的自动配置，根据 Spring Boot 自动配置类的命名规范可以找到它的自动配置类 ThymeleafAutoConfiguration，有需要的话可以自行覆盖里面的配置，也能找到对应的配置参数类 ThymeleafProperties，它的源码如下：

```
@ConfigurationProperties(prefix = "spring.thymeleaf")
public class ThymeleafProperties {

    private static final Charset DEFAULT_ENCODING = StandardCharsets.UTF_8;

    public static final String DEFAULT_PREFIX = "classpath:/templates/";

      public static final String DEFAULT_SUFFIX = ".html";

    ...
```

配置参数的前缀为 spring.thymeleaf，默认的模板引擎目录是 src/main/resources/templates，默认的模板文件后缀为 .html，默认的编码为 UTF-8，还有其他更多的参数，比如是否检查模板、是否开启缓存等，都可以自行配置。

笔者这里在默认模板目录下创建了一个 test.html 模板文件：

```
<!DOCTYPE html>
<html lang="en" xmlns:th="http://www.thymeleaf.org">
<head>
    <meta charset="utf-8"/>
```

```
    <title>test</title>
    <link rel="stylesheet" type="text/css" th:href="@{/css/test.css}"/>
    <script th:src="@{/js/test.js}"></script>
</head>
<body>

<a id="content" th:utext="${content}"
   class="red" onclick="changeColor()"></a>

</body>
</html>
```

使用 Thymeleaf 模板引擎的语法来引入需要渲染的内容：

（1）在模板文件头部使用 <html lang=” en” xmlns:th=” http://www.thymeleaf.org” > 定义模板文件，使用 th 作为模板语法前缀。

（2）使用 @{} 引入静态资源文件。

（3）使用 ${} 引入 Spring MVC 绑定的请求数据。

然后新建一个 HTTP 请求，返回模板页面路径名称，如下面的示例所示。

```
@GetMapping(value = "/test/{content}")
public String test(@PathVariable("content") String content) {
    request.setAttribute("content", content);
    return "test";
}
```

启动应用并访问 /test 页面，结果如下图所示。

正常获取值，并能通过单击操作来更换颜色，引入的静态资源也都生效了。

6.10 异常处理

6.10.1 默认的异常处理

Spring Boot 提供了一个默认的 /error 映射处理所有的异常错误，并提供了一个默认的

Whitelabel 错误页面，如下图所示。

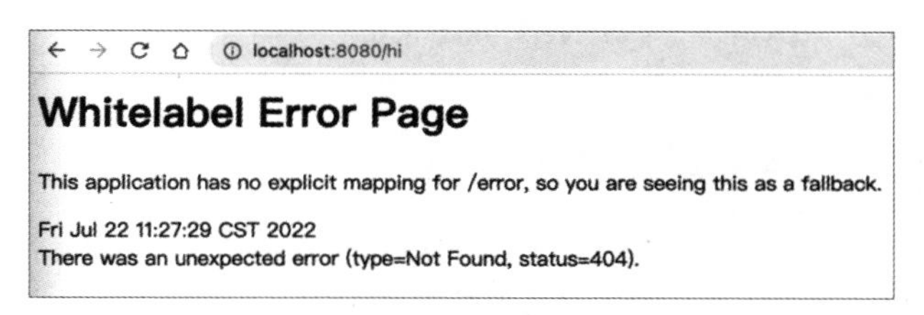

对应的错误处理器实现类为 BasicErrorController，它实现了 ErrorController 接口，类结构如下图所示。

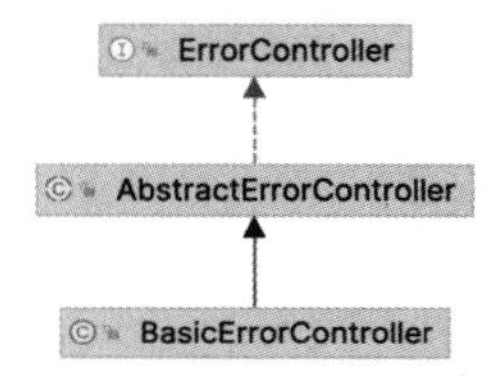

该错误处理器可以通过 server.error 系列参数配置一些错误信息，比如更改默认的 /error 错误请求，或者包含更多指定的错误信息，或者禁用默认的错误页面等。

如果要完全替换默认的错误机制，则可以扩展 BasicErrorController 类，还可以实现 ErrorController 接口并注册该类型的 Bean 或者添加 ErrorAttributes 类型的 Bean 以使用现有的机制来替换相关内容。

6.10.2 自定义全局异常

很多时候在项目中不想使用默认的错误机制，需要自定义全局异常，这时一般使用 @ControllerAdvice+@ExceptionHandler 注解实现全局异常处理，比如下面的配置示例：

```
@ControllerAdvice
public class GlobalExceptionHandler {

    @ResponseBody
    @ExceptionHandler(value = { Exception.class })
    public ResponseEntity<?> handleException(HttpServletRequest request, Throwable ex) {
        return new ResponseEntity<>("global exception", HttpStatus.OK);
    }

}
```

这个 @ExceptionHandler 注解用于指定要处理的异常，value 可以指定多个异常类，在 @ControllerAdvice 类中也可以指定多个 @ExceptionHandler 处理方法。

6.10.3 自定义异常状态码页面

如果想定制自定义错误码的错误页面，则可以在任何一个静态目录下创建 /error 目录，并创建对应的状态码静态模板文件，如下面的文件目录结构所示。

```
src/
 +- main/
     +- resources/
         +- public/
             +- error/
             |   +- 404.html
             |   +- 500.html
```

这里创建了 404.html 和 500.html 两个状态码模板文件，当发生 404、500 错误时会跳转到该目录下的页面。当访问一个不存在的路径时，浏览器返回到自定义的 404 页面，如下图所示。

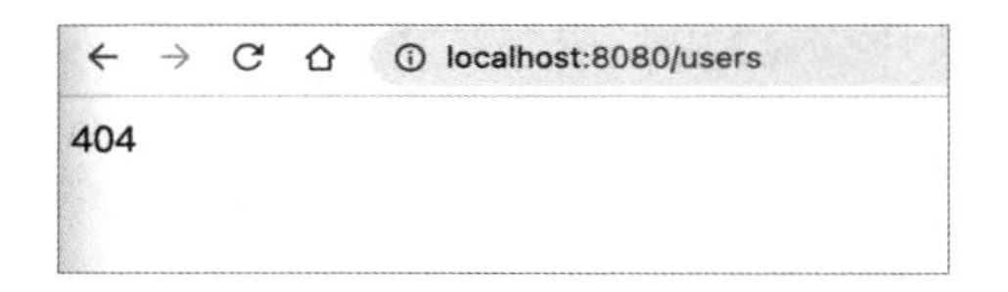

6.11 参数校验

6.11.1 概述

和配置参数校验一样，Spring Boot 同样支持基于 JSR-303 规范对接口参数进行校验，比如常用的 hibernate-validator 就是基于 JSR-303 规范实现的依赖：

```
<dependency>
  <groupId>org.hibernate.validator</groupId>
  <artifactId>hibernate-validator</artifactId>
</dependency>
```

Spring Boot 应用无须手动添加 hibernate-validator 依赖，可以使用导入 spring-boot-starter-validation 启动器的方式：

```xml
<dependency>
    <groupId>org.springframework.boot</groupId>
    <artifactId>spring-boot-starter-validation</artifactId>
</dependency>
```

这样做更全面、更专业，它除了会涵盖 hibernate-validator 依赖，还会导入 Spring Boot 参数校验所需的全部依赖组件，如下面的依赖关系图所示。

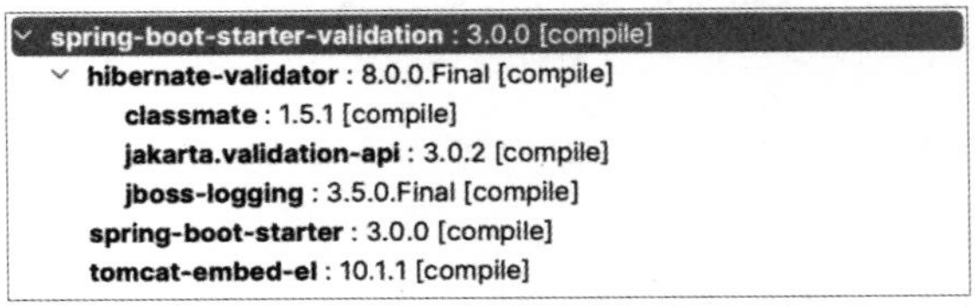

```
spring-boot-starter-validation : 3.0.0 [compile]
  hibernate-validator : 8.0.0.Final [compile]
    classmate : 1.5.1 [compile]
    jakarta.validation-api : 3.0.2 [compile]
    jboss-logging : 3.5.0.Final [compile]
  spring-boot-starter : 3.0.0 [compile]
  tomcat-embed-el : 10.1.1 [compile]
```

6.11.2 约束注解

校验接口参数需要用到基于 J2EE 规范的 javax.validation 包下的约束注解，现在包名已经更改为 jakarta.validation，因为早在几年前 Java EE 已经正式更名为 Jakarta，所以，所有相关的名称都变了。

如下图所示，可以看到该包下的全部约束注解。

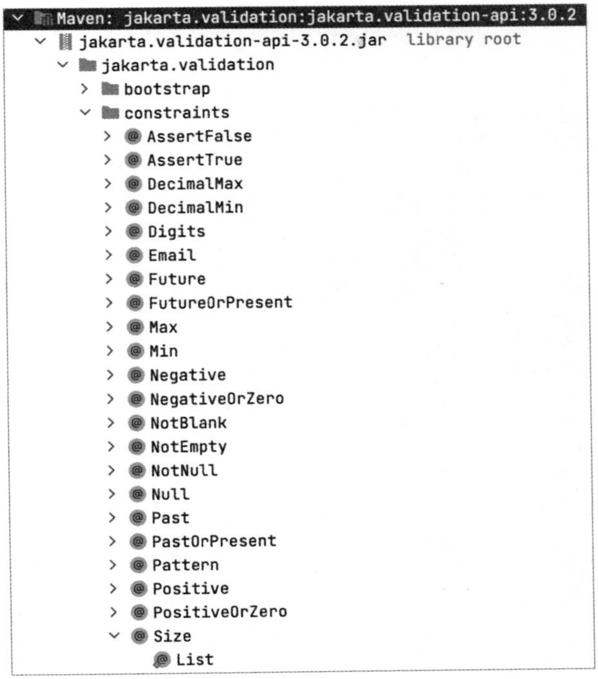

```
Maven: jakarta.validation:jakarta.validation-api:3.0.2
  jakarta.validation-api-3.0.2.jar  library root
    jakarta.validation
      bootstrap
      constraints
        AssertFalse
        AssertTrue
        DecimalMax
        DecimalMin
        Digits
        Email
        Future
        FutureOrPresent
        Max
        Min
        Negative
        NegativeOrZero
        NotBlank
        NotEmpty
        NotNull
        Null
        Past
        PastOrPresent
        Pattern
        Positive
        PositiveOrZero
        Size
          List
```

一看这些约束注解的名称就知道它们的作用，这里不再赘述，它们既可以用在方法参数上，也可以用在参数类成员变量上。

6.11.3 参数校验示例

导入spring-boot-starter-validation启动器之后，可以使用@Valid或者@Validated注解来校验参数，@Valid是Jakarta规范的注解，@Validated是Spring中支持的JSR-303注解，它其实就是@Valid注解的变体。

一般推荐使用Spring中的@Validated注解，功能比@Valid注解要强大，比如它支持分组效验等，常用的做法是：

- 校验接口参数类时，将@Validated注解放在方法参数类之前，然后将约束注解放在类成员变量上面。
- 校验接口单个参数时，将@Validated注解放在当前Controller类上面，然后将约束注解放在方法参数类之前。

比如在参数类上使用各种约束注解，示例如下：

```
@Data
@NoArgsConstructor
public class User {

    public User(String userName, Integer age) {
        this.userName = userName;
        this.age = age;
    }

    private Long id;

    @NotNull
    @Size(min = 5, max = 10)
    private String userName;

    @NotNull
    private Integer age;

    private String address;

    private String memo;


}
```

这里定义了用户名和年龄不能为空，且用户名长度在 5 ～ 10 之间。

校验该参数类时需要在接口方法参数类之前放置 @Validated 注解，校验单个参数时就在 Controller 类上面放置 @Validated 注解，示例如下：

```
@RestController
@Validated
public class ResponseBodyController {

    @CrossOrigin
    @GetMapping(value = "/user/json/{userId}", produces = MediaType.APPLICATION_JSON_VALUE)
    public ResponseEntity getJsonUserInfo(@PathVariable("userId") @Size(min = 5, max = 8)
String userId) {
        User user = new User("Java 技术栈 ", 18);
        user.setId(Long.valueOf(userId));
        return new ResponseEntity(user, HttpStatus.OK);
    }

    @PostMapping(value = "/user/save")
    public ResponseEntity saveUser(@RequestBody @Validated User user) {
        user.setId(RandomUtils.nextLong());
        return new ResponseEntity(user, HttpStatus.OK);
    }


}
```

如果要返回具体的参数错误，则需要拦截对应的 MethodArgumentNotValidException 异常，并做对应的返回处理，示例如下：

```
@ResponseBody
@ExceptionHandler(value = {MethodArgumentNotValidException.class})
public ResponseEntity handleMethodArgumentNotValidException (MethodArgumentNotValid-
Exception ex) {
    BindingResult bindingResult = ex.getBindingResult();
    StringBuilder sb = new StringBuilder(" 参数校验失败 :");
    for (FieldError fieldError : bindingResult.getFieldErrors()) {
        sb.append(fieldError.getField()).append(": ").append(fieldError.getDefaultMessage()).
append(", ");
    }
    String msg = sb.toString();
    return new ResponseEntity(msg, HttpStatus.OK);
}
```

重启应用，调用保存用户接口测试一下，结果如下图所示。

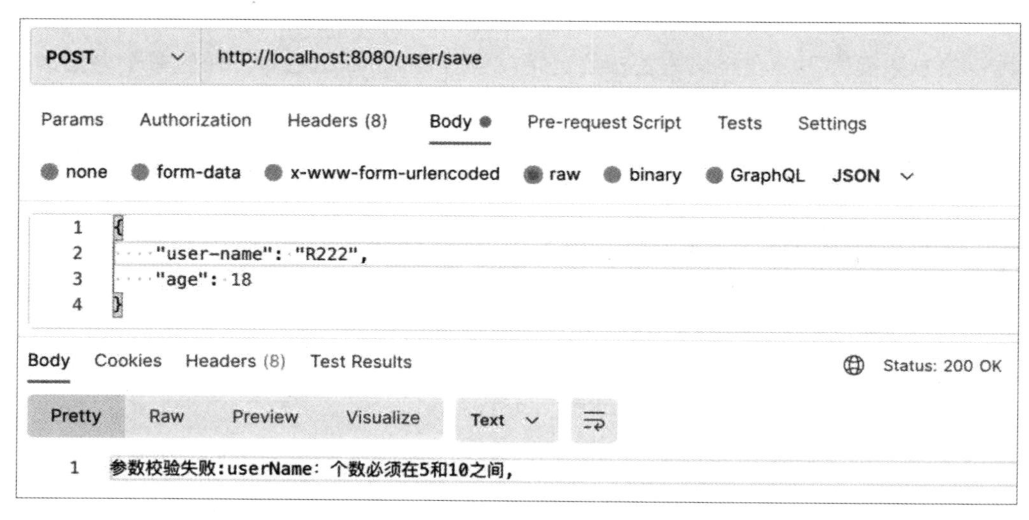

如果参数不符合条件就会返回对应的错误。

> 以上仅为异常测试，实际项目中应把接口响应信息封装成带有 code、msg、data 的标准化 JSON 格式的数据返回，然后将错误信息放在 msg 字段中。

6.12 国际化

6.12.1 概述

Spring Boot 支持本地化的消息，也就是国际化，如果应用需要满足不同语言偏好的用户，这就很有用。默认情况下，Spring Boot 会在类路径的根目录下搜索消息资源包是否存在。

Spring Boot 中已经对国际化做了自动配置，国际化自动配置类如下：

> org.springframework.boot.autoconfigure.context.MessageSourceAutoConfiguration

它同样被注册在新的 org.springframework.boot.autoconfigure.AutoConfiguration.imports 自动配置文件中，该自动配置类的核心源码如下：

```
@AutoConfiguration
@ConditionalOnMissingBean(name = AbstractApplicationContext.MESSAGE_SOURCE_BEAN_NAME,
search = SearchStrategy.CURRENT)
@AutoConfigureOrder(Ordered.HIGHEST_PRECEDENCE)
```

```
@Conditional(ResourceBundleCondition.class)
@EnableConfigurationProperties
public class MessageSourceAutoConfiguration {

    private static final Resource[] NO_RESOURCES = {};

    @Bean
    @ConfigurationProperties(prefix = "spring.messages")
    public MessageSourceProperties messageSourceProperties() {
        return new MessageSourceProperties();
    }

    @Bean
    public MessageSource messageSource(MessageSourceProperties properties) {
        ResourceBundleMessageSource messageSource = new ResourceBundleMessageSource();
        if (StringUtils.hasText(properties.getBasename())) {
            messageSource.setBasenames(StringUtils
                    .commaDelimitedListToStringArray(StringUtils.trimAllWhitespace
(properties.getBasename())));
        }
        if (properties.getEncoding() != null) {
            messageSource.setDefaultEncoding(properties.getEncoding().name());
        }
        messageSource.setFallbackToSystemLocale(properties.isFallbackToSystemLocale());
        Duration cacheDuration = properties.getCacheDuration();
        if (cacheDuration != null) {
            messageSource.setCacheMillis(cacheDuration.toMillis());
        }
        messageSource.setAlwaysUseMessageFormat(properties.isAlwaysUseMessageFormat());
        messageSource.setUseCodeAsDefaultMessage(properties.isUseCodeAsDefaultMessage());
        return messageSource;
    }

    ...

}
```

主要是注册了一个 MessageSource 实例，通过一系列 spring.messages.* 参数可以完成自动配置，查看参数绑定类 MessageSourceProperties 的源码，有以下主要几个重要的配置参数：

```
private String basename = "messages";

private Charset encoding = StandardCharsets.UTF_8;
```

```
@DurationUnit(ChronoUnit.SECONDS)
private Duration cacheDuration;

private boolean fallbackToSystemLocale = true;
```

这几个参数的说明如下：

- basename：指定要扫描的国际化文件名，默认为 messages，即在 resources 资源目录下建立 messages_xx.properties 资源文件，可以通过逗号指定多个，如果不指定包名，则默认从 classpath 根目录下搜索。
- encoding：默认的编码为 UTF-8。
- cacheDuration：国际化资源文件被加载后的缓存时间，默认单位为秒，如果不指定则默认为永久缓存。
- fallbackToSystemLocale：找不到当前语言的资源文件时是否降级为当前操作系统的语言对应的资源文件，默认为 true，比如在中国则加载 messages_zh_CN.properties，如果为 false 则加载系统默认的文件，如 messages.properties。

6.12.2 自动国际化

了解了国际化的自动配置和参数绑定类的作用，本节就可以进行国际化应用实践了，本节将介绍如何使用 Spring Boot 和 Thymeleaf 模板引擎实现页面国际化。

应用国际化支持在 Spring Boot 自动配置核心包中，引入 Spring Boot Web 包即可，不需要额外引入其他依赖包，因为本节使用 Thymeleaf 作为模板页面来显示国际化消息，所以还需要引入 Thymeleaf 的依赖：

```
<dependency>
    <groupId>org.springframework.boot</groupId>
    <artifactId>spring-boot-starter-web</artifactId>
</dependency>

<dependency>
    <groupId>org.springframework.boot</groupId>
    <artifactId>spring-boot-starter-thymeleaf</artifactId>
</dependency>
```

然后配置国际化资源文件的搜索路径：

```
spring:
  messages:
    basename: i18n/common, i18n/index
fallback-to-system-locale: false
```

配置参数说明如下：

- basename：这里指定了多个国际化资源文件，在 common 资源文件中定义了公共的消息，在 index 资源文件中定义了首页的消息，让页面可以读取多个资源文件的消息。
- fallback-to-system-locale：设置为 false 表示如果找不到当前语言的文件就使用默认的资源文件，这里默认的资源文件是 index.properties，里面定义为英文，即除了中文，其他都使用默认的英文资源文件。

然后在资源根路径下创建 i18n 国际化资源目录，用来统一管理国际化资源文件，以及创建对应的国际化资源文件，如下图所示。

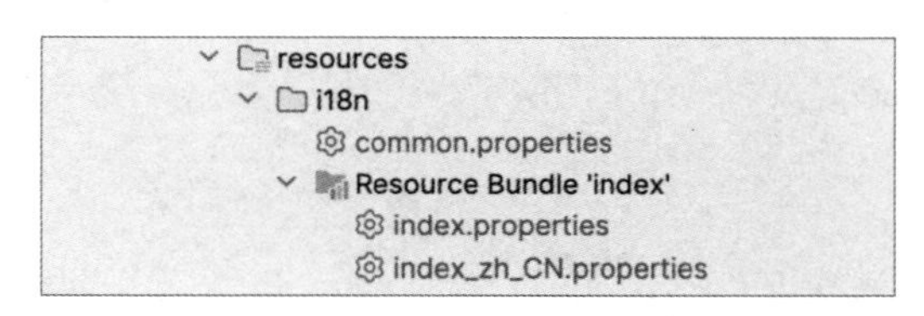

common.properties文件的内容如下：

```
brand=Java 技术栈
```

index.properties 文件的内容如下：

```
index.hi=hello
index.welcome=welcome
```

index_zh_CN.properties 文件的内容如下：

```
index.hi= 你好
index.welcome= 欢迎光临
```

创建 Thymeleaf 模板文件，引用国际化资源文件中的消息：

```
<!DOCTYPE html>
<html lang="en" xmlns:th="http://www.thymeleaf.org">
<head>
    <meta charset="utf-8"/>
```

```
    <title>index</title>
</head>
<body>

<label th:text="#{brand}"></label>
<label th:text="#{index.hi}"></label>
<label th:text="#{index.welcome}"></label>

</body>
</html>
```

创建首页接口：

```
@GetMapping(value = "/index")
public String index() {
    return "index";
}
```

然后在浏览器上访问 index 页面，如下图所示。

因为笔者的计算机是中文环境，所以页面显示了 index_zh_CN.properties 资源文件中的消息内容，这就是页面的自动国际化。

6.12.3 切换国际化

大部分场景下，页面需要支持手动切换语言，比如很多网站会在页面顶部或者底部设置切换语言的链接，这时之前的自动国际化配置就不符合要求了。

这时需要注册一个 LocaleResolver 区域解析器和区域拦截器：

```
@Slf4j
@Configuration
public class WebConfig implements WebMvcConfigurer {

    /**
     * Locale 默认设置为英文
     * @return
     */
```

```
    @Bean
    public LocaleResolver localeResolver() {
        SessionLocaleResolver sessionLocaleResolver = new SessionLocaleResolver();
        sessionLocaleResolver.setDefaultLocale(Locale.US);
        return sessionLocaleResolver;
    }

    @Override
    public void addInterceptors(InterceptorRegistry registry) {
        registry.addInterceptor(localeChangeInterceptor());
    }

    /**
     * 切换语言拦截器，通过 url?lang=zh_CN 形式进行切换
     * @return
     */
    private LocaleChangeInterceptor localeChangeInterceptor() {
        LocaleChangeInterceptor localeChangeInterceptor = new LocaleChangeInterceptor();
        localeChangeInterceptor.setParamName("lang");
        return localeChangeInterceptor;
    }

}
```

配置类说明如下：

- 首先注册了一个 LocaleResolver 实例，LocaleResolver 接口有许多实现，比如可以通过 Session、Cookie、Accept-Language Header 或者一个固定的值来判断当前的语言环境，上面使用的是 SessionLocaleResolver（基于 Session）的方式来判断的，并设置默认区域语言为英文。
- 然后注册了一个 LocaleChangeInterceptor 实例，并且设置了切换语言的参数名为 lang，即可以通过传递区域语言参数 url?lang=zh_CN 的形式进行区域语言的切换。

现在再在浏览器上访问 index 页面，如下图所示。

页面显示了默认的英文区域的内容，前面的中文是因为加载了 common 公共资源文件中的中文内容，common 是公共资源内容，没有做国际化资源配置。

现在再通过传递中文区域来访问 index页面，如下图所示。

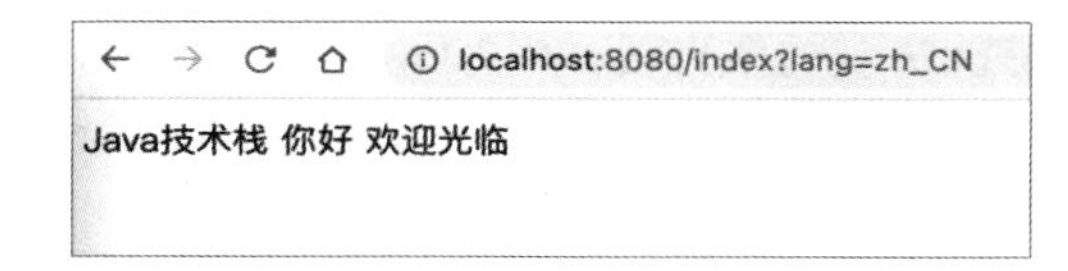

可以看到语言切换成功了，然后传递一个不存在的区域语言，如下图所示。

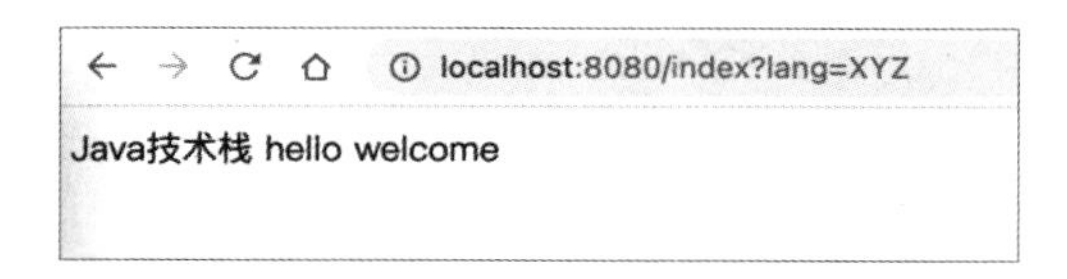

可以看到显示的是默认的英文区域的内容了。

6.13 分布式会话

传统的会话机制保存在应用服务器端，会话共享需要借助 Servlet 容器来实现，比如 Tomcat 提供了 Session 复制的功能，简单的集群复制没有什么问题，但是，如果集群节点过多，就可能会因网络延迟等问题造成会话不能及时同步，进而给应用带来性能影响，严重的问题会影响用户体验。

Spring Session 让集群会话变得轻而易举，它是 Spring 提供的一个用于管理用户会话的框架，可以将 Session 数据存储到第三方存储中并替换容器中的 HttpSession 机制，而无须绑定特定的应用容器。

Spring Boot 为 Spring Session 及各种数据存储提供了自动配置，支持 Servlet Web 应用和 Reactive Web 应用。以 Servlet Web 应用为例，可以自动配置基于以下组件的 Session 存储：

- Redis。
- JDBC。
- Hazelcast。
- MongoDB。

如果类路径中存在以上单个 Spring Session 模块，则 Spring Boot 会自动使用该存储实现，如果有多个实现，则 Spring Boot 按以上列举的组件顺序来选择，如果一个都没有，则使用默认的 Session 机制。

Session 自动配置类为 org.springframework.boot.autoconfigure.session.SessionAutoConfiguration，它同样被注册在新的 org.springframework.boot.autoconfigure.AutoConfiguration.imports 自动配置文件中，Spring Boot 自动配置后，就不再需要使用 @Enable*HttpSession 注解了，比如基于 Redis 方式

实现会话则需要使用 @EnableRedisHttpSession 注解，现在都不需要了。

Session 参数绑定类为 SessionProperties，通过一系列 spring.session.* 参数配置 Session 会话，每个存储也都有自己的扩展实现，比如 Redis 参数绑定类为 RedisSessionProperties，通过一系列 spring.session.redis.* 可以扩展 Redis 会话存储配置。

本节就以 Redis 存储会话为例，演示如何使用 Spring Boot 集成 Redis 实现分布式的会话。

> Redis 的安装、集成及具体介绍请参考第 7章。

首先添加集成 Redis 及会话相关的依赖：

```
<dependency>
    <groupId>org.springframework.session</groupId>
    <artifactId>spring-session-data-redis</artifactId>
</dependency>

<dependency>
    <groupId>org.springframework.boot</groupId>
    <artifactId>spring-boot-starter-data-redis</artifactId>
</dependency>
```

然后添加以下 Redis 会话及连接配置：

```
spring:
  session:
    timeout: 30s
  redis:
    host: localhost
port: 6379
```

Spring Boot 集成 Spring Session 后，可以使用 spring.session.timeout 参数设置超时配置，如果没有设置，就会取 server.servlet.session.timeout 设置的值（默认为 30 分钟），如果两个参数都未设置，则为后者的默认值（30 分钟），这个逻辑可以在 RedisSessionConfiguration 配置类中找到，如下图所示。

```
PropertyMapper map = PropertyMapper.get().alwaysApplyingWhenNonNull();
return (sessionRepository) -> {
    map.from(sessionProperties
            .determineTimeout(() -> serverProperties.getServlet().getSession().getTimeout()))
            .to(sessionRepository::setDefaultMaxInactiveInterval);
    map.from(redisSessionProperties::getNamespace).to(sessionRepository::setRedisKeyNamespace);
    map.from(redisSessionProperties::getFlushMode).to(sessionRepository::setFlushMode);
    map.from(redisSessionProperties::getSaveMode).to(sessionRepository::setSaveMode);
};
```

然后新建几个基础接口用于测试会话：

```
@Slf4j
@RequiredArgsConstructor
@Controller
public class IndexController {

    private final HttpSession httpSession;

    /**
     * 登录页面
     * @return
     */
    @ResponseBody
    @RequestMapping("/login")
    public String login() {
        return "login page.";
    }

    /**
     * 登录请求
     * @param username
     * @return
     */
    @RequestMapping("/login/submit")
    public String loginSubmit(@RequestParam("username") String username) {
        if (StringUtils.isNotBlank(username)) {
            httpSession.setAttribute("username", username);
            return "/index";
        }
        return "/login";
    }

    /**
     * 首页
     * @return
     */
    @ResponseBody
    @RequestMapping("/index")
    public String index() {
        log.info("session id: {}", httpSession.getId());
        return "index page.";
    }
```

```
    /**
     * 退出登录
     * @return
     */
    @RequestMapping("/logout")
    public String logout() {
        httpSession.invalidate();
        return "/login";
    }

}
```

这里包括登录页面、首页、登录请求、退出登录接口，可以像往常一样正常注入并使用HttpSession，它会自动存储到Redis中。

> **过滤器、Web相关配置代码略，请查看笔者提供的Spring Boot最佳实践项目仓库中的spring-boot-session模块的完整源代码。**

笔者这里同时启动了两个应用实例：

- 8080。
- 8081。

在没有登录的情况下，访问两个实例的/index首页接口都会跳转到登录页面，如下图所示。

然后在8080实例上发起登录请求：

```
http://localhost:8080/login/submit?username=test
```

这时Redis中应该有会话数据了，再次访问两个实例的/index首页接口，都成功跳转到了首页，如下图所示。

这就说明分布式会话生效了，然后验证是否在Redis中存储了会话：

```
127.0.0.1:6379> keys *
1) "spring:session:sessions:bbb3dc40-7d77-426f-b079-bdd55533061f"
127.0.0.1:6379> ttl spring:session:sessions:bbb3dc40-7d77-426f-b079-bdd55533061f
(integer) 24
```

已经在 Redis 中查询到了一个会话数据，过期时间是配置的 30 秒，在笔者查询时还有 24 秒过期，并且每次访问应用的任何接口都可以进行会话续期。

24 秒后再访问两个实例的 /index 首页接口，如下图所示。

因为会话过期了，所以又跳转到了登录页面。

除了会话自动过期，还可以通过主动退出登录让 Redis 中的会话过期，具体见上面的 /logout 退出登录方法，当退出登录后，再去 Redis 中查询，会话数据会马上消失。

另外，Redis 中存储的 Session ID 是应用的 HttpSession 中的 Session ID，如下图所示。

```
127.0.0.1:6379> keys *
1) "spring:session:sessions:bc680e11-6866-47b3-9081-932ccc3f5eec"
127.0.0.1:6379>
session.IndexController   : session id: bc680e11-6866-47b3-9081-932ccc3f5eec
session.IndexController   : session id: bc680e11-6866-47b3-9081-932ccc3f5eec
session.IndexController   : session id: bc680e11-6866-47b3-9081-932ccc3f5eec
```

通过读取请求头 Cookie 中的 SESSION值实现会话，如下图所示。

```
▾ Request Headers                                    View source
  Accept: text/html,application/xhtml+xml,application/xml;q=0.9,image/avif,image/webp,image/apng,*/*;q=0.8,application/signed-exchange;v=b3;
  q=0.9
  Accept-Encoding: gzip, deflate, br
  Accept-Language: zh-CN,zh;q=0.9,en;q=0.8,zh-TW;q=0.7,de;q=0.6,da;q=0.5
  Cache-Control: no-cache
  Connection: keep-alive
  Cookie: JSESSIONID=74A7663CC73EF1ABCA01003257C01AAF; SESSION=YmM2ODBlMTEtNjg2Ni00N2IzLTkwODEtOTMyY2NjM2Y1ZWVj
  Host: localhost:8080
  Pragma: no-cache
  sec-ch-ua: "Chromium";v="106", "Google Chrome";v="106", "Not;A=Brand";v="99"
  sec-ch-ua-mobile: ?0
  sec-ch-ua-platform: "macOS"
  Sec-Fetch-Dest: document
  Sec-Fetch-Mode: navigate
```

Cookie 中传递了 JSESSIONID 和 SESSION 两个值，Spring Session 取的是 SESSION 的值，同一个客户端的多个应用会分配同一个 SESSION 值，这样就能实现分布式的会话，如果不是基于浏览器的多个应用之间的相互调用，则只需手动传输 SESSION 值即可。

可以发现这里的 SESSION 值和应用、Redis 中的 Session ID 值并不一样，这是因为这个 SESSION 值是经过 BASE64 编码后返回的，SessionAutoConfiguration 自动配置类中注册了一个 DefaultCookieSerializer（默认 Cookie 序列化器），默认的 Cookie 名称为 SESSION 并启用了 BASE64 编码，如下图所示。

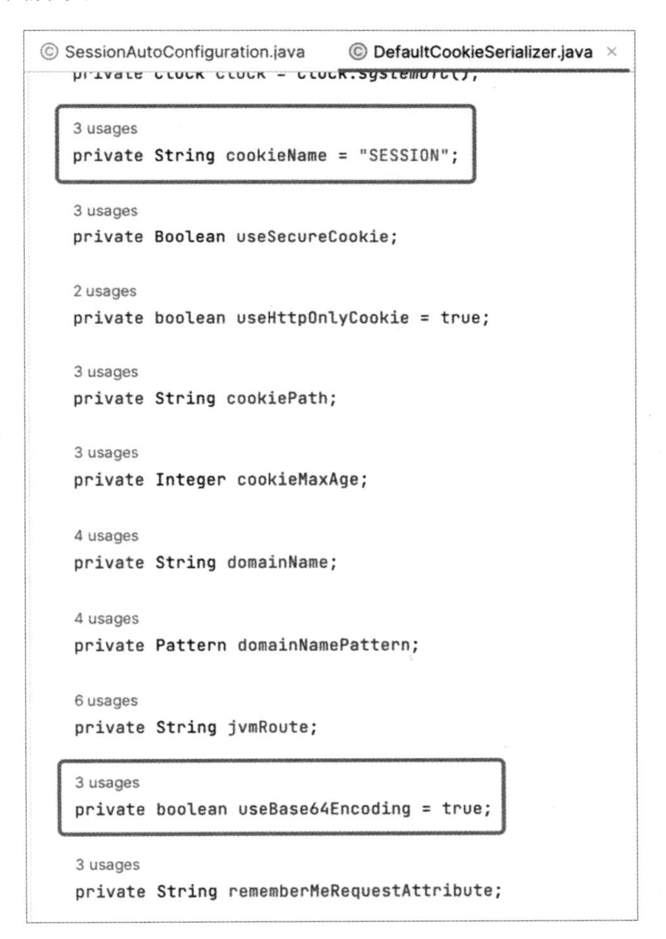

如果希望这个 SESSION 值和服务器中的保持一致，则可以自定义一个 DefaultCookieSerializer 实例，覆盖自动配置注册的默认实例，然后设置该实例不使用 BASE64 编码即可。

以上演示了 Redis 的分布式会话的使用方式，其他存储组件也是同样的道理。

6.14 跨域

要了解跨域，先要说一说同源策略，同源策略是由 Netscape 公司提出的一个著名的安全策略，所有支持 JavaScript 的浏览器都会使用这个策略。所谓同源是指，域名、协议、端口相同，当页面在执行一个脚本时会检查访问的资源是否同源，如果非同源，那么在请求数据时，浏览器会在控制台中报一个异常，提示拒绝访问。

所以，只要协议、域名、端口有任何一个不同，就被当作跨域，跨域情况下浏览器是不能执行其他域名网站的脚本的，这是由浏览器的同源策略造成的，是浏览器施加的安全限制。

实际工作开发中经常会有跨域的情况，因为公司会有很多项目，也会有很多域名，各个项目或者网站之间需要相互调用对方的资源，避免不了跨域请求。

Spring MVC 框架本身是支持跨域请求的，在需要跨域的方法上使用 @CrossOrigin 注解即可：

```
@RestController
public class ResponseBodyController {

    @CrossOrigin
    @GetMapping(value = "/user/json/{userId}", produces = MediaType.APPLICATION_JSON_VALUE)
    public ResponseEntity getJsonUserInfo(@PathVariable("userId") String userId) {
        User user = new User("Java 技术栈 ", 18);
        user.setId(Long.valueOf(userId));
        return new ResponseEntity(user, HttpStatus.OK);
    }

}
```

默认情况下 @CrossOrigin 支持：

- 所有的请求来源。
- 所有的 HTTP 头。
- 所有的 HTTP 方法。

如果有需要，则可以在 @CrossOrigin 注解中指定跨域的详细参数，除了作用在指定方法上，@CrossOrigin 注解还可以作用在类上：

```
@CrossOrigin(origins = "https://javastack.cn", maxAge = 3600)
@RestController
public class ResponseBodyController {
```

```
    ...
```

这样所有方法都能实现跨域支持，当然也可以同时作用于类和方法上，比如公共的跨域参数在类注解上指定，其他的跨域参数在各自接口方法上指定。

如果不想在类或者方法上指定，也可以在WebMvcConfigurer配置类中全局添加跨域注册的功能：

```
@Configuration
public class WebConfig implements WebMvcConfigurer {


    @Override
    public void addCorsMappings(CorsRegistry registry) {
        registry.addMapping("/user/**")
            .allowedMethods("GET", "POST")
            .allowedOrigins("https://javastack.cn")
            .allowCredentials(true).maxAge(3600);
    }


}
```

如果应用需要全局统一配置跨域请求，则可以考虑使用最后这种方式，但又回到6.2.5节的情况，实际项目中Spring Boot应用不可能直接对外，需要使用Nginx之类的负载均衡组件，所以会由运维人员统一在Nginx上配置跨域。

6.15　安全性

6.15.1　默认的安全机制

如果存在Spring Security类库依赖，Spring Boot就会对Web应用实现默认的安全保护，包括对Actuator Endpoints端点的保护，具体见第12章Spring Boot监控与报警。

加入Spring Security启动器依赖：

```
<dependency>
    <groupId>org.springframework.boot</groupId>
    <artifactId>spring-boot-starter-security</artifactId>
</dependency>
```

加入 Spring Security 启动器依赖后重启应用，再访问应用的 /test/** 接口，如下图所示。

如果没有登录授权，那么 Spring Boot 默认会跳到登录页面，需要输入默认的用户信息才能访问。默认的用户名为 user，密码则在应用启动时生成并输出到日志中：

```
Using generated security password: fff9585f-741b-4a58-bf0b-53ce924a20c2
```

Spring Security 自动配置类都被注册在新的 org.springframework.boot.autoconfigure.AutoConfiguration.imports 自动配置文件中，主要包括以下几个自动配置类：

- SecurityAutoConfiguration。
- UserDetailsServiceAutoConfiguration。

SecurityAutoConfiguration 是 Spring Security 框架的主要自动配置类，源码如下：

```
@AutoConfiguration
@ConditionalOnClass(DefaultAuthenticationEventPublisher.class)
@EnableConfigurationProperties(SecurityProperties.class)
@Import({ SpringBootWebSecurityConfiguration.class, SecurityDataConfiguration.class })
public class SecurityAutoConfiguration {

	@Bean
	@ConditionalOnMissingBean(AuthenticationEventPublisher.class)
	public DefaultAuthenticationEventPublisher authenticationEventPublisher
(ApplicationEventPublisher publisher) {
		return new DefaultAuthenticationEventPublisher(publisher);
	}

}
```

它注册了一个 DefaultAuthenticationEventPublisher 实例，用于发布授权成功、失败事件，在类上还导入了一些配置类，最主要的就是 SpringBootWebSecurityConfiguration 配置类，它的源码如下：

```
@Configuration(proxyBeanMethods = false)
@ConditionalOnWebApplication(type = Type.SERVLET)
class SpringBootWebSecurityConfiguration {

    /**
     * The default configuration for web security. It relies on Spring Security's
     * content-negotiation strategy to determine what sort of authentication to use. If
     * the user specifies their own {@link SecurityFilterChain} bean, this will back-off
     * completely and the users should specify all the bits that they want to configure as
     * part of the custom security configuration.
     */
    @Configuration(proxyBeanMethods = false)
    @ConditionalOnDefaultWebSecurity
    static class SecurityFilterChainConfiguration {

        @Bean
        @Order(SecurityProperties.BASIC_AUTH_ORDER)
        SecurityFilterChain defaultSecurityFilterChain(HttpSecurity http) throws Exception {
            http.authorizeHttpRequests().anyRequest().authenticated();
            http.formLogin();
            http.httpBasic();
            return http.build();
        }

    }

    ...

}
```

它主要注册了一个 SecurityFilterChain，并提供了默认的授权机制，即默认所有请求都需要授权，并且支持 formLogin 和 httpBasic 两种授权方式，基于请求中的 Accept 请求头的值决定使用哪种方式。

UserDetailsServiceAutoConfiguration 自动配置类提供了 Spring Security 默认授权用户的自动配置，它的源码如下：

```
@AutoConfiguration
@ConditionalOnClass(AuthenticationManager.class)
@ConditionalOnBean(ObjectPostProcessor.class)
@ConditionalOnMissingBean(
        value = { AuthenticationManager.class, AuthenticationProvider.class,
                UserDetailsService.class, AuthenticationManagerResolver.class },
        type = { "org.springframework.security.oauth2.jwt.JwtDecoder",
```

```
                "org.springframework.security.oauth2.server.resource.introspection.
OpaqueTokenIntrospector",
                "org.springframework.security.oauth2.client.registration.
ClientRegistrationRepository",
                "org.springframework.security.saml2.provider.service.registration.
RelyingPartyRegistrationRepository" })
public class UserDetailsServiceAutoConfiguration {

    ...

    @Bean
    @Lazy
    public InMemoryUserDetailsManager inMemoryUserDetailsManager(SecurityProperties properties,
            ObjectProvider<PasswordEncoder> passwordEncoder) {
        SecurityProperties.User user = properties.getUser();
        List<String> roles = user.getRoles();
        return new InMemoryUserDetailsManager(
                User.withUsername(user.getName()).password(getOrDeducePassword(user,
passwordEncoder.getIfAvailable()))
                        .roles(StringUtils.toStringArray(roles)).build());
    }

    ...

}
```

它注册了一个InMemoryUserDetailsManager，提供了单个用户的授权管理器，对应的SecurityProperties参数类的源码如下：

```
@ConfigurationProperties(prefix = "spring.security")
public class SecurityProperties {

    ...

    public static class User {

        /**
         * Default user name.
         */
        private String name = "user";

        /**
         * Password for the default user name.
```

```
         */
        private String password = UUID.randomUUID().toString();

        /**
         * Granted roles for the default user name.
         */
        private List<String> roles = new ArrayList<>();

        private boolean passwordGenerated = true;

        ...

    }

}
```

默认授权用户为：

- name：user。
- password：通过 UUID生成的随机值。

用户类提供了默认的用户名和密码，默认的密码是生成的 UUID随机值，可以通过 spring.security.user.* 来配置并替换默认的用户信息，例如：

```
spring:
  security:
    user:
      name: root
      password: root
```

默认的用户信息和明文密码只供开发阶段使用，如果应用部署到生产环境中，则建议做好严格的安全配置。

> Spring Security 同样支持 Actuator Endpoints 端点的默认安全保护机制，具体见 12.2 节。

6.15.2　自定义安全机制

如果不想使用 Spring Boot 自动配置的默认机制，那么可以通过注册一个 SecurityFilterChain 类和 UserDetailsService 类到 Spring 容器中实现自定义的安全机制，Spring Boot 中的默认安全机制就不会自动配置。

自定义安全机制的示例代码如下：

```
@Configuration
public class SecurityConfig {

    @Bean
    public SecurityFilterChain securityFilterChain(HttpSecurity http) throws Exception {
        return http.authorizeHttpRequests((authorize) -> {
                    authorize.requestMatchers("/test/**").hasRole("TEST")
                            .requestMatchers("/**").permitAll();
                })
                .logout().logoutSuccessUrl("/")
                .and().formLogin(withDefaults())
                .build();
    }

    @Bean
    public UserDetailsService userDetailsService() {
        InMemoryUserDetailsManager manager = new InMemoryUserDetailsManager();
        manager.createUser(User.withUsername("test").password("{noop}test")
                .roles("ADMIN", "TEST").build());
        manager.createUser(User.withUsername("root").password("{noop}root")
                .roles("ADMIN").build());
        return manager;
    }

}
```

这里注册了一个 SecurityFilterChain 实例，首先配置了以 /test/** 开头的 URL 需要拥有 TEST 角色才能访问，然后其他所有 URL 都能正常访问，并且提供了 formLogin 授权登录方式及注销后跳转首页功能，默认注销地址为 /logout，注销成功后跳转到 / 首页。

注册的 UserDetailsService 实例创建了两个用户，只有 test 用户才拥有 TEST 角色，然后重启应用，只有输入拥有 TEST 角色的 test 用户才能访问 /test/** 接口，输入其他用户会返回 403 错误。

先使用 root 用户登录并访问 /test/** 接口，接口会返回 403 错误页面，如下图所示。

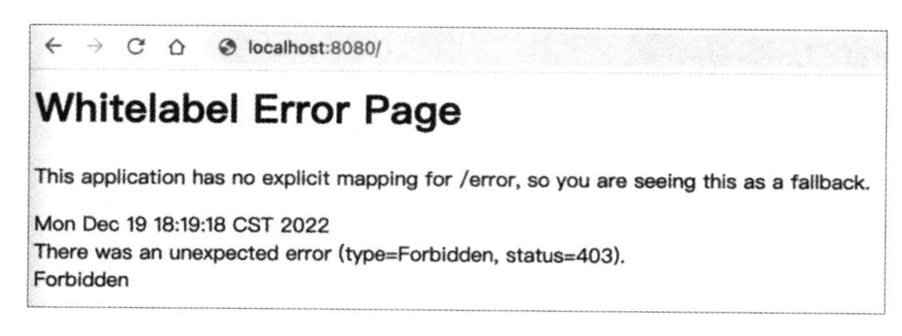

然后访问 /logout 注销登录接口，如下图所示。

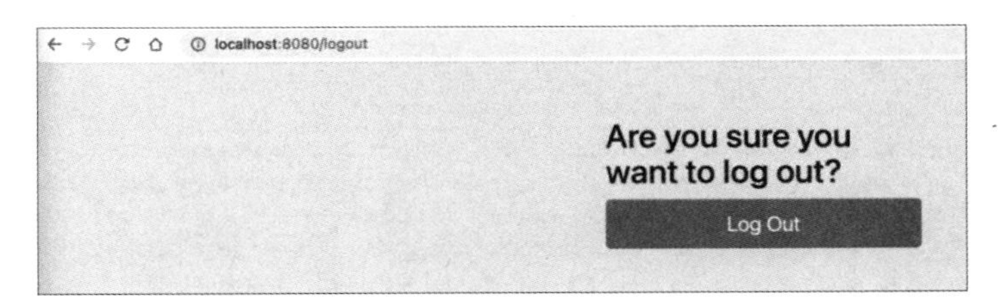

注销登录后再次以 test 用户登录并访问 /test/**接口，如下图所示。

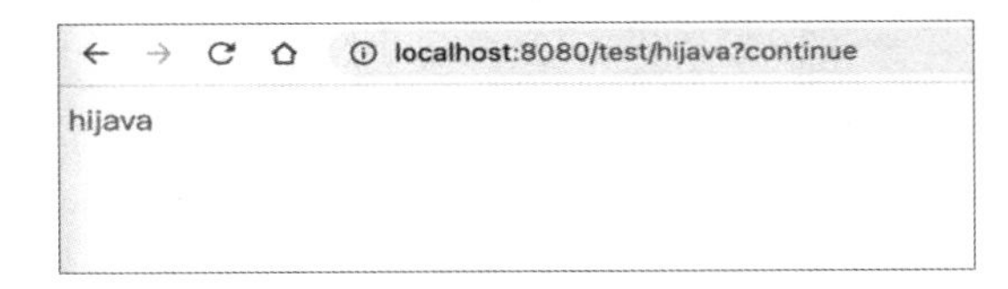

因为 test 用户拥有 TEST 角色，所以授权访问成功。

6.16　REST 服务调用

在 Spring Boot 中实现 REST 服务调用有以下两种方式：

- RestTemplate（SERVLET）。
- WebClient（REACTIVE）。

本节将分别介绍这两种调用方式。

6.16.1　RestTemplate（Servlet）

1. 实现 REST 服务调用

RestTemplate 是 Spring 框架中实现 REST 服务调用最基础的方式，因为 RestTemplate 是基于 Servlet 的 Web 应用，所以需要加入 spring-boot-starter-web 启动器依赖：

```
<dependency>
    <groupId>org.springframework.boot</groupId>
    <artifactId>spring-boot-starter-web</artifactId>
</dependency>
```

根据 Spring Boot 自动配置规则可以知道，RestTemplate 的自动配置类为 RestTemplateAuto-

Configuration，它同样被注册在新的 org.springframework.boot.autoconfigure.AutoConfiguration. imports 自动配置文件中，如下图所示。

```
org.springframework.boot.autoconfigure.AutoConfiguration.imports ×
119  org.springframework.boot.autoconfigure.transaction.TransactionAutoConfiguration
120  org.springframework.boot.autoconfigure.transaction.jta.JtaAutoConfiguration
121  org.springframework.boot.autoconfigure.validation.ValidationAutoConfiguration
122  org.springframework.boot.autoconfigure.web.client.RestTemplateAutoConfiguration
123  org.springframework.boot.autoconfigure.web.embedded.EmbeddedWebServerFactoryCust
124  org.springframework.boot.autoconfigure.web.reactive.HttpHandlerAutoConfiguration
125  org.springframework.boot.autoconfigure.web.reactive.ReactiveMultipartAutoConfigu
126  org.springframework.boot.autoconfigure.web.reactive.ReactiveWebServerFactoryAuto
127  org.springframework.boot.autoconfigure.web.reactive.WebFluxAutoConfiguration
```

因为 RestTemplate 一般要根据应用自身情况自定义配置后才能使用，所以 Spring Boot 没有提供任何默认的 RestTemplate 注册实例，但是通过 RestTemplateAutoConfiguration 自动配置提供了 RestTemplateBuilder 的默认注册实例，通过它可以创建 RestTemplate 实例。

RestTemplateBuilder 提供了许多有用的方法，可用于快速配置 RestTemplate，比如笔者这里注册了一个简单的 RestTemplate 实例，如下面的示例所示。

```
@Bean
public RestTemplate defaultRestTemplate(RestTemplateBuilder restTemplateBuilder) {
    return restTemplateBuilder.build();
}
```

添加一个测试接口，然后注入并使用 RestTemplate，如下面的示例所示。

```
@Slf4j
@RestController
@RequiredArgsConstructor
public class CallRestController {

    public static final String GET_USERINFO_URL = "http://localhost:8080/user/json/{uid}";

    private final RestTemplate restTemplate;

    @GetMapping("/restTemplate/{uid}")
    public User restTemplate(@PathVariable("uid") String uid) {
        return this.restTemplate.getForObject(GET_USERINFO_URL, User.class, uid);
    }

}
```

RestTemplate 的用法很简单，这里直接调用自身应用的获取用户信息接口，传入要调用的URL、反序列化的类型、接口请求参数，然后发起调用测试，如下图所示。

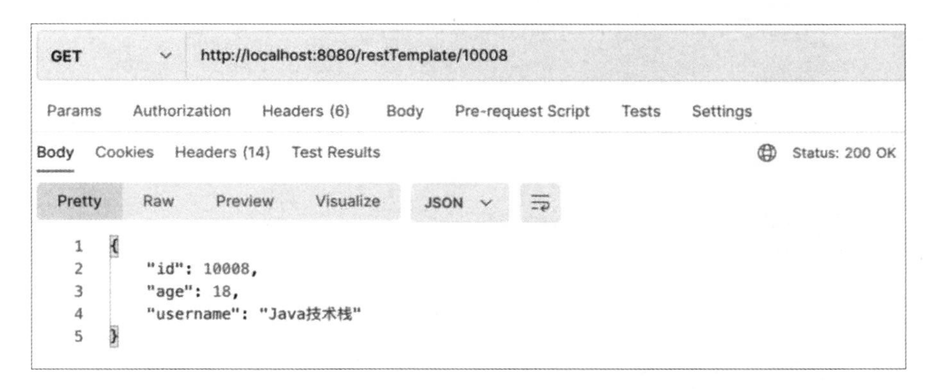

数据返回成功。RestTemplate 还有更多的 API 及用法，读者可以自行研究。

2. 自定义 RestTemplate

一般的自定义 RestTemplate 可以通过 RestTemplateBuilder 的方法实现，示例如下：

```
@Bean
public RestTemplate defaultRestTemplate(RestTemplateBuilder restTemplateBuilder) {
    return restTemplateBuilder
            .setConnectTimeout(Duration.ofSeconds(5))
            .setReadTimeout(Duration.ofSeconds(5))
            .basicAuthentication("test", "test")
            .build();
}
```

这里设置了请求和读取超时时间，以及 Basic 认证信息。

如果要实现其他附加的自定义配置，则可以实现并注册一个 RestTemplateCustomizer 接口的实例，它会应用到 RestTemplateBuilder 实例及由它构建的 RestTemplate 实例上，如下面的示例所示。

```
@Slf4j
@Component
public class CustomRestTemplateCustomizer implements RestTemplateCustomizer {

    @Override
    public void customize(RestTemplate restTemplate) {
        HttpRoutePlanner routePlanner = new CustomRoutePlanner(new HttpHost("proxy.
javastack.cn"));
```

```
        RequestConfig requestConfig = RequestConfig.custom().build();
        HttpClient httpClient = HttpClientBuilder.create()
                .setDefaultRequestConfig(requestConfig)
                .setRoutePlanner(routePlanner).build();
        restTemplate.setRequestFactory(new HttpComponentsClientHttpRequestFactory (httpClient));
    }

    static class CustomRoutePlanner extends DefaultProxyRoutePlanner {

        CustomRoutePlanner(HttpHost proxy) {
            super(proxy);
        }

        @Override
        protected HttpHost determineProxy(HttpHost target, HttpContext context) throws
HttpException {
            log.info("hostName is {}", target.getHostName());
            if ("localhost".equals(target.getHostName())) {
                return null;
            }
            return super.determineProxy(target, context);
        }

    }

}
```

上面使用了 httpclient5 作为连接客户端，所以需要引入 httpclient5 相关的依赖：

```
<dependency>
    <groupId>org.apache.httpcomponents.client5</groupId>
    <artifactId>httpclient5</artifactId>
</dependency>

<dependency>
    <groupId>org.apache.httpcomponents</groupId>
    <artifactId>httpcore</artifactId>
</dependency>
```

注册这个 CustomRestTemplateCustomizer 实例后，所有 localhost 之外的 URL 调用都会经过指定的代理。

除此之外，还可以自定义 RestTemplateBuilder 实例，以覆盖自动配置的 RestTemplateBuilder 实例，

不过一般很少会使用，默认提供的机制完全可以满足日常需要。

3. 底层实现客户端

RestTemplate 的类结构如下图所示。

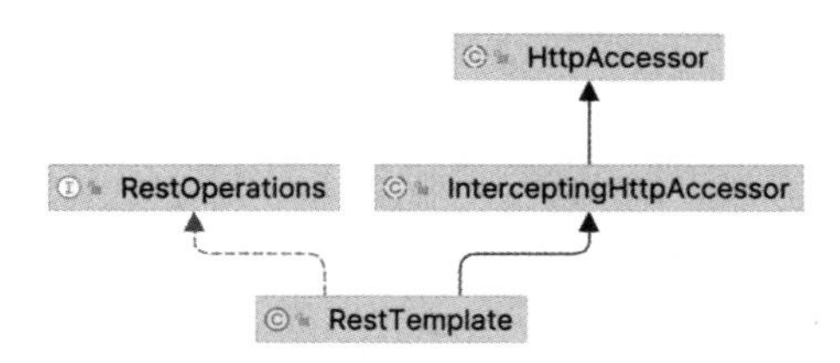

HttpAccessor 是顶层的抽象类，RestOperations 接口则定义了一系列的常用的 RESTful 操作方法，RestTemplate 便是该接口默认和唯一的实现。

HttpAccessor 默认实例化了一个简单的 SimpleClientHttpRequestFactory 工厂类，如 HttpAccessor 类的源码所示。

```
public abstract class HttpAccessor {

    /** Logger available to subclasses. */
    protected final Log logger = HttpLogging.forLogName(getClass());

    private ClientHttpRequestFactory requestFactory = new SimpleClientHttpRequestFactory();

    private final List<ClientHttpRequestInitializer> clientHttpRequestInitializers = new
ArrayList<>();

    ...

}
```

ClientHttpRequestFactory 是用来创建 HTTP 连接的工厂接口，它包含了几个主流的客户端连接工厂实现类，如下图所示。

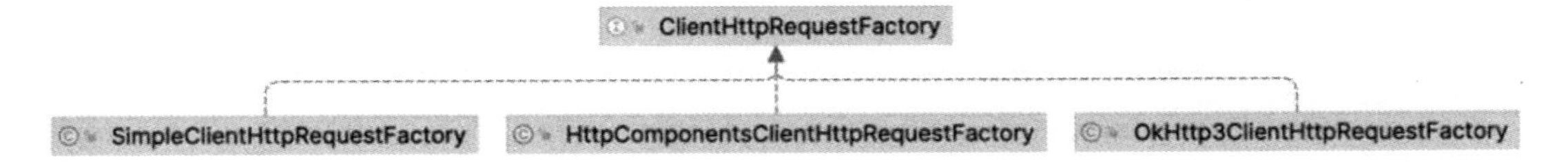

默认使用的是 SimpleClientHttpRequestFactory工厂类，其内部使用的是 JDK 自带的 URLConnection 类创建的连接，如该类中创建连接方法的源码所示。

```
public class SimpleClientHttpRequestFactory implements ClientHttpRequestFactory {

    ...

    /**
     * Opens and returns a connection to the given URL.
     * <p>The default implementation uses the given {@linkplain #setProxy(java.net.
Proxy) proxy} -
     * if any - to open a connection.
     * @param url the URL to open a connection to
     * @param proxy the proxy to use, may be {@code null}
     * @return the opened connection
     * @throws IOException in case of I/O errors
     */
    protected HttpURLConnection openConnection(URL url, @Nullable Proxy proxy) throws
IOException {
        URLConnection urlConnection = (proxy != null ? url.openConnection(proxy) :
url.openConnection());
        if (!(urlConnection instanceof HttpURLConnection)) {
            throw new IllegalStateException(
                    "HttpURLConnection required for [" + url + "] but got: " + urlConnection);
        }
        return (HttpURLConnection) urlConnection;
    }

    ...

}
```

RestTemplate 中的构造器和 setRequestFactory 方法都可以用来切换 ClientHttpRequestFactory，比如在自定义 RestTemplate 示例中，就实现了 HttpComponentsClientHttpRequestFactory 工厂类的切换，它其实就是 httpclient5 连接客户端的实现。

6.16.2 WebClient（Reactive）

1. 实现 REST 服务调用

如果使用的是 Spring WebFlux 基于 Reactive 的 Web 应用，则可以使用 WebClient 来调用远程 REST 服务。与 RestTemplate 相比，WebClient 客户端的功能更强大，并且是完全基于响应式的。

首先加入 spring-boot-starter-webflux 启动器依赖：

```
<dependency>
    <groupId>org.springframework.boot</groupId>
    <artifactId>spring-boot-starter-webflux</artifactId>
</dependency>
```

根据 Spring Boot 自动配置规则可以知道，WebClient 的自动配置类为 WebClientAutoConfiguration，它同样被注册在新的 org.springframework.boot.autoconfigure. AutoConfiguration.imports 自动配置文件中，如下图所示。

```
org.springframework.boot.autoconfigure.AutoConfiguration.imports ×
126 ork.boot.autoconfigure.web.reactive.ReactiveWebServerFactoryAutoConfiguration
127 ork.boot.autoconfigure.web.reactive.WebFluxAutoConfiguration
128 ork.boot.autoconfigure.web.reactive.WebSessionIdResolverAutoConfiguration
129 ork.boot.autoconfigure.web.reactive.error.ErrorWebFluxAutoConfiguration
130 ork.boot.autoconfigure.web.reactive.function.client.ClientHttpConnectorAutoConfiguration
131 ork.boot.autoconfigure.web.reactive.function.client.WebClientAutoConfiguration
132 ork.boot.autoconfigure.web.servlet.DispatcherServletAutoConfiguration
133 ork.boot.autoconfigure.web.servlet.ServletWebServerFactoryAutoConfiguration
```

Spring Boot 在该自动配置类中注册了一个 WebClient.Builder 接口默认实现类 DefaultWebClientBuilder 的实例，官方推荐使用它来创建 WebClient 客户端实例，比如笔者这里注册了一个简单的 WebClient 实例，如下面的示例所示。

```
@Bean
public WebClient webClient(WebClient.Builder webClientBuilder) {
    return webClientBuilder.build();
}
```

添加一个测试接口，然后注入 WebClient 就能使用了，如下面的示例所示。

```
@Slf4j
@RestController
@RequiredArgsConstructor
public class CallRestController {

    public static final String GET_USERINFO_URL = "http://localhost:8080/user/json/{uid}";

    private final WebClient webClient;

    @GetMapping("/webClient/{uid}")
    public Mono<User> webClient(@PathVariable("uid") String uid) {
        return this.webClient.get().uri(GET_USERINFO_URL, uid)
```

```
            .retrieve().bodyToMono(User.class);
    }

}
```

WebClient 的用法也很简单，这里直接调用上一节 Servlet 应用的获取用户信息接口，传入要调用的 URL、反序列化的类型、接口请求参数，然后发起调用测试，如下图所示。

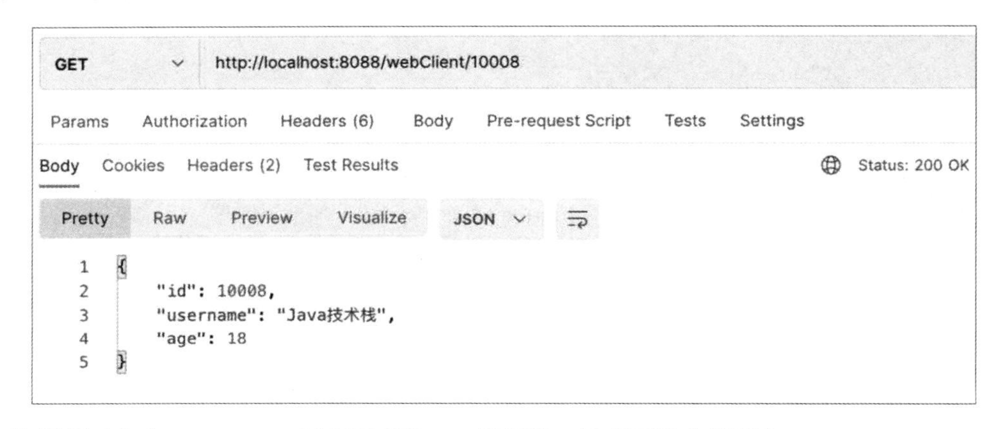

数据返回成功。WebClient 还有更多的 API 及用法，读者可以自行研究。

另外，本节介绍的是基于 Webflux 的应用，所以默认使用的是 Netty 嵌入式容器，如下面的启动日志所示。

```
.   ____          _            __ _ _
/\\ / ___'_ __ _ _(_)_ __  __ _ \ \ \ \
( ( )\___ | '_ | '_| | '_ \/ _` | \ \ \ \
\\/  ___)| |_)| | | | | || (_| |  ) ) ) )
'  |____| .__|_| |_|_| |_\__, | / / / /
=========|_|==============|___/=/_/_/_/
:: Spring Boot ::                (v3.0.0)

...
... o.s.b.web.embedded.netty.NettyWebServer  : Netty started on port 8088
...
```

2. 自定义 WebClient

一般的自定义 WebClient 可以通过 WebClient.Builder 的方法来实现，示例如下：

```
@Bean
public WebClient webClient(WebClient.Builder webClientBuilder) {
    HttpClient httpClient = HttpClient.create()
            .tcpConfiguration(client ->
                    client.option(ChannelOption.CONNECT_TIMEOUT_MILLIS, 3)
                            .doOnConnected(conn -> {
                                conn.addHandlerLast(new ReadTimeoutHandler(3000));
                                conn.addHandlerLast(new WriteTimeoutHandler(3000));
                            })
            );
    ReactorClientHttpConnector connector = new ReactorClientHttpConnector(httpClient);
    return webClientBuilder.clientConnector(connector).build();
}
```

这里创建了一个基于 Reactor Netty 的 ReactorClientHttpConnector 连接器，通过它来设置连接、读写超时时间，通过自定义 ClientHttpConnector 的方式来完全控制 Client 端的配置，它会改变 WebClient.Builder 注册实例的默认配置。所以，如果需要使用这种方式构建多个不同配置的 WebClient 客户端，则可以在 WebClient.Builder 被修改前调用其 clone 方法复制一个新实例再进行构建。

> 如果使用的是 Jetty 和 Reactor Netty 响应式服务器，则可以通过自定义 ReactorResourceFactory、JettyResourceFactory 的方式来覆盖资源配置，它会同时作用于 Server 端和 Client 端。

另外，和 RestTemplate 一样，WebClient 也提供了附加的自定义配置，实现并注册一个 WebClientCustomizer 类型的 Bean 即可。

最后，还可以通过使用 WebClient.create() 方法来手动创建 WebClient 实例，此时，自动配置中的 WebClient 就会自动取消注册。同时，它也不会应用 WebClientCustomizer 配置。

3. 底层实现客户端

前面使用了 ClientHttpConnector 连接器，它是 HTTP 客户端的底层抽象驱动接口，用于驱动 WebClient 客户端连接到服务器，并提供发送 ClientHttpRequest 和接收 ClientHttpResponse 所需的所有支撑。

Spring Boot 会自动检测应用要使用的 ClientHttpConnector 连接器类型，并通过它来驱动 WebClient 客户端。ClientHttpConnector 的类结构如下图所示。

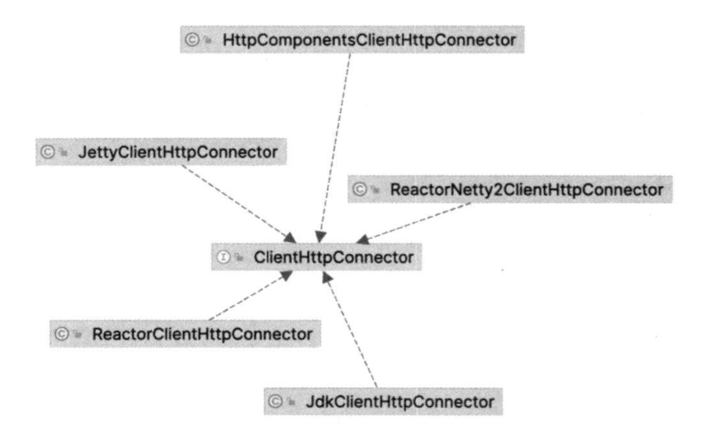

具体使用哪种 ClientHttpConnector 连接器，就要看应用使用了哪个客户端，Spring Boot 目前支持以下几种类型的客户端：

- Reactor Netty（默认）。
- Jetty RS Client。
- Apache HttpClient。
- JDK HttpClient。

DefaultWebClientBuilder#build 方法调用的 DefaultWebClientBuilder#initConnector 初始化连接器方法的实现源码如下所示。

```
final class DefaultWebClientBuilder implements WebClient.Builder {

    ...

    static {
        ClassLoader loader = DefaultWebClientBuilder.class.getClassLoader();
        reactorNettyClientPresent = ClassUtils.isPresent("reactor.netty.http.client.
HttpClient", loader);
        reactorNetty2ClientPresent = ClassUtils.isPresent("reactor.netty5.http.client.
HttpClient", loader);
        jettyClientPresent = ClassUtils.isPresent("org.eclipse.jetty.client.HttpClient",
loader);
        httpComponentsClientPresent =
                ClassUtils.isPresent("org.apache.hc.client5.http.impl.async.
CloseableHttpAsyncClient", loader) &&
```

```
                    ClassUtils.isPresent("org.apache.hc.core5.reactive.
ReactiveDataConsumer", loader);
    }

    @Override
    public WebClient build() {
        ClientHttpConnector connectorToUse =
                (this.connector != null ? this.connector : initConnector());
       ...
    }

    private ClientHttpConnector initConnector() {
        if (reactorNettyClientPresent) {
            return new ReactorClientHttpConnector();
        }
        else if (reactorNetty2ClientPresent) {
            return new ReactorNetty2ClientHttpConnector();
        }
        else if (jettyClientPresent) {
            return new JettyClientHttpConnector();
        }
        else if (httpComponentsClientPresent) {
            return new HttpComponentsClientHttpConnector();
        }
        else {
            return new JdkClientHttpConnector();
        }
    }

    ...

}
```

上述源码的含义是判断当前应用使用的是哪种响应式服务器配套的客户端，如果使用的是 spring-boot-starter-webflux 启动器，则它默认使用的是 Reactor Netty，再根据使用的 Reactor Netty 版本来创建对应的 ClientHttpConnector 连接器。

> Spring Boot 为 Reactor Netty 带来了服务端和客户端的实现，如果使用了其他的响应式服务器，则建议使用相同的 Server 和 Client 技术，这样就能在 Server 和 Client 之间共享 HTTP 资源。

第 7 章 Spring Boot 数据访问

数据库几乎是所有 Java 应用都需要用到的技术，一般项目中主要有关系型数据库和非关系型数据库（NoSQL）的应用，而在 Spring Boot 框架中提供了对主流数据库的自动配置，可以直接拿来使用。本章介绍关系型数据库的使用方式，包括嵌入式数据库、数据源、连接池、事务管理、JdbcTemplate、Spring Data JPA 等常用的数据库技术，以及对常用框架 MyBatis、MyBatis-Plus 的集成使用，另外还会介绍几个常用的非关系型数据库的安装、集成和简单应用，主要包括 Redis、MongoDB、Elasticsearch。

7.1 概述

Spring Boot 支持关系型数据库相关组件的自动配置，包括数据库、连接池、事务等的自动配置，还提供了 JdbcTemplate 及 Spring Data JPA 数据库技术，极大地简化了关系型数据库的使用难度，帮助开发者快速集成和操作关系型数据库。

Spring Boot 还支持 H2、HSQL、Derby 等嵌入式数据库的自动配置，以及 MongoDB、Neo4J、Elasticsearch、Redis、GemFire or Geode、Cassandra、Couchbase、Couchbase、LDAP、InfluxDB 等常用的 NoSQL 数据库的自动配置。

对于其他不支持的 NoSQL，可以查找官网有没有提供对应的 Starter，如果没有就需要自己配置了。本章主要以常用的 Redis、MongoDB、Elasticsearch 为例来介绍 Spring Boot 是如何集成和使用的。

7.2　嵌入式数据库

嵌入式数据库相当于“内存形式的关系型数据库”，它在应用中只有一个jar文件，而没有其他的外部环境，所有数据都存储在内存中，也不能持久化数据，应用关闭就会丢弃所有数据。

嵌入式数据库最大的好处就是足够轻量，可以集成在应用中一起打包发布，不需要连接额外的数据库服务器，这样就可以快速存储少量的结构化数据，省去了大量的连接资源消耗。

使用嵌入式数据库不需要指定数据库URL的连接地址，只需要导入对应数据库的依赖即可，比如：

```
<dependency>
    <groupId>org.springframework.boot</groupId>
    <artifactId>spring-boot-starter-data-jpa</artifactId>
</dependency>
<dependency>
    <groupId>org.hsqldb</groupId>
    <artifactId>hsqldb</artifactId>
    <scope>runtime</scope>
</dependency>
```

嵌入式数据库在实际工作中很少使用，一般可用于演示Demo项目，以及单元测试时需要临时数据库的场景，因为不能持久化数据，所以也不可能将嵌入式数据库用于实际生产项目，实际生产项目中还是要使用MySQL等专业的持久化关系型数据库。

7.3　数据源

7.3.1　概述

Java是通过javax.sql.DataSource数据源接口来连接关系型数据库的，它提供了连接和处理关系型数据库的一系列标准方法和规范。

Spring Boot支持数据库数据源的自动配置，只需要导入spring-boot-starter-data-jdbc启动器依赖及对应的数据库驱动依赖即可，如下面的依赖配置：

```
<dependency>
    <groupId>org.springframework.boot</groupId>
    <artifactId>spring-boot-starter-data-jdbc</artifactId>
```

```
</dependency>
<dependency>
    <groupId>mysql</groupId>
    <artifactId>mysql-connector-java</artifactId>
</dependency>
```

spring-boot-starter-data-jdbc 的相关依赖如下图所示。

```
∨ spring-boot-starter-data-jdbc : 3.0.0 [compile]
    ∨ spring-boot-starter-jdbc : 3.0.0 [compile]
        > HikariCP : 5.0.1 [compile]
          spring-boot-starter : 3.0.0 [compile]
        > spring-jdbc : 6.0.2 [compile]
    ∨ spring-data-jdbc : 3.0.0 [compile]
          slf4j-api : 2.0.4 [compile]
        > spring-beans : 6.0.2 [compile]
        > spring-context : 6.0.2 [compile]
        > spring-core : 6.0.2 [compile]
        > spring-data-commons : 3.0.0 [compile]
        > spring-data-relational : 3.0.0 [compile]
          spring-jdbc : 6.0.2 [compile]
        > spring-tx : 6.0.2 [compile]
```

spring-boot-starter-data-jdbc 包含一个默认的 HikariCP 连接池，以及连接 JDBC 需要的相关依赖，但是，具体要使用的数据库驱动是需要额外导入的，上面笔者使用的是 MySQL 的连接驱动，如果需要换成其他数据库，那么替换这个驱动依赖即可。

根据 Spring Boot 的自动配置规则可以知道，数据源的自动配置类为 DataSourceAutoConfiguration，和别的自动配置类一样，在 Spring Boot 3.0.0 中，它已经被移到了新的 org.springframework. boot. autoconfigure.AutoConfiguration.imports 自动配置文件中，数据源的自动配置类如下列核心源码所示。

```
@AutoConfiguration(before = SqlInitializationAutoConfiguration.class)
@ConditionalOnClass({ DataSource.class, EmbeddedDatabaseType.class })
@ConditionalOnMissingBean(type = "io.r2dbc.spi.ConnectionFactory")
@EnableConfigurationProperties(DataSourceProperties.class)
@Import(DataSourcePoolMetadataProvidersConfiguration.class)
public class DataSourceAutoConfiguration {

    @Configuration(proxyBeanMethods = false)
    @Conditional(EmbeddedDatabaseCondition.class)
    @ConditionalOnMissingBean({ DataSource.class, XADataSource.class })
    @Import(EmbeddedDataSourceConfiguration.class)
    protected static class EmbeddedDatabaseConfiguration {
```

```
    }

    @Configuration(proxyBeanMethods = false)
    @Conditional(PooledDataSourceCondition.class)
    @ConditionalOnMissingBean({ DataSource.class, XADataSource.class })
    @Import({ DataSourceConfiguration.Hikari.class, DataSourceConfiguration.Tomcat.class,
            DataSourceConfiguration.Dbcp2.class, DataSourceConfiguration.OracleUcp.class,
            DataSourceConfiguration.Generic.class, DataSourceJmxConfiguration.class })
    protected static class PooledDataSourceConfiguration {

    }

    ...

}
```

如数据源自动配置类使用的 @ConditionalOnClass 注解所示，如果类路径下有以下两个类，就会完成自动配置：

- DataSource.class。
- EmbeddedDatabaseType.class。

优先配置外部化关系型数据库的 DataSource 数据源，再尝试配置嵌入式数据库，如果没有自定义的数据源，就会通过导入的连接池配置类进行默认配置。

如自动配置类上的注解所示，数据源绑定配置类为 DataSourceProperties，通过一系列的 spring.datasource.* 数据源配置参数即可完成自动配置。

```
spring:
  datasource:
#    driver-class-name: com.mysql.cj.jdbc.Driver
    url: jdbc:mysql://localhost:3306/javastack
    username: root
    password: 12345678
```

数据库驱动类参数 driver-class-name 可以不用指定，主流的数据库 Spring Boot 都可以进行自动推断，更多的配置参数可以参考 DataSourceProperties 参数类，另外还有特定于具体连接池的参数，比如 spring.datasource.hikari.* 等。

应用启动后可以看到连接池完成初始化的相关日志：

```
HikariPool-1 - Starting...
```

```
HikariPool-1 - Added connection com.mysql.cj.jdbc.ConnectionImpl@1376883
HikariPool-1 - Start completed.
```

7.3.2 自定义数据源

如果想自定义数据源，则直接注册一个 DataSource 数据源实现类的 Bean 即可，根据 Spring Boot 自动配置类使用的条件注解的配置规则，如果应用已注册相关数据源实例，那么数据源自动配置类中的相关连接池的 DataSource 数据源就不会再自动配置了，如下面的示例所示。

```
@Configuration(proxyBeanMethods = false)
public class DsConfig {

    @Bean
    @ConfigurationProperties(prefix = "spring.datasource")
    public DataSource dataSource() {
        return new XxxDataSource();
    }

}
```

还可以自定义多个数据源，如下面的示例所示。

```
@Configuration
public class DsConfig {

    @Primary
    @Bean
    @ConfigurationProperties("spring.datasource.xx.one")
    public DataSource dataSource1() {
        return new XxxDataSource();
    }

    @Bean
    @ConfigurationProperties("spring.datasource.xx.two")
    public DataSource dataSource2() {
        return new XxxDataSource();
    }

}
```

需要注意的是：

如果应用同时出现多个相同类型的 Bean，则需要使用 @Primary 注解定义一个主 Bean，否则在根据类型注入 DataSource 的地方就会产生 Bean 注入冲突错误。

自定义多个数据源一般用于小型项目读写分离的场景，然后通过实现抽象类的 AbstractRoutingDataSource#determineCurrentLookupKey 方法决定选择哪个数据源，在大型项目中一般会通过第三方的数据库中间件技术来实现。

7.4 连接池

7.4.1 概述

上一节介绍了 Spring Boot 默认使用的是 HikariCP 连接池，Spring Boot 推荐和优先选择使用 HikariCP 连接池，因为它性能好、并发高，所以如果使用的是以下两个官方的启动器依赖：

- spring-boot-starter-jdbc。
- spring-boot-starter-data-jpa。

那么它们默认使用的就是 HikariCP，再回到上面的 DataSourceAutoConfiguration 数据源自动配置类源码，进入其 PooledDataSourceConfiguration 配置类导入的 DataSourceConfiguration 类的核心源码，如下所示。

```
abstract class DataSourceConfiguration {

    ...

    /**
     * Hikari DataSource configuration.
     */
    @Configuration(proxyBeanMethods = false)
    @ConditionalOnClass(HikariDataSource.class)
    @ConditionalOnMissingBean(DataSource.class)
    @ConditionalOnProperty(name = "spring.datasource.type", havingValue =
"com.zaxxer.hikari.HikariDataSource",
            matchIfMissing = true)
    static class Hikari {

        @Bean
        @ConfigurationProperties(prefix = "spring.datasource.hikari")
        HikariDataSource dataSource(DataSourceProperties properties) {
            HikariDataSource dataSource = createDataSource(properties, HikariDataSource.class);
```

```
            if (StringUtils.hasText(properties.getName())) {
                dataSource.setPoolName(properties.getName());
            }
            return dataSource;
        }

    }

    ...

}
```

Spring Boot 在这里提供了常用的几个连接池的自动配置，比如 HikariCP、Tomcat、DBCP2 等，因为默认导入的是 HikariCP 连接池依赖，所以默认使用的是 HikariCP 连接池，如果没有 HikariCP 连接池依赖，则根据类路径下的包按顺序选择对应的连接池，选择连接池的顺序如下图所示。

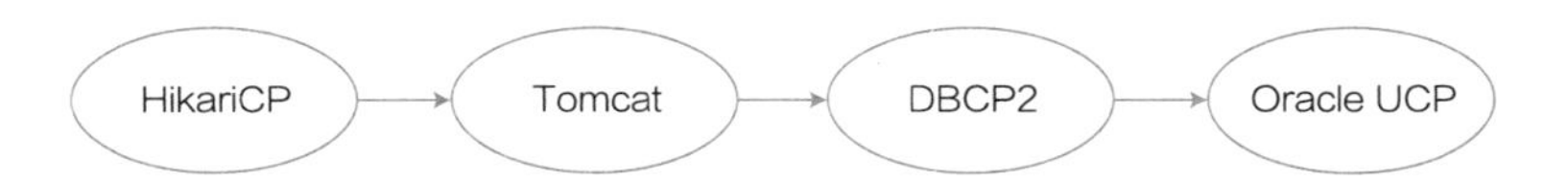

如果以上连接池依赖都不存在，则使用非连接池的通用方案，即生成并返回一个对应数据库的默认 DataSource 数据源实例，如下面的 DataSourceConfiguration.Generic 配置类的源码所示。

```
abstract class DataSourceConfiguration {

    ...

    @Configuration(proxyBeanMethods = false)
    @ConditionalOnMissingBean(DataSource.class)
    @ConditionalOnProperty(name = "spring.datasource.type")
    static class Generic {

        @Bean
        DataSource dataSource(DataSourceProperties properties) {
            return properties.initializeDataSourceBuilder().build();
        }

    }

    ...

}
```

如果不想通过自动检测方式使用连接池，也可以通过 spring.datasource.type 参数指定具体要使用的连接池，如以下配置示例所示。

```
spring:
  datasource:
    type: org.apache.commons.dbcp2.BasicDataSource
```

7.4.2　使用 Druid 连接池

在国内，笔者推荐使用阿里开源的 Druid 连接池，这是一个为监控而生的连接池，它提供了强大的监控和扩展功能，对于那些需要监控 SQL 执行性能的项目就显得尤为重要。

Druid 连接池也提供了一站式集成的 Starter：

```
<dependency>
    <groupId>com.alibaba</groupId>
    <artifactId>druid-spring-boot-starter</artifactId>
    <version>1.2.11</version>
</dependency>
```

不过它不是 Spring Boot 官方提供的，是 Druid 自行提供的，这个从 Starter 的命名也可以看出来，集成后再重启应用，如下面的启动日志所示。

```
com.alibaba.druid.pool.DruidDataSource    : {dataSource-1} inited
```

从启动日志可以发现已经切换到 Druid 连接池了，该 Starter 提供了 Druid 连接池的自动配置，对应的自动配置类为 DruidDataSourceAutoConfigure，如果不存在 DataSource 数据源的实例，则创建一个 Druid 自己的数据源，下面是它的核心源码：

```
@Configuration
@ConditionalOnClass(DruidDataSource.class)
@AutoConfigureBefore(DataSourceAutoConfiguration.class)
@EnableConfigurationProperties({DruidStatProperties.class, DataSourceProperties.class})
@Import({DruidSpringAopConfiguration.class,
    DruidStatViewServletConfiguration.class,
    DruidWebStatFilterConfiguration.class,
    DruidFilterConfiguration.class})
public class DruidDataSourceAutoConfigure {
    private static final Logger LOGGER = LoggerFactory.getLogger
(DruidDataSourceAutoConfigure.class);

    @Bean(initMethod = "init")
```

```
    @ConditionalOnMissingBean
    public DataSource dataSource() {
        LOGGER.info("Init DruidDataSource");
        return new DruidDataSourceWrapper();
    }
}

@ConfigurationProperties("spring.datasource.druid")
public class DruidDataSourceWrapper extends DruidDataSource implements InitializingBean {

    @Autowired
    private DataSourceProperties basicProperties;

    @Override
    public void afterPropertiesSet() throws Exception {
        //if not found prefix 'spring.datasource.druid' jdbc properties , 'spring.datasource'
prefix jdbc properties will be used.
        if (super.getUsername() == null) {
            super.setUsername(basicProperties.determineUsername());
        }
        if (super.getPassword() == null) {
            super.setPassword(basicProperties.determinePassword());
        }
        if (super.getUrl() == null) {
            super.setUrl(basicProperties.determineUrl());
        }
        if (super.getDriverClassName() == null) {
            super.setDriverClassName(basicProperties.getDriverClassName());
        }
    }

    ...

}
```

可以使用 spring.datasource.druid.* 参数完成一系列自动配置，同时注入 Spring Boot 的数据源通用配置参数类 DataSourceProperties，如果 Druid 没有配置数据源参数（username、password、url、driverClassName），则再取 Spring Boot 通用的数据源配置参数，这也是为什么集成 Druid 之后什么都不用修改就可以正常使用的原因。

7.5　数据库初始化

Spring Boot 可以在应用启动时初始化数据库，包括 DDL、DML 操作，但在 Spring Boot 2.5 中，又把数据库初始化逻辑重构了，最新的自动配置类为 SqlInitializationAutoConfiguration，这个类同样注册在新的自动配置文件中，以下是 SqlInitializationAutoConfiguration 自动配置类的源码：

```
@AutoConfiguration
@EnableConfigurationProperties(SqlInitializationProperties.class)
@Import({ DatabaseInitializationDependencyConfigurer.class, R2dbcInitializationConfiguration.class,
        DataSourceInitializationConfiguration.class })
@ConditionalOnProperty(prefix = "spring.sql.init", name = "enabled", matchIfMissing = true)
@Conditional(SqlInitializationModeCondition.class)
public class SqlInitializationAutoConfiguration {

    static class SqlInitializationModeCondition extends NoneNestedConditions {

        SqlInitializationModeCondition() {
            super(ConfigurationPhase.PARSE_CONFIGURATION);
        }

        @ConditionalOnProperty(prefix = "spring.sql.init", name = "mode", havingValue = "never")
        static class ModeIsNever {

        }

    }

}
```

对应的参数绑定类为 SqlInitializationProperties，可以通过一系列 spring.sql.init.* 参数配置数据库的初始化，如下面的示例所示。

```
spring:
  datasource:
    ...
  sql:
    init:
      mode: ALWAYS
      continueOnError: true
      schemaLocations:
        - classpath:sql/create_t_user.sql
```

```
    dataLocations:
      - classpath:sql/insert_t_user.sql
```

配置参数说明如下：

- mode：数据库初始化模式。
- continueOnError：初始化错误时是否继续启动。
- schemaLocations：需要执行初始化的数据库 DDL 脚本。
- dataLocations：需要执行初始化的数据库 DML 脚本。

初始化模式（mode）支持的选项可以参考 DatabaseInitializationMode 枚举类，如下所示。

- ALWAYS。
- EMBEDDED（默认）。
- NEVER。

因为默认为 EMBEDDED，即只对嵌入式数据库自动初始化，因此如果使用外部关系型数据库，那么首次启动时需要切换为 ALWAYS，成功初始化之后需要关闭，不然会重复插入数据。如果有唯一主键就会产生冲突导致启动失败，虽然可以通过 continueOnError 参数设置发生错误后继续启动，但始终会影响应用的启动速度。

在资源目录下分别创建两个脚本，Spring Boot 启动时就会运行这些脚本，脚本内容如下所示。

sql/create_t_user.sql：

```
DROP TABLE IF EXISTS `t_user`;
CREATE TABLE `t_user` (
  `id` int NOT NULL AUTO_INCREMENT,
  `username` varchar(50) CHARACTER SET utf8mb4 COLLATE utf8mb4_general_ci NOT NULL,
  `phone` varchar(20) CHARACTER SET utf8mb4 COLLATE utf8mb4_general_ci NOT NULL,
  `create_time` datetime NOT NULL,
  `status` int NOT NULL DEFAULT '1' COMMENT '-1：禁用，1：正常，0：初始 ',
  PRIMARY KEY (`id`) USING BTREE,
  UNIQUE KEY `idx_username` (`username`) USING BTREE
) ENGINE=InnoDB AUTO_INCREMENT=1 DEFAULT CHARSET=utf8mb4 COLLATE=utf8mb4_general_ci;

sql/insert_t_user.sql：

BEGIN;
INSERT INTO `t_user` VALUES (1, 'Lily', '18800000001', NOW(), 1);
INSERT INTO `t_user` VALUES (2, 'Jod', '18800000002', NOW(), 1);
```

```
INSERT INTO `t_user` VALUES (3, 'Jack', '18800000003', NOW(), 1);
INSERT INTO `t_user` VALUES (4, 'Sherry', '18800000004', NOW(), 1);
INSERT INTO `t_user` VALUES (5, 'James', '18800000005', NOW(), 1);
COMMIT;
```

然后启动应用，验证表是否创建，数据是否插入，如下所示。

```
mysql> desc t_user;
+-------------+-------------+------+-----+---------+----------------+
| Field       | Type        | Null | Key | Default | Extra          |
+-------------+-------------+------+-----+---------+----------------+
| id          | int         | NO   | PRI | NULL    | auto_increment |
| username    | varchar(50) | NO   | UNI | NULL    |                |
| phone       | varchar(20) | NO   |     | NULL    |                |
| create_time | datetime    | NO   |     | NULL    |                |
| status      | int         | NO   |     | 1       |                |
+-------------+-------------+------+-----+---------+----------------+
5 rows in set (0.00 sec)
```

可以看到表在 Spring Boot 应用启动后成功创建了，数据也成功插入了。

这就是 Spring Boot 数据库初始化的目的，实际项目中可能不被允许这样做，任何 SQL 脚本的执行都需要运维人员操作或经授权后处理，但在某些场景下可能会很有用，比如非正式项目中需要进行的数据库操作演示、测试等场景，后面的 JdbcTemplate 章节就需要用到数据库临时操作。

7.6　事务管理

7.6.1　概述

Spring Boot 提供了事务的自动配置，spring-boot-starter-data-jdbc 启动器会引入事务相关的包，相关的事务自动配置类有以下几个：

- TransactionAutoConfiguration。
- DataSourceTransactionManagerAutoConfiguration。

它们同样注册在新的自动配置文件中，前者是事务相关的自动配置，后者是事务管理器的自动配置，对应的参数绑定类为 TransactionProperties，通过一系列 spring.transaction.* 参数可以自定义一些事务配置。

如果没有自定义的事务需求，那么使用默认自动配置的事务就相当简单，只需要在代理方法上使用 @Transactional 事务注解即可，比如下面的示例：

```
@RequiredArgsConstructor
@Service
public class UserDaoImpl implements UserDao {

    public final JdbcTemplate jdbcTemplate;

    @Transactional
    @Override
    public void update() {
        jdbcTemplate.execute("update t_user set username = 'Petty' where id = 1");
        jdbcTemplate.execute("update t_user set username = 'Yoga' where id = 2");
        throw new RuntimeException("test exception");
    }

}
```

这里连续执行了两条更新语句，再手动抛出一个异常，用于测试事务回滚，当异常抛出后事务就会回滚，然后把异常代码注释后，两条语句正常更新。

7.6.2 事务失效的场景

前面的事务测试为什么要单独创建一个 DAO 实现类呢？这是为了能启用 Spring 中的代理，使用 @Transactional 事务注解必须注意下面几个失效的场景。

1. 数据库引擎不支持事务

这里以 MySQL 为例，MyISAM 引擎是不支持事务操作的，一般要支持事务都会使用 InnoDB 引擎，根据 MySQL 的官方文档说明，从 MySQL 5.5.5 开始的默认存储引擎是 InnoDB，之前默认的都是 MyISAM，所以这一点要值得注意，如果底层引擎不支持事务，那么再怎么设置也没有用。

2. 没有被 Spring 管理

示例如下：

```
// @Service
public class OrderServiceImpl implements OrderService {

    @Transactional
```

```
    public void updateOrder(Order order) {
        // update order
    }

}
```

如果此时把 @Service 注解注释掉，那么这个类就不会被加载成一个 Bean，这个类就不会被 Spring 管理了，事务自然就失效了。

3. 方法不是 public 的

@Transactional 注解只能用于 public 的方法上，否则事务不会生效，如果要用在非 public 的方法上，则可以开启基于 AspectJ 框架的静态代理模式。

4. 发生自身调用

示例如下：

```
@Service
public class OrderServiceImpl implements OrderService {

    public void update(Order order) {
        updateOrder(order);
    }

    @Transactional
    public void updateOrder(Order order) {
        // update order
    }

}
```

update 方法上面没有加 @Transactional 注解，如果调用有 @Transactional 注解的 updateOrder 方法，那么 updateOrder 方法上的事务还可以生效吗？读者可以先想一想，后面会揭晓答案。

再来看下面这个例子：

```
@Service
public class OrderServiceImpl implements OrderService {

    @Transactional
    public void update(Order order) {
        updateOrder(order);
```

```
    }

    @Transactional(propagation = Propagation.REQUIRES_NEW)
    public void updateOrder(Order order) {
        // update order
    }

}
```

这次在 update 方法上加了 @Transactional，如果在 updateOrder 上加了 REQUIRES_NEW 新开启一个事务，那么新开启的事务可以生效吗？

这两个例子中的事务都不会生效，因为它们发生了自身调用，就调用了该类自己的方法，而没有经过 Spring 的代理类，默认只有调用外部代理类的方法，事务才会生效，这也是老生常谈的问题了。

这个问题的解决方案之一就是在事务所在的类中注入自己，用注入的对象再调用另外一个方法，这个不太优雅，在 Spring 中可以在当前线程中暴露并获取当前代理类，通过在启动类上添加以下注解来启用暴露代理类，如下面的示例所示。

```
@EnableAspectJAutoProxy(exposeProxy = true)
```

然后通过以下代码获取当前代理类，并调用代理类的事务方法：

```
((OrderService) AopContext.currentProxy()).updateOrder();
```

Spring 默认只有调用 Spring 代理类的 public 方法，事务才能生效。

5. 没有配置事务管理器

如果没有配置以下 DataSourceTransactionManager 数据源事务管理器，那么事务也不会生效：

```
@Bean
public PlatformTransactionManager transactionManager(DataSource dataSource) {
    return new DataSourceTransactionManager(dataSource);
}
```

但在 Spring Boot 中只要引入了 spring-boot-starter-data-jdbc 启动器依赖就会自动配置 DataSourceTransactionManager 数据源事务管理器，所以 Spring Boot 框架不存在这个问题，但在传统的 Spring 框架中需要注意。

6. 设置了不支持事务

示例如下：

```
@Service
public class OrderServiceImpl implements OrderService {

    @Transactional
    public void update(Order order) {
        updateOrder(order);
    }

    @Transactional(propagation = Propagation.NOT_SUPPORTED)
    public void updateOrder(Order order) {
        // update order
    }

}
```

这里的 Propagation.NOT_SUPPORTED 表示当前方法不以事务方式运行，当前若存在事务则挂起，这就是主动不支持以事务方式运行了。

7. 异常没有被抛出

示例如下：

```
// @Service
public class OrderServiceImpl implements OrderService {

    @Transactional
    public void updateOrder(Order order) {
        try {
            // update order
        } catch {

        }
    }

}
```

这个方法把异常给捕获了，但没有抛出来，所以事务不会回滚，只有捕捉到异常事务才会生效。

8. 异常类型不匹配

示例如下：

```
// @Service
public class OrderServiceImpl implements OrderService {

    @Transactional
    public void updateOrder(Order order) {
        try {
            // update order
        } catch {
            throw new Exception("更新错误");
        }
    }

}
```

因为 Spring 默认回滚的是 RuntimeException 异常，和程序抛出的 Exception 异常不匹配，所以事务也是不生效的。如果要触发默认 RuntimeException 之外异常的回滚，则需要在 @Transactional 事务注解上指定异常类，示例如下：

```
@Transactional(rollbackFor = Exception.class)
```

本节总结了使用 @Transactional 注解导致事务失效的几个常见场景，如果 @Transactional 事务不生效，则可以根据这几种情形排查一下，其实次数最多的也就是发生自身调用、异常被捕获、异常抛出类型不匹配这几种场景。

7.7 JdbcTemplate

7.7.1 数据库操作

JDBC 的全称为 Java Database Connectivity，即 Java 数据库连接，是 Java 语言规范应用访问数据库的接口，从而不需要关心底层特定数据库的细节，提供了如查询、更新、删除数据库中的数据等方法。

JdbcTemplate 即 JDBC 模板，是 Spring 用于简化 JDBC 操作的包装模板类，只需要导入 spring-boot-starter-data-jdbc 启动器依赖即可，它在新的自动配置文件中同样提供了对 JdbcTemplateAutoConfiguration 自动配置类的注册，以下是该自动配置类的源码：

```
@AutoConfiguration(after = DataSourceAutoConfiguration.class)
@ConditionalOnClass({ DataSource.class, JdbcTemplate.class })
@ConditionalOnSingleCandidate(DataSource.class)
@EnableConfigurationProperties(JdbcProperties.class)
@Import({ DatabaseInitializationDependencyConfigurer.class, JdbcTemplateConfiguration.class,
        NamedParameterJdbcTemplateConfiguration.class })
public class JdbcTemplateAutoConfiguration {

}

@Configuration(proxyBeanMethods = false)
@ConditionalOnMissingBean(JdbcOperations.class)
class JdbcTemplateConfiguration {

    @Bean
    @Primary
    JdbcTemplate jdbcTemplate(DataSource dataSource, JdbcProperties properties) {
        JdbcTemplate jdbcTemplate = new JdbcTemplate(dataSource);
        JdbcProperties.Template template = properties.getTemplate();
        jdbcTemplate.setFetchSize(template.getFetchSize());
        jdbcTemplate.setMaxRows(template.getMaxRows());
        if (template.getQueryTimeout() != null) {
            jdbcTemplate.setQueryTimeout((int) template.getQueryTimeout().getSeconds());
        }
        return jdbcTemplate;
    }

}
```

JdbcTemplateAutoConfiguration 自动配置类中导入了 JdbcTemplateConfiguration 配置类，它提供了默认的 JdbcTemplate 实例，在应用中直接注入就能使用，示例如下：

```
@Slf4j
@RequiredArgsConstructor
@SpringBootApplication
public class Application {
```

```
    public final JdbcTemplate jdbcTemplate;

    public static void main(String[] args) {
        SpringApplication.run(Application.class);
    }

    @Bean
    public CommandLineRunner commandLineRunner() {
        return (args) -> {
            String username = jdbcTemplate.queryForObject("select username from t_user
where id = 2",
                    String.class);
            log.info("query username is : {}", username);
        };
    }


}
```

这里的逻辑是，在应用启动之后，在数据库中查询 id=2 的 username 字段的值并输出日志。

7.7.2 自定义 JdbcTemplate

如果默认注册的 JdbcTemplate 实例不符合要求，则可以自定义 JdbcTemplate，或者通过一系列 spring.jdbc.template.* 参数进行定制，如下面的示例所示。

```
spring:
  jdbc:
    template:
      max-rows: 3
```

这里对 JdbcTemplate 进行了定制，最多可返回 3 条记录，然后新增一个查询操作测试一下，如下面的示例所示。

```
@Bean
public CommandLineRunner commandLineRunner() {
    return (args) -> {
        ...
```

```
        List<Map<String, Object>> list = jdbcTemplate.queryForList("select id from t_user");
        log.info("total list: {}", list.size());
    };
}
```

重启应用会输出以下日志：

```
query username is : Jod
total list: 3
```

虽然用户表中有超过 3 条的记录，但是 JdbcTemplate 只返回了最多 3 条记录。更多的 JdbcTemplate 定制参数可以参考 JdbcProperties#Template 参数配置类。

7.8 Spring Data JPA

7.8.1 概述

很多人可能对 JPA 相关技术有误解，下面先对相关技术术语做一个梳理。

1. JPA

JPA 其实不是指具体框架，它是 Java Persistence API 的简称，是 Java 提供的持久化 API，它为 POJO 提供持久化的标准规范，可以把 Java 对象映射为数据库中的记录。

2. Hibernate

JPA 是持久化规范，而 Hibernate、TopLink、OpenJPA 等都是 JPA 这种标准规范的实现，Hibernate 则是 Java 最流行的 JPA 规范的实现框架。

3. Spring Data JPA

Spring Data 系列项目为数据访问提供基于 Spring 的统一编程模型，而其中之一的 Spring Data JPA 则是 Spring 为了简化 JPA 的使用而推出的一个项目，底层使用的是 Hibernate 框架，可以将 Spring Data JPA 理解为对 Hibernate 更上一层的封装。

三者的关系如下图所示。

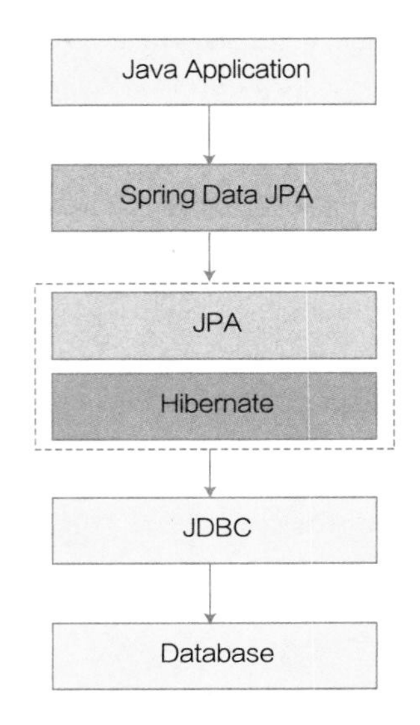

Spring Data JPA 框架是对 JPA 顶层的封装而已，底层使用的是基于 JPA 规范的 Hibernate 框架，最后通过 JDBC 连接底层数据库。

7.8.2 数据库操作

Spring Data JPA 框架简化了 JPA 的复杂操作，Spring Boot 又提供了 spring-boot-starter-data- jpa 启动器，自动配置了 JPA 相关组件，让开发者可以更加快速地集成和使用 JPA。

首先在应用中导入该启动器依赖：

```
<dependency>
    <groupId>org.springframework.boot</groupId>
    <artifactId>spring-boot-starter-data-jpa</artifactId>
</dependency>
```

然后查看它导入的相关依赖，如下图所示。

```
∨ spring-boot-starter-data-jpa : 3.0.0 [compile]
    > hibernate-core : 6.1.5.Final [compile]
    > spring-aspects : 6.0.2 [compile]
    > spring-boot-starter-aop : 3.0.0 [compile]
    ∨ spring-boot-starter-jdbc : 3.0.0 [compile]
        > HikariCP : 5.0.1 [compile]
          spring-boot-starter : 3.0.0 [compile]
        > spring-jdbc : 6.0.2 [compile]
    ∨ spring-data-jpa : 3.0.0 [compile]
          jakarta.annotation-api : 2.1.1 [compile]
          slf4j-api : 2.0.4 [compile]
          spring-aop : 6.0.2 [compile]
          spring-beans : 6.0.2 [compile]
          spring-context : 6.0.2 [compile]
          spring-core : 6.0.2 [compile]
        > spring-data-commons : 3.0.0 [compile]
        > spring-orm : 6.0.2 [compile]
        > spring-tx : 6.0.2 [compile]
```

spring-boot-starter-data-jpa 包含了 JPA 访问数据库所需要的所有依赖，另外，在新的自动配置文件中注册了以下 JPA 自动配置类：

- JpaRepositoriesAutoConfiguration。
- HibernateJpaAutoConfiguration。

对应的参数绑定类为 JpaProperties 和 HibernateProperties，通过一系列 spring.jpa.* 和 spring.jpa.hibernate.* 参数可以自定义 JPA 配置，比如以下常用的两个配置：

```
spring:
  jpa:
    hibernate:
      ddl-auto: update
    show-sql: true
```

这里指定了 ddl-auto 为 update 更新模式，并且输出 SQL 语句日志，方便调试 SQL 问题，后面的章节中会看到输出的 SQL 语句日志。

传统的 JPA 实体需要在 persistence.xml 文件中指定，而在 Spring Boot 中就没有必要，实体类上面只需要一个 @Entity 注解，启动类上有 @SpringBootApplication 或者 @EnableAutoConfiguration 注解就能自动扫描所有 @Entity 实体类，如下面的用户表的实体类所示。

```
@Data
@Entity(name = "t_user")
```

```
public class UserDO implements Serializable {

    @Id
    @GeneratedValue(strategy = GenerationType.IDENTITY)
    private Long id;

    @Column(nullable = false, unique = true)
    private String username;

    @Column(nullable = false)
    private String phone;

    @Column(name = "create_time", nullable = false)
    private Date createTime;

    @Column(nullable = false)
    private Integer status;

}
```

如果自动配置注解扫描不到，则可以使用 @EntityScan 注解指定需要扫描的实体类所在的包，或者直接指定实体类。

Spring Data JPA 的数据库操作被称为 repositories，Repository 是所有 repositories 的顶级接口，它是一个空接口，官方还提供了几个常用的 Repositorie，如下图所示。

```
Repository (org.springframework.data.repository)
    RevisionRepository (org.springframework.data.rep
    RxJava3CrudRepository (org.springframework.data.
  > PagingAndSortingRepository (org.springframework.
    ReactiveCrudRepository (org.springframework.data
  v CrudRepository (org.springframework.data.reposit
      SimpleJdbcRepository (org.springframework.data
    > ListCrudRepository (org.springframework.data.r
    > JpaRepository (org.springframework.data.jpa.re
      UserRepository (cn.javastack.springboot.jpa.re
    ReactiveSortingRepository (org.springframework.d
    RxJava3SortingRepository (org.springframework.da
    CoroutineCrudRepository (org.springframework.data.
    CoroutineSortingRepository (org.springframework.da
```

这些 Repositorie 接口一般可以满足常用的需求，项目中的数据库操作可以根据需要选择对应的 Repository 继承 / 扩展即可，比如下面的创建一个用户 UserRepository 的示例：

```
public interface UserRepository extends CrudRepository<UserDO, Long> {

}
```

UserRepository 接口继承了 CrudRepository 接口，CrudRepository 接口又继承自 Repository 接口，CrudRepository 接口提供了常用的增删改查操作方法，如果有更复杂的操作，则在自己的 Repository 接口中再额外定义即可，如下图所示。

```
CrudRepository
    count(): long
    delete(T): void
    deleteAll(): void
    deleteAll(Iterable<? extends T>): void
    deleteAllById(Iterable<? extends ID>): void
    deleteById(ID): void
    existsById(ID): boolean
    findAll(): Iterable<T>
    findAllById(Iterable<ID>): Iterable<T>
    findById(ID): Optional<T>
    save(S): S
    saveAll(Iterable<S>): Iterable<S>
```

UserRepository 接口定义后也会被自动扫描到，可以直接拿来注入使用，如果自动配置注解扫描不到，则可以使用 @EnableJpaRepositories 注解指定 Repository 所在的包，或者直接指定 Repository 类。

这里为了方便测试，直接在 UserController 控制器类中注入 UserRepository，并调用其 findById 方法查询返回用户表的某条记录，如以下示例所示。

```
@RequiredArgsConstructor
@RestController
public class UserController {

    private final UserRepository userRepository;

    @GetMapping("/user/info/{id}")
    public UserDO getUserInfo(@PathVariable("id") long id){
        UserDO userDO = userRepository.findById(id).orElseGet(null);
        return userDO;
    }

}
```

重启应用，再调用查询用户接口，如下图所示。

结果正常返回，控制台也会打印相关 SQL 语句：

```
Hibernate: select u1_0.id,u1_0.create_time,u1_0.phone,u1_0.status,u1_0.username from
t_user u1_0 where u1_0.id=?
```

Spring Data JPA 的使用绝不仅限于此，一般的应用使用 Spring Data JPA 即可满足需求，还可以有效提升效率。如果是复杂的业务系统，涉及大量复杂的 SQL 语句，则建议使用 MyBatis 框架。MyBatis 是把 SQL 语句配置在 XML 文件中，编写灵活，易于调试和对 SQL 调优，另外加上 JPA 有一定的使用门槛，需要团队成员熟练掌握，所以国内大部分公司主要还是使用 MyBatis 框架。

7.9 MyBatis

7.9.1 概述

国内一般都使用 MyBatis 框架，可以直接写 SQL，上手容易，维护也方便。

Spring Boot 官方没有提供 MyBatis Starter，但 MyBatis 官方提供了 mybatis-spring-boot-starter 一站式启动器，如以下依赖配置所示。

```
<dependency>
    <groupId>org.mybatis.spring.boot</groupId>
    <artifactId>mybatis-spring-boot-starter</artifactId>
```

```
    <version>3.0.1</version>
</dependency>
```

在 Spring Boot 3.0.0 发布不久后，MyBatis Starter 也进行了兼容性适配，Spring Boot 3.0.0 对应的 mybatis-spring-boot-starter 版本为 3.x，它导入了使用 MyBatis 所需的所有依赖，如下图所示。

```
∨ mybatis-spring-boot-starter : 3.0.1 [compile]
     mybatis : 3.5.11 [compile]
     mybatis-spring : 3.0.1 [compile]
   > mybatis-spring-boot-autoconfigure : 3.0.1 [compile]
   > spring-boot-starter : 3.0.0 [compile]
   > spring-boot-starter-jdbc : 3.0.0 [compile]
```

Starter 包含 Spring Boot 使用 MyBatis 时所需的自动配置，对应的自动配置类为 MybatisAutoConfiguration，它代替了传统 Spring 项目集成 MyBatis 所需的 XML 配置文件，自动配置类中主要注册了一个 SqlSessionFactory 和 SqlSessionTemplate 实例，对应的参数绑定配置类为 MybatisProperties，通过一系列 mybatis.* 参数可以自定义 MyBatis 配置。

7.9.2　数据库操作

本节使用 Spring Boot 集成 MyBatis 实现一个简单的查询操作，使用 MyBatis 框架一般需要以下两个组件：

- *Mapper.java：数据库映射接口，提供对应的数据库操作方法。
- *Mapper.xml：数据库 SQL 映射配置文件，提供数据库映射接口对应执行的 SQL。

首先导入 mybatis-spring-boot-starter 启动器依赖，然后在应用配置文件中添加 MyBatis 配置参数，如下面常用的配置示例所示。

```
mybatis:
  mapper-locations: classpath:mapper/*.xml
  type-aliases-package: cn.javastack.springboot.mybatis.entity
```

配置参数说明如下：

- mapper-locations：指定 *.xml SQL 配置文件的路径。
- type-aliases-package：指定数据库实体类的全包路径，方便在 SQL 配置文件中使用实体类时不用写包名。

然后定义一个 *Mapper 映射接口，接口需要使用 @Mapper 注解修饰，如以下 UserMapper 类的代码所示。

```
@Mapper
public interface UserMapper {

    UserDO findById(@Param("id") long id);

}
```

UserDO 实体类的代码如下所示。

```
@Data
public class UserDO implements Serializable {

    private static final long serialVersionUID = 1L;

    private Integer id;

    private String username;

    private String phone;

    private Date createTime;

    private Integer status;

}
```

然后在资源目录下新建对应的 mapper/UserMapper.xml 配置文件：

```
<?xml version="1.0" encoding="UTF-8"?>
<!DOCTYPE mapper
        PUBLIC "-//mybatis.org//DTD Mapper 3.0//EN"
        "http://mybatis.org/dtd/mybatis-3-mapper.dtd">
<mapper namespace="cn.javastack.springboot.mybatis.mapper.UserMapper">

    <resultMap id="BaseResultMap" type="UserDO">
            <id property="id" column="id" jdbcType="INTEGER"/>
            <result property="username" column="username" jdbcType="VARCHAR"/>
            <result property="phone" column="phone" jdbcType="VARCHAR"/>
            <result property="createTime" column="create_time" jdbcType="TIMESTAMP"/>
            <result property="status" column="status" jdbcType="INTEGER"/>
    </resultMap>
```

```
    <sql id="Base_Column_List">
        id,username,phone,create_time,status
    </sql>

    <select id="findById" parameterType="long" resultMap="BaseResultMap">
        select
            <include refid="Base_Column_List" />
        from t_user
        where id = #{id}
    </select>

</mapper>
```

XML 配置说明如下：

- namespace：定义命名空间，即 Mapper 映射接口的全路径，比如 cn.javastack.springboot.mybatis.mapper.UserMapper。
- resultMap：定义结果映射集，将数据库表字段映射为 Java 实体类。
- sql：定义通用的 SQL 语句，可包含在其他 SQL 语句中。
- select：定义查询语句，id 要和 Mapper 映射接口中的方法一致。

如果 *Mapper 在启动类所在的子包中，那么 MyBatis 自动配置会自动扫描该注解修饰的接口，应用可以直接注入使用，否则需要使用 @MapperScan 注解实现自定义的 Mapper 扫描。

这里为了方便测试，直接在 UserController 控制器类中注入 UserMapper，并调用其 findById 方法查询并返回用户表的某条记录，如以下代码所示。

```
@RequiredArgsConstructor
@RestController
public class UserController {

    private final UserMapper userMapper;

    @GetMapping("/user/info/{id}")
    public UserDO getUserInfo(@PathVariable("id") long id) {
        UserDO userDO = userMapper.findById(id);
        return userDO;
    }

}
```

启动应用测试一下，如下图所示。

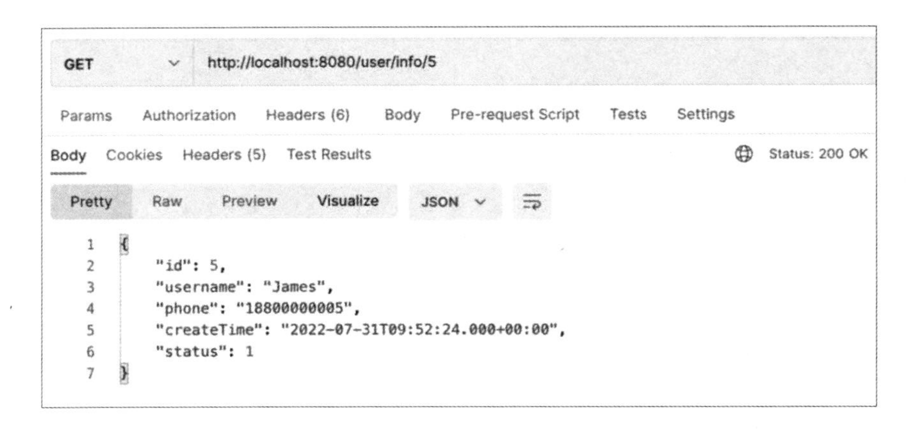

7.10 MyBatis-Plus

7.10.1 概述

MyBatis-Plus 是针对 MyBatis 的开源增强框架，它只是增强，而不改变 MyBatis 框架本身，在 MyBatis 框架的基础上提供了许多人性化的功能，比如通用的 Mapper、Service 接口，可以节省大量基本的 CRUD 操作，还有更多丰富的实用特性，它是使用 MyBatis 框架的最佳搭档。

开源地址为链接 7。

本节使用 Spring Boot 集成 MyBatis-Plus，并实现上一节使用 MyBatis 实现的同样的功能。

> 在笔者撰写本章及后续复核期间，MyBatis-Plus 最新只支持 Spring Boot 2.x，使用最新的 Spring Boot 3.0.0 会出现兼容性问题，所以本章基于 Spring Boot 2.7.0 进行实战演示。最新的 MyBatis-Plus 版本只是没有兼容 Spring Boot 3.x 而已，框架的底层架构和 API 是不会有大的改动的，所以不必担心。

MyBatis-Plus 官方也提供了 Starter 一站式启动器，如下面的依赖配置所示。

```
<dependency>
    <groupId>com.baomidou</groupId>
    <artifactId>mybatis-plus-boot-starter</artifactId>
    <version>${mybatis-plus-boot-starter.version}</version>
</dependency>
```

导入 MyBatis-Plus 的依赖，它会自动引入使用 MyBatis 框架所需的所有依赖，如下图所示。

```
∨ mybatis-plus-boot-starter : 3.5.2 [compile]
    ∨ mybatis-plus : 3.5.2 [compile]
        > kotlin-stdlib-jdk8 : 1.7.21 [compile]
        > mybatis-plus-extension : 3.5.2 [compile]
    > spring-boot-autoconfigure : 3.0.0 [compile]
    > spring-boot-starter-jdbc : 3.0.0 [compile]
```

对应的自动配置类为 MybatisPlusAutoConfiguration，它其实就是代替了官方的 MyBatis Starter 及 MybatisAutoConfiguration 自动配置类，注册了一个 SqlSessionFactory 实例并对其进行定制，如下面的核心源码所示。

```
@Configuration(proxyBeanMethods = false)
@ConditionalOnClass({SqlSessionFactory.class, SqlSessionFactoryBean.class})
@ConditionalOnSingleCandidate(DataSource.class)
@EnableConfigurationProperties(MybatisPlusProperties.class)
@AutoConfigureAfter({DataSourceAutoConfiguration.class, MybatisPlusLanguageDriverAuto-
Configuration.class})
public class MybatisPlusAutoConfiguration implements InitializingBean {

    ...

    @Bean
    @ConditionalOnMissingBean
    public SqlSessionFactory sqlSessionFactory(DataSource dataSource) throws Exception {
        // TODO 使用 MybatisSqlSessionFactoryBean 而不是 SqlSessionFactoryBean
        MybatisSqlSessionFactoryBean factory = new MybatisSqlSessionFactoryBean();

        ...

        // TODO 修改源码支持定义 TransactionFactory
        this.getBeanThen(TransactionFactory.class, factory::setTransactionFactory);

        // TODO 对源码做了一定的修改（因为源码适配了老旧的 Mybatis 版本，但我们不需要适配）
        Class<? extends LanguageDriver> defaultLanguageDriver = this.properties.
getDefaultScriptingLanguageDriver();
        if (!ObjectUtils.isEmpty(this.languageDrivers)) {
            factory.setScriptingLanguageDrivers(this.languageDrivers);
        }
        Optional.ofNullable(defaultLanguageDriver).ifPresent(factory::
setDefaultScriptingLanguageDriver);
```

```
        applySqlSessionFactoryBeanCustomizers(factory);

        // TODO 此处必为非 NULL
        GlobalConfig globalConfig = this.properties.getGlobalConfig();
        // TODO 注入填充器
        this.getBeanThen(MetaObjectHandler.class, globalConfig::setMetaObjectHandler);
        // TODO 注入主键生成器
        this.getBeansThen(IKeyGenerator.class, i -> globalConfig.getDbConfig().
setKeyGenerators(i));
        // TODO 注入 SQL 注入器
        this.getBeanThen(ISqlInjector.class, globalConfig::setSqlInjector);
        // TODO 注入 ID 生成器
        this.getBeanThen(IdentifierGenerator.class, globalConfig::setIdentifierGenerator);
        // TODO 设置 GlobalConfig 到 MybatisSqlSessionFactoryBean
        factory.setGlobalConfig(globalConfig);
        return factory.getObject();
    }

    ...

}
```

对应的参数绑定类为 MybatisPlusProperties，通过一系列 mybatis-plus.* 参数可以对 Mybatis-Plus 进行定制，而 MyBatis 的自动配置参数为 mybatis.*，mybatis-plus.* 继承了 mybatis.* 的部分原生参数并进行了增强。

7.10.2 通用数据库操作

在应用配置文件中添加对应的 MyBatis-Plus 配置，如以下配置所示。

```
mybatis-plus:
  mapper-locations: classpath:mapper/*.xml
  type-aliases-package: cn.javastack.springboot.mybatis.entity
```

参数和 MyBatis一致，只需要把 mybatis 前缀改为 mybatis-plus 即可。

与 MyBatis 不同的是，MyBatis-Plus 会自动加载 mapper 资源目录下的 XML 映射文件，如果使用默认目录，则这里的 mapper-locations 可以不用定义，如下面的参数绑定类的源码所示。

```
@ConfigurationProperties(
    prefix = "mybatis-plus"
)
```

```
public class MybatisPlusProperties {
    private static final ResourcePatternResolver resourceResolver = new PathMatching-
ResourcePatternResolver();
    private String configLocation;
    private String[] mapperLocations = new String[]{"classpath*:/mapper/**/*.xml"};

    ...

}
```

还有一点与 MyBatis 框架不同的是，因为引入的不是 MyBatis 官方的 Starter，所以 MyBatis-Plus 中的 Mapper 映射类不能自动扫描注册，需要使用 @MapperScan 注解手动开启扫描，如下面的配置代码所示。

```
@MapperScan(basePackages = {"cn.javastack.springboot.mybatisplus.mapper"})
@SpringBootApplication
public class Application {

    public static void main(String[] args) {
        SpringApplication.run(Application.class);
    }

}
```

新建一个 UserMapper 接口，如下面的示例代码所示。

```
public interface UserMapper extends BaseMapper<UserDO> {

}
```

这里继承了 MyBatis-Plus 提供的通用 BaseMapper 接口，它提供了 DAO 层常用的 CRUD 操作，UserMapper 可以直接注入使用。

MyBatis-Plus 实体类与 MyBatis 也有不同，因为 MyBatis-Plus 不需要 XML 配置，所以需要使用一系列注解完成实体类字段的映射，MyBatis-Plus 提供了一系列注解对实体类进行定义、填充、约束等，如以下 UserDO 实体类的示例代码所示。

```
@TableName("t_user")
@Data
public class UserDO implements Serializable {

    @TableField()
```

```
    private static final long serialVersionUID = 1L;

    @TableId(type = IdType.AUTO)
    private Integer id;

    @TableField(insertStrategy = FieldStrategy.NOT_NULL)
    private String username;

    @TableField(insertStrategy = FieldStrategy.NOT_NULL)
    private String phone;

    @TableField(insertStrategy = FieldStrategy.NOT_NULL, fill = FieldFill.INSERT)
    private LocalDateTime createTime;

    @TableField(insertStrategy = FieldStrategy.NOT_NULL)
    private Integer status;

}
```

新建一个 UserService 接口，如以下示例代码所示。

```
public interface UserService extends IService<UserDO> {

}
```

这里继承了 MyBatis-Plus 提供的通用 IService 接口，它提供了 Serivice 层常用的 CRUD 操作，然后创建一个实现类实现该接口，并继承 MyBatis-Plus 提供的通用 ServiceImpl 实现，如以下示例代码所示。

```
@Slf4j
@RequiredArgsConstructor
@Service
public class UserServiceImpl
        extends ServiceImpl<UserMapper, UserDO>
        implements UserService {

}
```

然后直接在 UserController 控制器类中注入 UserService，并调用 IService 通用接口中的 getById 方法查询并返回用户表的某条记录，如以下示例代码所示。

```
@RequiredArgsConstructor
@RestController
```

```
public class UserController {

    private final UserService userService;

    @GetMapping("/user/info/{id}")
    public UserDO getUserInfoById(@PathVariable("id") long id) {
        UserDO userDO = userService.getById(id);
        return userDO;
    }


}
```

在 MyBatis 原生框架中，一个简单的操作也要写一套相关的代码，包括 Mapper 方法、SQL 映射语句、Service 接口方法及实现方法，而 MyBatis-Plus 提供的通用 Mapper、通用 IService、通用 ServiceImpl 实现可以节省大量的冗余代码。

7.10.3 自定义数据库操作

如果通用的接口不符合要求，也可以自定义数据库操作，包括直接编写 SQL 映射文件，或者使用 MyBatis-Plus 提供的 QueryWrapper 和 LambdaQueryWrapper、UpdateWrapper 实现无 XML 化操作。

下面实现了除通用接口外的其他两种处理方式：

```
@Slf4j
@RequiredArgsConstructor
@Service
public class UserServiceImpl
        extends ServiceImpl<UserMapper, UserDO> implements UserService {

    private final UserMapper userMapper;

    @Override
    public UserDO getByUsername(String username, int type) {
        if (type == 0) {
            // XML
            log.info("query from xml");
            return userMapper.selectByUsername(username);
        } else {
            // QueryWrapper
            log.info("query from wrapper");
            LambdaQueryWrapper<UserDO> queryWrapper = new LambdaQueryWrapper();
```

```
            queryWrapper.eq(UserDO::getUsername, username);
            queryWrapper.eq(UserDO::getStatus, 1);
            return userMapper.selectOne(queryWrapper);
        }

    }

}
```

其他基本代码略，完全版本见仓库源码。

第一种方式就是 MyBatis 原生的 XML 映射文件的处理方式，根据方法名称查询并运行 XML 映射文件中的 SQL 语句，这里不再赘述。

第二种方式是 MyBatis-Plus 提供的条件构造器，无须 XML 文件，无须编写 SQL 语句，通过条件构造器同样可以实现对一些简单 SQL 语句的处理。

以上介绍的 MyBatis-Plus 的相关功能已经很强大了，MyBatis-Plus 还有更多实用的功能，比如主键自动生成策略、字段自动填充、分页功能等，这里不再赘述。总之，MyBatis 结合 MyBatis-Plus，效率翻倍。

7.11 Redis

7.11.1 概述

Redis 的全称为 REmote DIctionary Server，它是一种 key-value 形式的 NoSQL 内存数据库，它最大的特性是将所有数据都存储在内存中，所以读写速度快、性能好，它也支持将内存中的数据持久化到硬盘中，这样就不怕数据会丢失，兼顾性能和安全。

Spring Boot 为 Redis 的集成及使用提供了许多便利，只需要简单的配置，基于 RedisTemplate 模板就能“开箱即用”。

7.11.2 Redis 环境搭建

1. 安装 Redis

Redis 支持多种操作系统的安装方式（官方不支持 Windows），包括 Docker 镜像，这里以 CentOS 为例进行演示。

首先从 Redis 官网下载最新稳定版本包，下载地址为链接 8。

下载后解压并切换到根目录进行编译安装，参考命令如下：

```
$ tar -zxvf redis-7.0.5.tar.gz
$ cd redis-7.0.5
$ make
$ make install
```

最后在 src 目录下安装下面几个核心命令的可执行文件：

- redis-server（服务端程序）。
- redis-benchmark（基准性能测试）。
- redis-cli（客户端程序）。

2. 启动 Redis

运行 redis src 目录下的 redis-server 命令来启动 Redis 服务，参考命令如下：

```
$ ./redis-server ../redis.conf &
```

Redis 服务启动成功的画面如下所示。

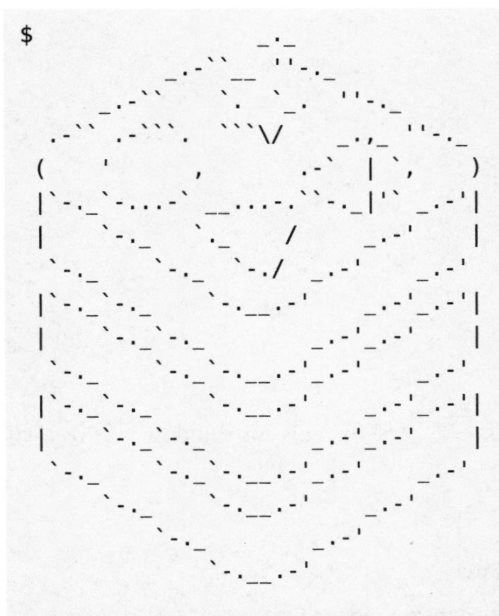

```
$
                          Redis 7.0.5 (00000000/0) 64 bit

                          Running in standalone mode
                          Port: 6379
                          PID: 62293

                                https://redis.io

... WARNING: The TCP backlog setting of 511 cannot be enforced because kern.ipc.
somaxconn is set to the lower value of 128.
```

```
... Server initialized
... Loading RDB produced by version 7.0.5
... RDB age 7 seconds
... RDB memory usage when created 1.04 Mb
... Done loading RDB, keys loaded: 0, keys expired: 0.
... DB loaded from disk: 0.000 seconds
... Ready to accept connections
```

上面显示了Redis启动的默认端口号（6379）、进程ID（62293），以及Redis使用的资源信息等。

3. 连接 Redis

运行 redis src 目录下的 redis-cli 命令来连接 Redis 服务，参考命令如下：

```
$ ./redis-cli
127.0.0.1:6379>
```

连接后可以使用 ping 命令测试 Redis 的连通性，参考命令如下：

```
$ ./redis-cli
127.0.0.1:6379> ping
PONG
```

PONG 表示连接 Redis 成功。

再进行一些简单的键值操作，参考命令如下：

```
127.0.0.1:6379> set test 123
OK
127.0.0.1:6379> get test
 "123"
127.0.0.1:6379>
```

设置值和查询都没问题，说明连接和操作都正常了。

4. 关闭 Redis

为了避免数据丢失，不要使用 kill -9 方式关闭 Redis，建议使用 redis-cli shutdown 命令关闭 Redis，参考命令如下：

```
./redis-cli shutdown
62293:M 04 Nov 2022 16:33:26.572 # User requested shutdown...
62293:M 04 Nov 2022 16:33:26.572 * Saving the final RDB snapshot before exiting.
62293:M 04 Nov 2022 16:33:26.593 * DB saved on disk
```

```
62293:M 04 Nov 2022 16:33:26.593 * Removing the pid file.
62293:M 04 Nov 2022 16:33:26.594 # Redis is now ready to exit, bye bye...
[1] + 62293 done ./redis-server ../redis.conf
```

如以上日志所示，Redis 会安全关闭。

7.11.3 Spring Boot 集成 Redis

1. 集成并连接 Redis

Spring Boot 提供了一个 spring-boot-starter-data-redis 启动器依赖，如以下依赖配置所示。

```
<dependency>
    <groupId>org.springframework.boot</groupId>
    <artifactId>spring-boot-starter-data-redis</artifactId>
</dependency>
```

该启动器依赖会导入 Spring Boot 集成 Redis 所需的所有依赖，如下图所示。

```
⌄ spring-boot-starter-data-redis : 3.0.0 [compile]
  > lettuce-core : 6.2.1.RELEASE [compile]
  > spring-boot-starter : 3.0.0 [compile]
  ⌄ spring-data-redis : 3.0.0 [compile]
      slf4j-api : 2.0.4 [compile]
    > spring-aop : 6.0.2 [compile]
    > spring-context-support : 6.0.2 [compile]
    > spring-data-keyvalue : 3.0.0 [compile]
    > spring-oxm : 6.0.2 [compile]
    > spring-tx : 6.0.2 [compile]
```

Spring Data Redis 提供了为 Spring 应用轻松配置和访问 Redis 的能力，spring-boot-starter- data-redis 启动器又在 Spring Data Redis 项目的基础上提供了 Redis 的自动配置功能，包括自动配置的 Lettuce 和 Jedis 客户端，如上图所示，默认客户端为 Lettuce。

Spring Boot在新的 org.springframework.boot.autoconfigure.AutoConfiguration.imports 自动配置文件中注册了几个 Redis 相关的自动配置类，具体如下：

- RedisAutoConfiguration。
- RedisReactiveAutoConfiguration。
- RedisRepositoriesAutoConfiguration。

比如自动配置类 RedisAutoConfiguration 对应的参数绑定类为 RedisProperties，通过一系列 spring.redis.* 参数可以自动配置 Redis，如下面的自动配置类的源码所示。

```
@AutoConfiguration
@ConditionalOnClass(RedisOperations.class)
@EnableConfigurationProperties(RedisProperties.class)
@Import({ LettuceConnectionConfiguration.class, JedisConnectionConfiguration.class })
public class RedisAutoConfiguration {

    @Bean
    @ConditionalOnMissingBean(name = "redisTemplate")
    @ConditionalOnSingleCandidate(RedisConnectionFactory.class)
    public RedisTemplate<Object, Object> redisTemplate(RedisConnectionFactory
redisConnectionFactory) {
        RedisTemplate<Object, Object> template = new RedisTemplate<>();
        template.setConnectionFactory(redisConnectionFactory);
        return template;
    }

    @Bean
    @ConditionalOnMissingBean
    @ConditionalOnSingleCandidate(RedisConnectionFactory.class)
    public StringRedisTemplate stringRedisTemplate(RedisConnectionFactory
redisConnectionFactory) {
        return new StringRedisTemplate(redisConnectionFactory);
    }

}
```

只需要在应用配置文件中添加以下 Redis 配置参数即可连接 Redis：

```
spring:
  data:
    redis:
      host: localhost
      port: 6379
      database: 0
```

> Spring Boot 3.0.0 新变化：
> RedisProperties 参数绑定类的前缀已经由 spring.redis.* 变更为 spring.data.redis.*。

Redis 支持 Sentinel（哨兵）、Cluster（集群）、Standalone（单机）部署方式，这里的配置默认是单机模式，如果 host、port、database 参数都是以上默认值，则可以不用配置。

从自动配置类中可以看到导入了 Lettuce 和 Jedis 两个客户端的配置类，Lettuce 是 Redis 内置默认支持的，切换为 Jedis 需要额外导入 Jedis 的依赖，并配合使用以下参数：

```
spring:
  redis:
client-type: jedis
```

2. 使用 RedisTemplate

自动配置类中还注册了以下两个默认的 Redis 模板：

- RedisTemplate。
- StringRedisTemplate。

这两个模板可以直接在应用中注入使用，它们都是线程安全的，"开箱即用"，可以在多个实例中重复使用，区别如下：

- 两者是处理不同 key-value 类型的数据操作模板，StringRedisTemplate 继承自 RedisTemplate，只能处理 <String, String> 的泛型类型，而 RedisTemplate 可以处理 <Object, Object> 所有泛型类型。
- StringRedisTemplate 默认使用 StringRedisSerializer 序列化器，RedisTemplate 默认使用 JDK 自带的序列化器。

因为 StringRedisTemplate 继承自 RedisTemplate，相当于要同时配置两个 RedisTemplate，所以需要配合 @ConditionalOnSingleCandidate 注解根据指定的类名注入。

RedisTemplate 模板还提供了很多通用的操作接口，用于对特定的数据类型或某些键进行操作，常用的接口如下表所示。

接口	描述
GeoOperations	地理空间操作
HashOperations	散列类型操作
HyperLogLogOperations	HyperLogLog 类型操作
ListOperations	列表操作
SetOperations	集合操作
ValueOperations	字符串操作
ZSetOperations	有序集合操作

新建一个接口注入 StringRedisTemplate 并使用，如以下示例代码所示。

```
@RequiredArgsConstructor
```

```
@RestController
public class RedisController {

    private final StringRedisTemplate stringRedisTemplate;

    @GetMapping("/redis/set")
    public String set(@RequestParam("name") String name, @RequestParam("value")
String value) {
        ValueOperations<String, String> valueOperations = stringRedisTemplate.opsForValue();
        valueOperations.set(name, value);
        return valueOperations.get(name);
    }

}
```

这里将请求参数名称（name）、参数值（value）作为 key-value 设置到 Redis 数据库中，最后返回从 Redis 数据库中取出来的 key 为 name 的值。

请求成功示例如下图所示。

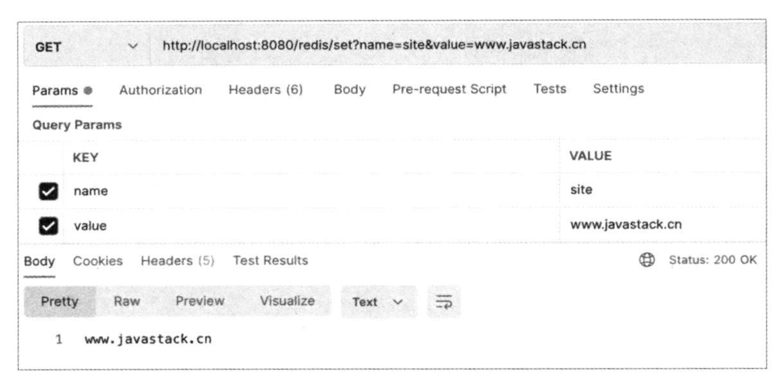

连接 Redis 控制台也可以获取对应 key 的值：

```
127.0.0.1:6379> get site
"www.javastack.cn"
```

每次使用 *Operations 时无须通过 RedisTemplate 获取，可以直接注入对应的 RedisTemplate 模板来使用 *Operations：

```
@Resource(name = "stringRedisTemplate")
private ValueOperations<String, String> valueOperations;
```

或者提前将其作为一个 Bean 注册到 Spring 容器中，然后直接注入使用：

```
@Bean
public ValueOperations<String, Object> valueOperations(RedisTemplate<String, Object>
redisTemplate) {
    return redisTemplate.opsForValue();
}

@Autowired
private ValueOperations<String, Object> valueOperations;
```

如果自动配置的 RedisTemplate 模板不符合要求，那么可以对其进行自定义，覆盖自动配置的 RedisTemplate 模板，如下面的示例所示。

```
@Configuration
public class RedisConfig {

    @Bean
    public RedisTemplate<String, Object> redisTemplate(RedisConnectionFactory factory) {
        RedisTemplate<String, Object> template = new RedisTemplate<>();
        template.setConnectionFactory(factory);

        StringRedisSerializer stringSerializer = new StringRedisSerializer();
        RedisSerializer jacksonSerializer = getJacksonSerializer();

        template.setKeySerializer(stringSerializer);
        template.setValueSerializer(jacksonSerializer);
        template.setHashKeySerializer(stringSerializer);
        template.setHashValueSerializer(jacksonSerializer);
        template.setEnableTransactionSupport(true);
        template.afterPropertiesSet();

        return template;
    }

    private RedisSerializer getJacksonSerializer() {
        ObjectMapper om = new ObjectMapper();
        om.setVisibility(PropertyAccessor.ALL, JsonAutoDetect.Visibility.ANY);
        om.activateDefaultTyping(LaissezFaireSubTypeValidator.instance,
                ObjectMapper.DefaultTyping.NON_FINAL);
        return new GenericJackson2JsonRedisSerializer(om);
    }

}
```

这里的 key 使用了 StringRedisSerializer 序列化器，value 使用了 GenericJackson2JsonRedisSerializer 序列化器。

Redis 序列化器的接口为 RedisSerializer，可以查看该接口的实现了解所有内置的序列化器，如下图所示。

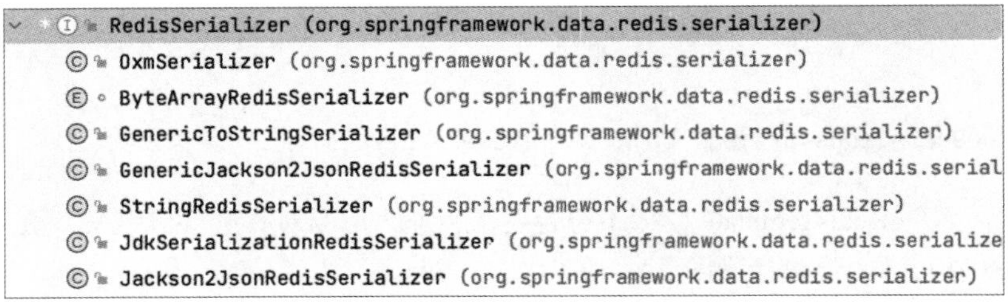

```
RedisSerializer (org.springframework.data.redis.serializer)
    OxmSerializer (org.springframework.data.redis.serializer)
    ByteArrayRedisSerializer (org.springframework.data.redis.serializer)
    GenericToStringSerializer (org.springframework.data.redis.serializer)
    GenericJackson2JsonRedisSerializer (org.springframework.data.redis.serial
    StringRedisSerializer (org.springframework.data.redis.serializer)
    JdkSerializationRedisSerializer (org.springframework.data.redis.serialize
    Jackson2JsonRedisSerializer (org.springframework.data.redis.serializer)
```

下表是常用的几个序列化器。

序列化器	说明
StringRedisSerializer	String/byte[] 转换器，速度快
JdkSerializationRedisSerializer	JDK 自带序列化器
OxmSerializer	XML 序列化器，占空间，速度慢
Jackson2JsonRedisSerializer	JSON 序列化器，需要定义 JavaType
GenericJackson2JsonRedisSerializer	JSON 序列化器，不需要定义 JavaType

如果只是简单的字符串类型，那么使用 StringRedisSerializer 即可，如果是存储对象，则推荐使用 JSON 序列化器，可以很方便地把 JSON 类型的数据组装成 Java 中的对象。

7.12 MongoDB

7.12.1 概述

MongoDB 是一个开源 NoSQL 文档数据库，它使用的是类似 JSON 的数据结构，面向集合存储，而不是传统的基于表的关系型数据库，它可以支持非常松散、复杂的数据类型，并且支持非常强大的类似于面向对象的查询语言，支持索引，几乎能完成关系型数据库单表查询的大部分功能，所以，MongoDB 是非关系型数据库中功能最丰富的，又最像关系型数据库的 NoSQL 数据库。

Spring Boot 为 MongoDB 的集成及使用提供了许多便利，只需要简单的配置，基于 MongoTemplate 模板或者 MongoRepository 存储库就能“开箱即用”，十分方便。

7.12.2 MongoDB 环境搭建

1. 安装 MongoDB

MongoDB 支持多种操作系统的安装方式，包括 Docker 镜像，这里以 CentOS 为例，使用 yum 包管理器的方式安装 MongoDB 社区免费版。

首先创建并配置 yum 包管理系统，参考命令如下：

```
sudo vi /etc/yum.repos.d/mongodb-org-6.0.repo
```

配置文件的内容如下：

```
[mongodb-org-6.0]
name=MongoDB Repository
baseurl=https://repo.mongodb.org/yum/redhat/$releasever/mongodb-org/6.0/x86_64/
gpgcheck=1
enabled=1
gpgkey=https://www.mongodb.org/static/pgp/server-6.0.asc
```

当然，也可以不用配置 yum 仓库，可以选择直接从 MongoDB 存储库下载 MongoDB，配置 yum 的目的是可以直接使用 yum 命令下载并安装最新的 MongoDB，参考命令如下：

```
sudo yum install -y mongodb-org
```

笔者这里下载的是 MongoDB 6.0.2，也可以指定要安装的具体版本，参考命令如下：

```
sudo yum install -y mongodb-org-6.0.2 mongodb-org-database-6.0.2 mongodb-org-server-6.0.2
mongodb-mongosh-6.0.2 mongodb-org-mongos-6.0.2 mongodb-org-tools-6.0.2
```

2. 启动 MongoDB

安装好 MongoDB 后，可以使用 systemd (systemctl) 和 System V Init (service) 的方式来管理 MongoDB 服务，下面以更强大的 systemd (systemctl) 服务为例进行演示。

启动 MongoDB 的参考命令如下：

```
sudo systemctl start mongod
```

关闭 MongoDB 的参考命令如下：

```
sudo systemctl stop mongod
```

重新启动 MongoDB 的参考命令如下：

```
sudo systemctl restart mongod
```

系统重启时自动启动 MongoDB 的参考命令如下：

```
sudo systemctl enable mongod
```

默认情况下，MongoDB 会使用 mongod 用户账户运行并使用以下默认目录：

- /var/lib/mongo（数据目录）。
- /var/log/mongodb（日志目录）。

这些目录在 MongoDB 安装时就创建好了。

MongoDB 启动之后，可以使用以下参考命令验证：

```
sudo systemctl status mongod
```

```
$ sudo systemctl status mongod
• mongod.service - MongoDB Database Server
Loaded: loaded (/usr/lib/systemd/system/mongod.service; enabled; vendor preset: disabled)
Active: active (running) ...
...
```

可以看到正在运行的 MongoDB 的状态信息。

3. 连接 MongoDB

使用 mongosh 命令连接 MongoDB，它会默认连接到本地的 27017 端口，参考命令如下：

```
$ mongosh
Current Mongosh Log ID:       6368b535d2a64709a8392ddf
Connecting to:                mongodb://127.0.0.1:27017/?directConnection=
true&serverSelectionTimeoutMS=2000&appName=mongosh+1.6.0
Using MongoDB:                6.0.2
Using Mongosh:                1.6.0
...
test>
```

默认连接到 test 数据库，使用 show dbs 显示所有数据库，参考命令如下：

```
test> show dbs;
admin   40.00 KiB
config  60.00 KiB
local   72.00 KiB
```

这三个数据库都是 MongoDB 预设的，可以使用 use 命令进行切换。

下面是一个基本的新增和查询操作，参考命令如下：

```
test> db.collection.insertOne({id: 1, name: 'java'})
{
  acknowledged: true,
  insertedId: ObjectId("6368b746495ae1aaf3ebce65")
}
test> db.collection.find({id: 1})
[ { _id: ObjectId("6368b746495ae1aaf3ebce65"), id: 1, name: 'java' } ]
```

新增和查询都没问题，说明连接和操作都正常了。

7.12.3 Spring Boot 集成 MongoDB

1. 集成并连接 MongoDB

Spring Boot 为使用 MongoDB 的两种模型都提供了 Starter 启动器：

- spring-boot-starter-data-mongodb。
- spring-boot-starter-data-mongodb-reactive。

本节使用第一种方式，先添加以下依赖配置：

```
<dependency>
    <groupId>org.springframework.boot</groupId>
    <artifactId>spring-boot-starter-data-mongodb</artifactId>
</dependency>
```

该启动器依赖会导入 Spring Boot 集成 MongoDB 所需的所有依赖，如下图所示。

```
∨ spring-boot-starter-data-mongodb : 3.0.0 [compile]
  > mongodb-driver-sync : 4.8.0 [compile]
  > spring-boot-starter : 3.0.0 [compile]
  ∨ spring-data-mongodb : 4.0.0 [compile]
      mongodb-driver-core : 4.8.0 [compile]
      slf4j-api : 2.0.4 [compile]
    > spring-beans : 6.0.2 [compile]
    > spring-context : 6.0.2 [compile]
      spring-core : 6.0.2 [compile]
    > spring-data-commons : 3.0.0 [compile]
    > spring-expression : 6.0.2 [compile]
    > spring-tx : 6.0.2 [compile]
```

这里面就包含了 MongoDB 驱动和 spring-data-mongodb 数据层访问依赖。Spring Boot 在新的 org.springframework.boot.autoconfigure.AutoConfiguration.imports 自动配置文件中注册了几个 MongoDB 相关的自动配置类，具体如以下所示。

- MongoDataAutoConfiguration。
- MongoDataAutoConfiguration。
- MongoReactiveDataAutoConfiguration。
- MongoRepositoriesAutoConfiguration。
- MongoAutoConfiguration。
- MongoReactiveAutoConfiguration。

对应的参数绑定为 MongoProperties，通过一系列 spring.data.mongodb.* 参数就能完成自动配置，比如这里添加最基本的连接参数，连接本地默认端口的 test 数据库：

```
spring:
  data:
    mongodb:
      uri: mongodb://localhost:27017/test
```

2. 使用 MongoTemplate

和其他数据库模板一样，Spring Boot 也提供了 MongoDB 的 MongoTemplate 模板，自动配置后，在应用中就可以直接注入。

MongoTemplate 实例就注册在 MongoDataAutoConfiguration 自动配置类中，如以下源码所示。

```
@AutoConfiguration(after = MongoAutoConfiguration.class)
@ConditionalOnClass({ MongoClient.class, MongoTemplate.class })
@EnableConfigurationProperties(MongoProperties.class)
@Import({ MongoDataConfiguration.class, MongoDatabaseFactoryConfiguration.class,
        MongoDatabaseFactoryDependentConfiguration.class })
public class MongoDataAutoConfiguration {

}

@Configuration(proxyBeanMethods = false)
@ConditionalOnBean(MongoDatabaseFactory.class)
class MongoDatabaseFactoryDependentConfiguration {

    ...
```

```
    @Bean
    @ConditionalOnMissingBean(MongoOperations.class)
    MongoTemplate mongoTemplate(MongoDatabaseFactory factory, MongoConverter converter) {
        return new MongoTemplate(factory, converter);
    }

    ...

}
```

下面添加一个 MongoTemplate 模板的使用测试接口，示例如下：

```
@RequiredArgsConstructor
@RestController
public class MongoController {

    public static final String COLLECTION_NAME = "javastack";

    private final MongoTemplate mongoTemplate;

    @RequestMapping("/mongo/insert")
    public User insert(@RequestParam("name") String name, @RequestParam("sex") int sex) {
        // 新增
        User user = new User(RandomUtils.nextInt(), name, sex);
        mongoTemplate.insert(user, COLLECTION_NAME);

        // 查询
        Query query = new Query(Criteria.where("name").is(name));
        return mongoTemplate.findOne(query, User.class, COLLECTION_NAME);
    }

}
```

User 实体类如以下示例代码所示。

```
@NoArgsConstructor
@AllArgsConstructor
@Data
public class User {

    private long id;
```

```
    private String name;

    private int sex;

}
```

通过注入 MongoTemplate 模板就能实现 CRUD 操作，先插入数据，然后根据名称查询刚插入的数据。重启应用后再调用插入数据接口，测试结果如下图所示。

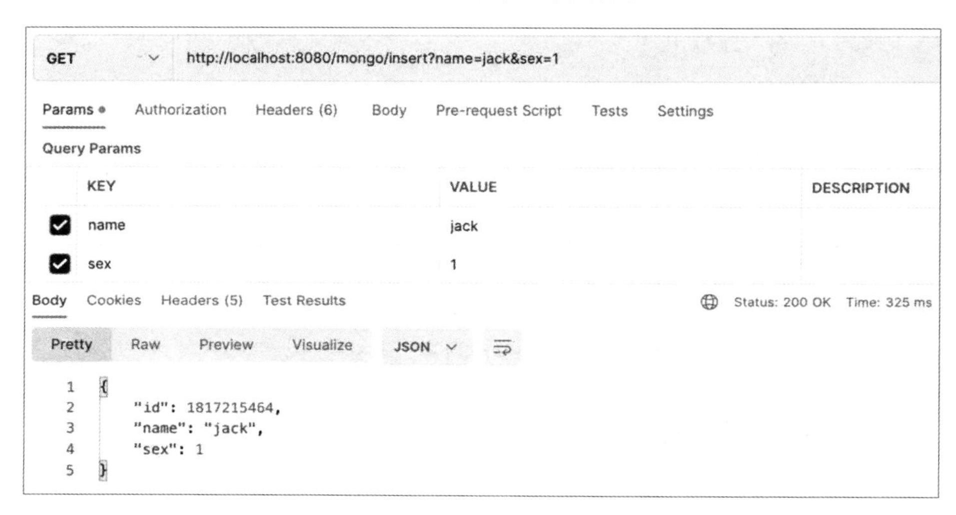

结果正常返回了，说明数据插入成功了，通过命令也能成功查询到数据，如以下参考命令所示。

```
test> db.javastack.findOne({name: 'jack'})
{
  _id: 1817215464,
  name: 'jack',
  sex: 1,
  _class: 'cn.javastack.mongodb.User'
}
```

3. 使用 MongoRepository

除了使用 MongoTemplate，Spring Data 还提供了对 MongoDB 的 Repositories 存储库的支持，可以像使用 JPA Repositories 一样操作 MongoDB 数据库，比如添加一个自定义的 UserRepository，示例如下：

```
public interface UserRepository extends MongoRepository<User, Long> {
```

```
    List<User> findByName(String name);

}
```

UserRepository 继承了 MongoRepository，MongoRepository 提供了针对 MongoDB 的丰富的 CRUD 操作方法，很多 CRUD 的数据库基本操作代码就不需要写了，类继承图如下图所示。

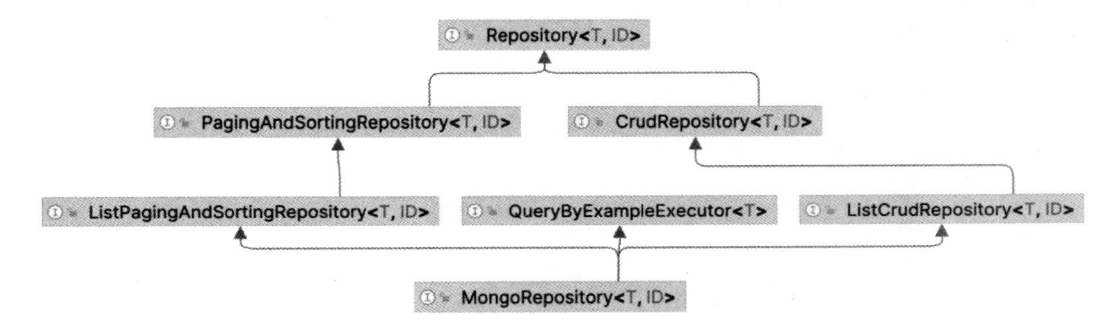

如果有需要，也可以按规范进行自定义，比如上面的 findByName 方法。这里再针对 MongoRepository 新增一个类似的接口进行测试，如以下测试代码所示。

```
@RequiredArgsConstructor
@RestController
public class MongoController {

    public static final String COLLECTION_NAME = "javastack";
    private final MongoTemplate mongoTemplate;

    private final UserRepository userRepository;

    ...

    @RequestMapping("/mongo/repo/insert")
    public User repoInsert(@RequestParam("name") String name, @RequestParam("sex")
int sex) {
        // 新增
        User user = new User(RandomUtils.nextInt(), name, sex);
        userRepository.save(user);

        // 查询
        return userRepository.findByName(name).get(0);
    }

}
```

MongoRepository 操作方法没有指定集合名称的参数，还需要在实体类上使用 @Document 注解指定集合名称，如以下示例代码所示。

```
@NoArgsConstructor
@AllArgsConstructor
@Data
@Document(collection = "javastack")
public class User {

    private long id;

    private String name;

    private int sex;

}
```

通过注入 UserRepository 存储库来实现 CRUD 操作，和之前的逻辑一样，先插入数据，再根据名称查询刚插入的数据，重启应用之后再调用插入数据接口，测试结果如下图所示。

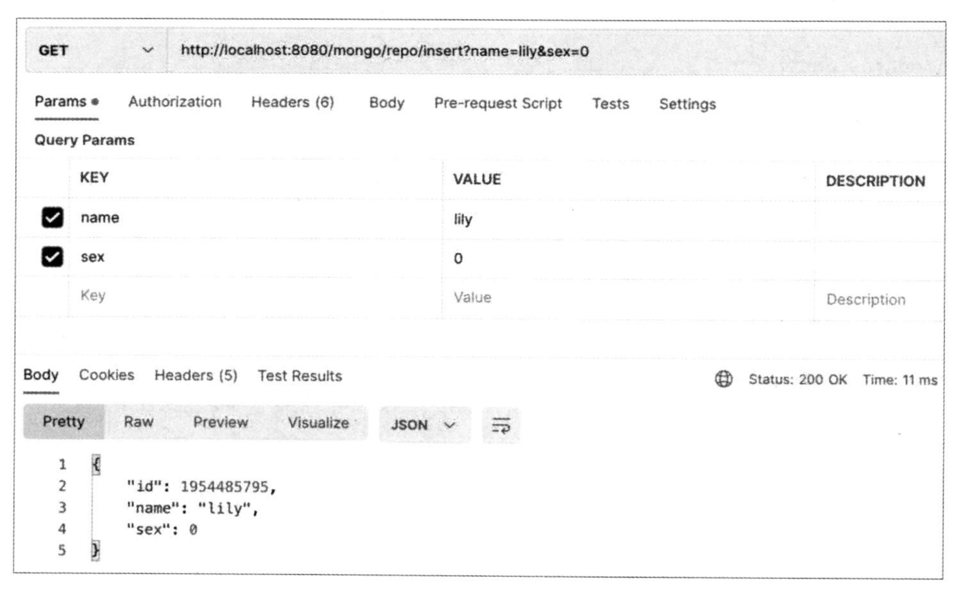

页面正常返回了，说明数据插入成功了，通过以下命令也能成功查询到数据：

```
test> db.javastack.findOne({name: 'lily'})
{
```

```
  _id: Long("1954485795"),
  name: 'lily',
  sex: 0,
  _class: 'cn.javastack.mongodb.User'
}
```

Spring Boot 集成和使用 MongoDB 的过程十分简单，添加依赖、连接配置即可以使用 Spring Boot“搞定”简单的 MongoDB 应用。

7.13 Elasticsearch

7.13.1 概述

Elasticsearch 是 Apache 开源的一个分布式、基于 RESTful 风格的数据分析和搜索引擎，由 Java 语言开发，它拥有很方便的搜索、分析和探索数据的能力，能够达到实时搜索，稳定、可靠、快速，安装使用方便，能使数据在生产环境中变得更有价值，它是目前最受企业欢迎的搜索引擎，其次是 Apache Solr（也基于 Lucene）。

Elasticsearch 中的数据是面向文档的，一个文档代表关系型数据库中的一行数据，它是对文档进行索引、检索、排序和过滤，而不是对行或者列数据，JSON 则作为文档的序列化格式。

下面这个 JSON 文档代表了一个 user 对象：

```
{
    "email":        "xxx@javastck.cn",
    "first_name":   "javastack",
    "last_name":    "admin",
    "info": {
        "wx":        "Java 技术栈 ",
        "sites": [ "javastack.cn", "www.javastack.cn" ]
    },
    "add_date":     "2022/11/08"
}
```

Elasticsearch 索引的两种概念如下：

- **索引（名词）：** 相当于关系型数据库中的一个数据库，是一个存储关系型文档的地方，索引（index）的复数词为 indices 或 indexes。
- **索引（动词）：** 指存储一个文档到一个索引中，类似于 SQL 语句中的 INSERT 关键词。

Spring Boot为Elasticsearch的集成及使用提供了许多便利，只需要简单的配置，基于ElasticsearchTemplate模板或者MongoRepository存储库就能“开箱即用”，十分方便。

7.13.2 Elasticsearch环境搭建

1. 安装Elasticsearch

Elasticsearch支持多种操作系统的安装方式，包括Docker镜像，这里以CentOS为例，使用yum包管理器的方式安装Elasticsearch。

首先创建一个yum管理系统配置文件，参考命令如下：

```
sudo vi /etc/yum.repos.d/elasticsearch.repo
```

配置文件的内容如下所示。

```
[elasticsearch]
name=Elasticsearch repository for 8.x packages
baseurl=https://artifacts.elastic.co/packages/8.x/yum
gpgcheck=1
gpgkey=https://artifacts.elastic.co/GPG-KEY-elasticsearch
enabled=0
autorefresh=1
type=rpm-md
```

当然，也可以不用配置yum仓库，选择直接从Elasticsearch存储库下载Elasticsearch，配置yum的目的是可以直接使用yum命令下载并安装最新的Elasticsearch，参考命令如下：

```
sudo yum install --enablerepo=elasticsearch elasticsearch
```

笔者这里下载的是Elasticsearch 8.5.0，安装过程中，默认启用并配置以下安全功能：

- 启用认证和授权功能，并生成一个默认密码的elastic超级用户。
- 为传输层和HTTP层生成TLS的证书和密钥，并使用这些密钥和证书启用并配置TLS。

默认生成的安全信息会显示在安装控制台上，如下所示。

```
-------------------- Security autoconfiguration information --------------------

Authentication and authorization are enabled.
TLS for the transport and HTTP layers is enabled and configured.

The generated password for the elastic built-in superuser is : nE-5Sq92L=+-0ofr7pTR
```

```
If this node should join an existing cluster, you can reconfigure this with
'/usr/share/elasticsearch/bin/elasticsearch-reconfigure-node --enrollment-token <token-here>'
after creating an enrollment token on your existing cluster.

You can complete the following actions at any time:

Reset the password of the elastic built-in superuser with
'/usr/share/elasticsearch/bin/elasticsearch-reset-password -u elastic'.

Generate an enrollment token for Kibana instances with
'/usr/share/elasticsearch/bin/elasticsearch-create-enrollment-token -s kibana'.

Generate an enrollment token for Elasticsearch nodes with
'/usr/share/elasticsearch/bin/elasticsearch-create-enrollment-token -s node'.

---------------------------------------------------------------------------------
```

比如默认的用户名和密码为 elastic/nE-5Sq92L=+-Oofr7pTR，上面也提供了修改密码的指令，如下所示。

```
./elasticsearch-reset-password -u elastic –interactive
```

默认重新生成随机密码，自定义修改密码需要设置 --interactive 参数，笔者修改密码为123456。

2. 启动 Elasticsearch

设置 Elasticsearch 开机自启动的参考命令如下：

```
sudo systemctl daemon-reload
sudo systemctl enable elasticsearch.service
```

启动 Elasticsearch 服务的参考命令如下：

```
sudo systemctl start elasticsearch.service
```

关闭 Elasticsearch 服务的参考命令如下：

```
sudo systemctl stop elasticsearch.service
```

验证 Elasticsearch 服务是否运行的参考命令如下：

```
curl --cacert $ES_HOME/config/certs/http_ca.crt -u elastic https://localhost:9200
```

```
Enter host password for user 'elastic':
{
```

```
  "name" : "...",
  "cluster_name" : "elasticsearch",
  "cluster_uuid" : "lE9ftJrLQ8CnksT0jxCf2Q",
  "version" : {
    "number" : "8.5.0",
    "build_flavor" : "default",
    "build_type" : "tar",
    "build_hash" : "c94b4700cda13820dad5aa74fae6db185ca5c304",
    "build_date" : "2022-10-24T16:54:16.433628434Z",
    "build_snapshot" : false,
    "lucene_version" : "9.4.1",
    "minimum_wire_compatibility_version" : "7.17.0",
    "minimum_index_compatibility_version" : "7.0.0"
  },
  "tagline" : "You Know, for Search"
}
```

或者直接在浏览器中访问 https://localhost:9200，然后输入用户名和密码也可以验证，如下图所示。


```
{
  "name"
  "cluster_name" : "elasticsearch",
  "cluster_uuid" : "lE9ftJrLQ8CnksT0jxCf2Q",
  "version" : {
    "number" : "8.5.0",
    "build_flavor" : "default",
    "build_type" : "tar",
    "build_hash" : "c94b4700cda13820dad5aa74fae6db185ca5c304",
    "build_date" : "2022-10-24T16:54:16.433628434Z",
    "build_snapshot" : false,
    "lucene_version" : "9.4.1",
    "minimum_wire_compatibility_version" : "7.17.0",
    "minimum_index_compatibility_version" : "7.0.0"
  },
  "tagline" : "You Know, for Search"
}
```


3. 连接 Elasticsearch

连接 Elasticsearch 后可以通过 Elasticsearch 提供的 Restful API 发起操作，比如查询 name=jack 的数据，参考命令如下：

```
curl -X GET "localhost:9200/javastack/_search?q=name:jack&pretty"
curl: (52) Empty reply from server
```

还可以安装一个可视化的开源工具 elasticsearch-head，开源地址为链接 9。

安装 elasticsearch-head 工具的参考命令如下：

```
git clone git://github.com/mobz/elasticsearch-head.git
cd elasticsearch-head
npm install
npm run start
```

然后需要设置 Elasticsearch API 以支持跨域，修改以下配置文件：

```
/config/elasticsearch.yml
```

在配置文件的最后添加跨域配置，参考配置的内容如下：

```
http.cors.enabled: true
http.cors.allow-origin: "*"
http.cors.allow-methods: "GET,PUT,POST,DELETE"
http.cors.allow-headers: Authorization,X-Requested-With,Content-Length,Content-Type
```

重启 Elasticsearch，之后就可以访问 elasticsearch-head 了，访问地址如下：

```
http://localhost:9100/?auth_user=elastic&auth_password=123456
```

效果如下图所示。

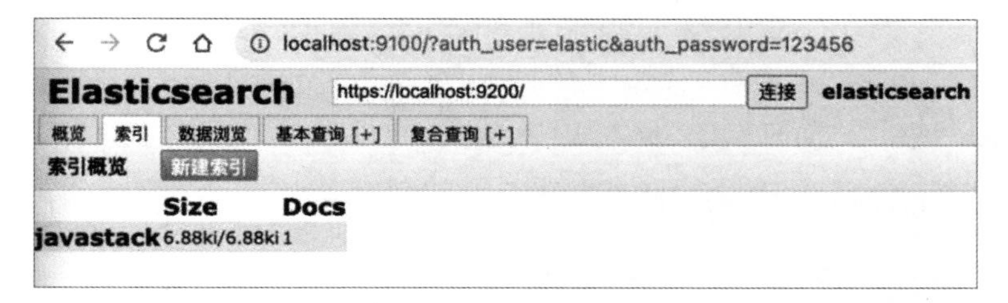

连接上 Elasticsearch 之后，笔者在这里添加了一个 javastack 索引，虽然后面的集成测试会用到，但这并不是必需的，应用在插入数据时也会自动创建。

7.13.3　Spring Boot 集成 Elasticsearch

1. 集成并连接 Elasticsearch

Spring Boot 支持以下多个 Elasticsearch 客户端：

- Elasticsearch 官方低级别的 REST 客户端。
- Elasticsearch 官方的 Java API 客户端。

- Spring Data Elasticsearch 提供的 ReactiveElasticsearchClient 客户端。

并且提供了一个专用的spring-boot-starter-data-elasticsearch一站式启动器，如以下依赖配置所示。

```
<dependency>
    <groupId>org.springframework.boot</groupId>
    <artifactId>spring-boot-starter-data-elasticsearch</artifactId>
</dependency>
```

该启动器依赖会导入 Spring Boot 集成 Elasticsearch 所需的所有依赖，如下图所示。

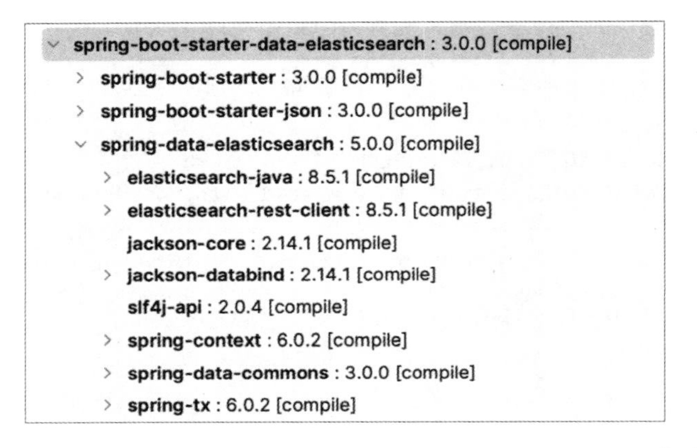

在新的 org.springframework.boot.autoconfigure.AutoConfiguration.imports 自动配置文件中注册了几个 Elasticsearch 相关的自动配置类，具体如下所示。

- ElasticsearchDataAutoConfiguration。
- ElasticsearchRepositoriesAutoConfiguration。
- ReactiveElasticsearchRepositoriesAutoConfiguration。
- ElasticsearchClientAutoConfiguration。
- ElasticsearchRestClientAutoConfiguration。
- ReactiveElasticsearchClientAutoConfiguration。

对应的参数绑定为 ElasticsearchProperties，通过一系列 spring.elasticsearch.* 参数就能完成自动配置。

这里添加最基本的连接参数，连接本地默认端口的 test 数据库，如以下配置示例所示。

```
spring:
  elasticsearch:
```

```
    uris: http://localhost:9200
    connection-timeout: 5s
    socket-timeout: 10s
    username: elastic
    password: 123456
```

因为 Elasticsearch 默认开启了 xpack 安全机制，所以只能通过 https 连接，而 Spring Boot 自动配置参数类 ElasticsearchProperties 暂时无法配置 xpack 安全信息，为了方便进行 http 测试，笔者这里先关闭 xpack 安全机制，修改配置如下：

```
xpack.security.http.ssl:
  enabled: false
  keystore.path: certs/http.p12
```

重启 Elasticsearch，随后即可通过 http 形式访问 API，应用也能顺利连接。

2. 使用 ElasticsearchTemplate

和其他数据库模板一样，Spring Boot 也提供了 Elasticsearch 的 ElasticsearchTemplate 模板，自动配置后，在应用中就可以直接注入。

ElasticsearchTemplate 就注册在 ElasticsearchDataAutoConfiguration 自动配置类中，如下面的源码所示。

```
@AutoConfiguration(
        after = { ElasticsearchClientAutoConfiguration.class, ReactiveElasticsearch-
ClientAutoConfiguration.class })
@ConditionalOnClass({ ElasticsearchTemplate.class })
@Import({ ElasticsearchDataConfiguration.BaseConfiguration.class,
        ElasticsearchDataConfiguration.JavaClientConfiguration.class,
        ElasticsearchDataConfiguration.ReactiveRestClientConfiguration.class })
public class ElasticsearchDataAutoConfiguration {

}

abstract class ElasticsearchDataConfiguration {

    ...

    @Configuration(proxyBeanMethods = false)
    @ConditionalOnClass(ElasticsearchClient.class)
```

```
    static class JavaClientConfiguration {

        @Bean
        @ConditionalOnMissingBean(value = ElasticsearchOperations.class, name =
"elasticsearchTemplate")
        @ConditionalOnBean(ElasticsearchClient.class)
        ElasticsearchTemplate elasticsearchTemplate(ElasticsearchClient client,
ElasticsearchConverter converter) {
            return new ElasticsearchTemplate(client, converter);
        }

    }

    ...

}
```

添加一个 ElasticsearchTemplate 模板的使用测试接口，如以下示例代码所示。

```
@RequiredArgsConstructor
@RestController
public class EsController {

    public static final String INDEX_JAVASTACK = "javastack";
    private final ElasticsearchTemplate elasticsearchTemplate;

    @RequestMapping("/es/insert")
    public User insert(@RequestParam("name") String name, @RequestParam("sex") int
sex) throws InterruptedException {
        // 新增
        User user = new User(RandomUtils.nextInt(), name, sex);
        IndexCoordinates indexCoordinates =  IndexCoordinates.of(INDEX_JAVASTACK);
        User save = elasticsearchTemplate.save(user, indexCoordinates);

        // 可能有延迟，休眠一秒再查询
        Thread.sleep(1000l);
        Query query = new CriteriaQuery(Criteria.where("name").is(name));
        return elasticsearchTemplate.searchOne(query, User.class, indexCoordinates).
getContent();
    }

}
```

User 实体类如以下示例代码所示。

```
@NoArgsConstructor
@AllArgsConstructor
@Data
public class User {

    private long id;

    private String name;

    private int sex;

}
```

通过注入 ElasticsearchTemplate 模板就能实现 CRUD 操作，先插入数据，再根据名称查询刚插入的数据，然后调用插入数据接口，测试结果如下图所示。

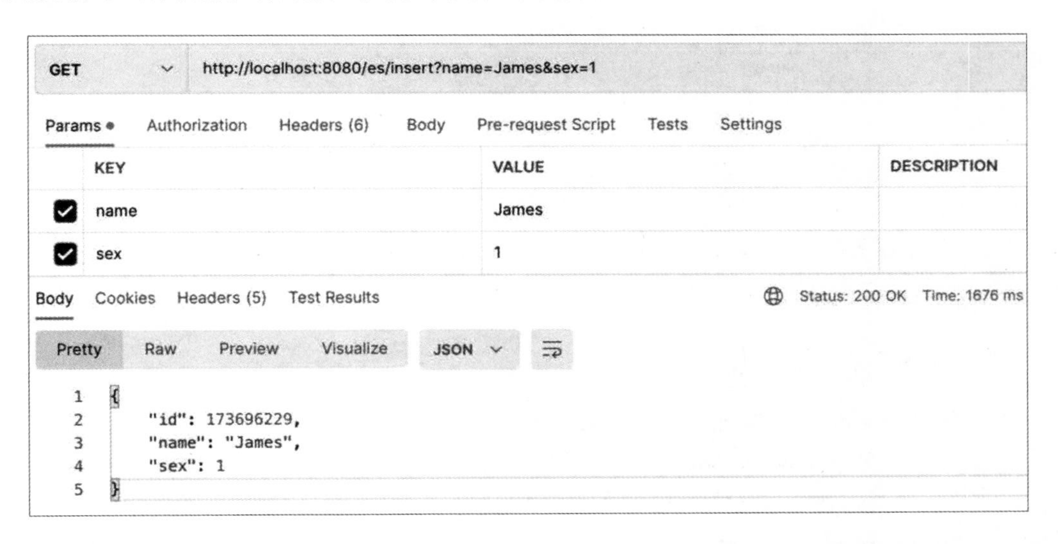

页面正常返回了，说明数据插入成功了。通过 elasticsearch-head 也能成功查询到数据，如下图所示。

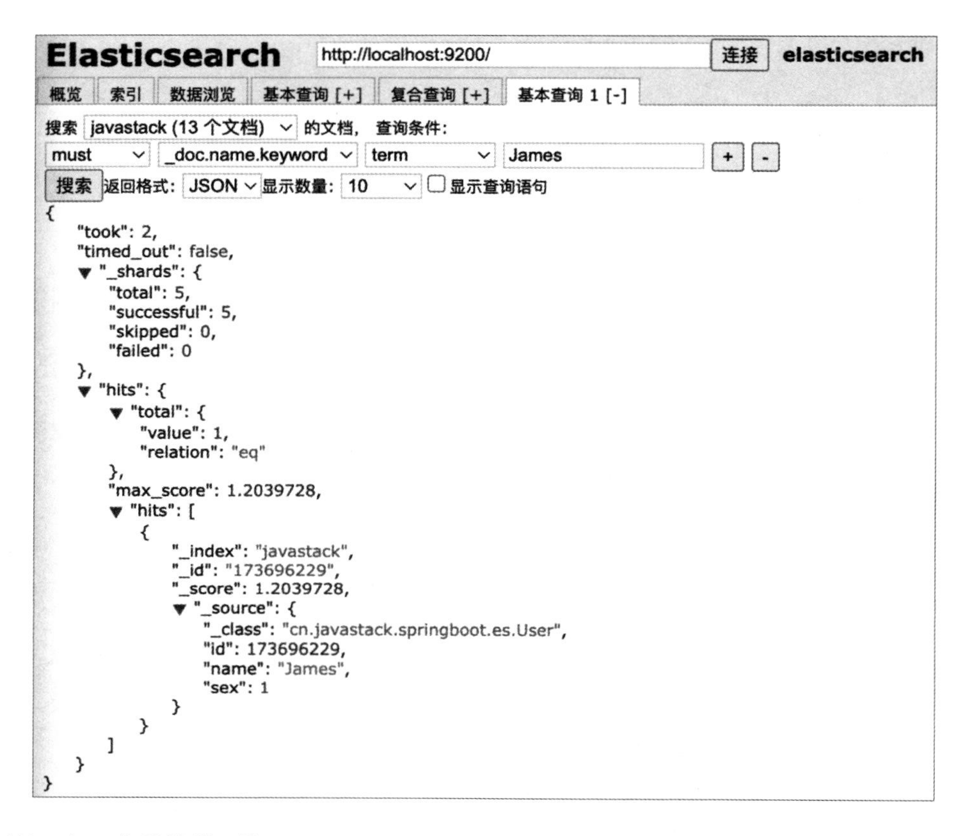

还是一如既往的简单，使用 Spring Boot 及 Elasticsearch 提供的 ElasticsearchTemplate 操作模板，可以轻松“搞定”一个 Elasticsearch 搜索应用。

3. 使用 ElasticsearchRepository

除了使用 ElasticsearchTemplate，Spring Data 还提供了对 Elasticsearch 的 Repositories 存储库的支持，可以像使用 JPA Repositories 一样操作 Elasticsearch 数据库，这里添加一个自定义的 UserRepository，示例如下：

```
public interface UserRepository extends ElasticsearchRepository<User, Long> {

    List<User> findByName(String name);

}
```

UserRepository 继承了 ElasticsearchRepository，ElasticsearchRepository 提供了针对 Elasticsearch

的丰富的CRUD操作方法，很多CRUD的数据库基本操作代码就不需要写了，类继承图如下图所示。

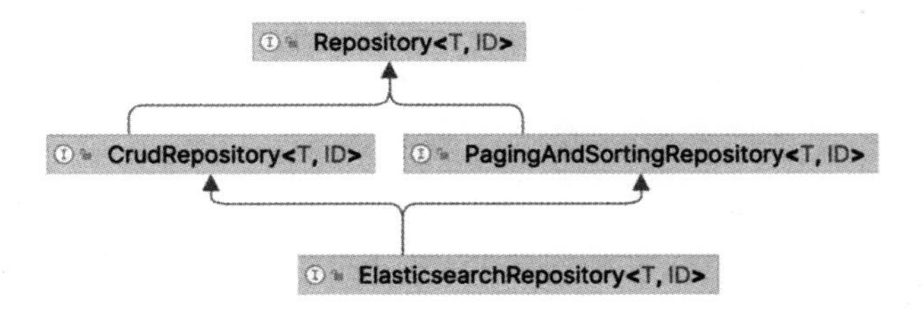

如果有需要，也可以按规范进行自定义，比如上面的findByName方法。这里针对ElasticsearchRepository再新增一个类似的接口进行测试，示例如下：

```
@RequiredArgsConstructor
@RestController
public class EsController {

    public static final String INDEX_JAVASTACK = "javastack";
    private final ElasticsearchTemplate elasticsearchTemplate;

    private final UserRepository userRepository;

    ...

    @RequestMapping("/es/repo/insert")
    public User repoInsert(@RequestParam("name") String name, @RequestParam("sex")
int sex) {
        // 新增
        User user = new User(RandomUtils.nextInt(), name, sex);
        userRepository.save(user);

        // 查询
        return userRepository.findByName(name).get(0);
    }

}
```

ElasticsearchRepository操作方法没有指定索引名称的参数，还需要在实体类上使用@Document注解指定索引名称，示例如下：

```
@NoArgsConstructor
@AllArgsConstructor
@Data
@Document(indexName = "javastack")
public class User {

    private long id;

    private String name;

    private int sex;

}
```

然后通过注入 UserRepository 存储库来实现 CRUD 操作，和之前的逻辑一样，先插入数据，再根据名称查询刚插入的数据，然后调用插入数据接口，测试结果如下图所示。

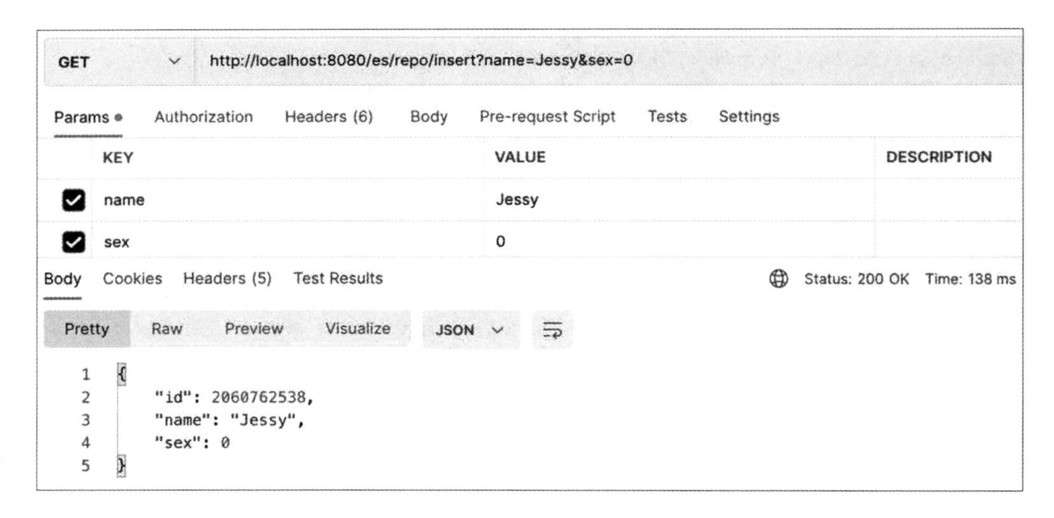

页面正常返回了，说明数据插入成功了，通过 elasticsearch-head 也能成功查询到数据，如下图所示。

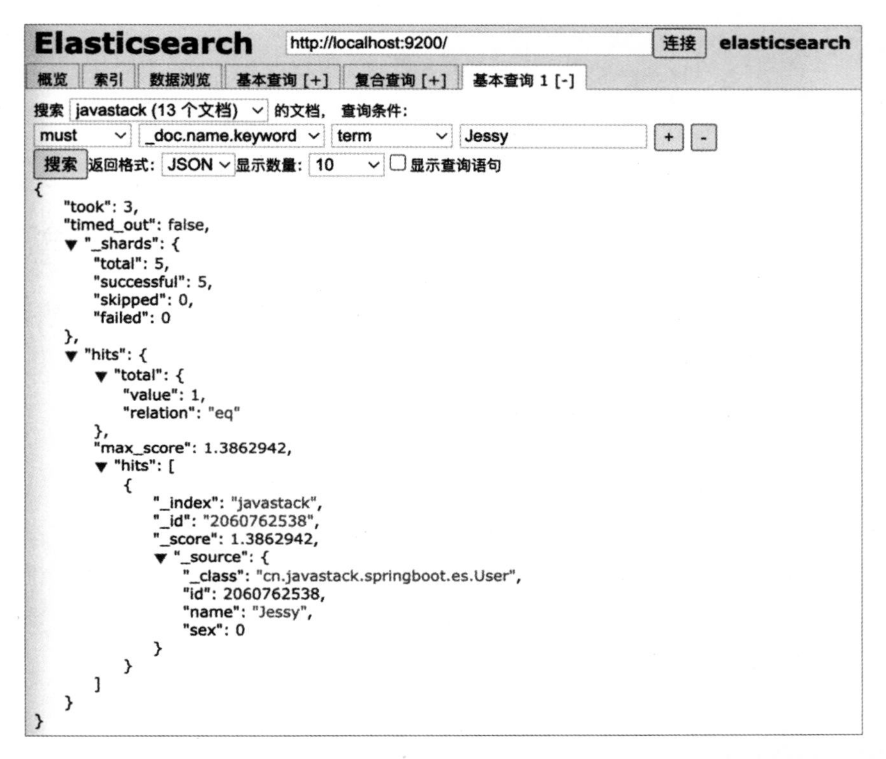

本节介绍了 Spring Boot 如何快速集成 Elasticsearch 并进行了实践，整个集成和使用过程十分简单，添加依赖、连接配置即可使用 Spring Boot “搞定” 简单的 Elasticsearch 应用。

第 8 章 Spring Boot 计划任务

计划任务是企业系统必不可少的部分，比如银行对账、定时发送报表、事务最终一致性补偿机制等都需要计划任务来辅助，Spring Boot 提供了对 Spring 计划任务及 Quartz 计划任务的官方支持，本章会介绍这两者的使用方式及相应的原理分析，包括基本概述、线程池工作流程、任务实现、自定义线程池配置等。

8.1 Spring 计划任务

8.1.1 概述

Spring Boot 实现计划任务不需要引入其他的依赖，它只会依赖 Spring 基础框架中的计划任务，Spring 3.1 开始支持计划任务，Spring Boot 只是在自动配置包中提供了对 Spring 计划任务的自动配置，其主要还是依赖于 Spring 中的 spring-context 包，所以只要任何一个 Spring Boot Starter 启动器包含 spring-context 包，它就能使用计划任务，比如 spring-boot-starter-web 一站式启动器。

Spring Boot 提供了对任务执行、任务调度的自动配置，对应的自动配置类如下所示。

- TaskExecutionAutoConfiguration：任务执行自动配置类。
- TaskSchedulingAutoConfiguration：任务调度自动配置类。

它们被注册在新的 org.springframework.boot.autoconfigure.AutoConfiguration.imports 自动配置文件中，TaskExecutionAutoConfiguration 自动配置类的源码如下所示。

```
@ConditionalOnClass(ThreadPoolTaskExecutor.class)
@AutoConfiguration
@EnableConfigurationProperties(TaskExecutionProperties.class)
public class TaskExecutionAutoConfiguration {

    /**
     * Bean name of the application {@link TaskExecutor}.
     */
    public static final String APPLICATION_TASK_EXECUTOR_BEAN_NAME = "applicationTaskExecu-
tor";

    @Bean
    @ConditionalOnMissingBean
    public TaskExecutorBuilder taskExecutorBuilder(TaskExecutionProperties properties,
            ObjectProvider<TaskExecutorCustomizer> taskExecutorCustomizers,
            ObjectProvider<TaskDecorator> taskDecorator) {
        TaskExecutionProperties.Pool pool = properties.getPool();
        TaskExecutorBuilder builder = new TaskExecutorBuilder();
        builder = builder.queueCapacity(pool.getQueueCapacity());
        builder = builder.corePoolSize(pool.getCoreSize());
        builder = builder.maxPoolSize(pool.getMaxSize());
        builder = builder.allowCoreThreadTimeOut(pool.isAllowCoreThreadTimeout());
        builder = builder.keepAlive(pool.getKeepAlive());
        Shutdown shutdown = properties.getShutdown();
        builder = builder.awaitTermination(shutdown.isAwaitTermination());
        builder = builder.awaitTerminationPeriod(shutdown.getAwaitTerminationPeriod());
        builder = builder.threadNamePrefix(properties.getThreadNamePrefix());
        builder = builder.customizers(taskExecutorCustomizers.orderedStream()::iterator);
        builder = builder.taskDecorator(taskDecorator.getIfUnique());
        return builder;
    }

    @Lazy
    @Bean(name = { APPLICATION_TASK_EXECUTOR_BEAN_NAME,
            AsyncAnnotationBeanPostProcessor.DEFAULT_TASK_EXECUTOR_BEAN_NAME })
    @ConditionalOnMissingBean(Executor.class)
    public ThreadPoolTaskExecutor applicationTaskExecutor(TaskExecutorBuilder builder) {
        return builder.build();
    }

}
```

如果当前上下文中没有任何 Executor 实例，那么它会自动注册一个默认的 ThreadPoolTaskExecutor

线程池任务执行器实例，它基于 JDK 并发包中的 Executor 接口实现了自己的任务执行器。同时注册了一个 TaskExecutorBuilder 实例，用来实现自定义的任务执行线程池，对应的参数绑定类为 TaskExecutionProperties 类，通过一系列 spring.task.execution.* 参数可以自定义默认任务执行的线程池配置。

TaskSchedulingAutoConfiguration 自动配置类的源码如下所示。

```
@ConditionalOnClass(ThreadPoolTaskScheduler.class)
@AutoConfiguration(after = TaskExecutionAutoConfiguration.class)
@EnableConfigurationProperties(TaskSchedulingProperties.class)
public class TaskSchedulingAutoConfiguration {

    @Bean
    @ConditionalOnBean(name = TaskManagementConfigUtils.SCHEDULED_ANNOTATION_PROCESSOR_BEAN_
NAME)
    @ConditionalOnMissingBean({ SchedulingConfigurer.class, TaskScheduler.class,
ScheduledExecutorService.class })
    public ThreadPoolTaskScheduler taskScheduler(TaskSchedulerBuilder builder) {
        return builder.build();
    }

    @Bean
    @ConditionalOnBean(name = TaskManagementConfigUtils.SCHEDULED_ANNOTATION_PROCESSOR_
BEAN_NAME)
    public static LazyInitializationExcludeFilter scheduledBeanLazyInitializationExcludeFilter() {
        return new ScheduledBeanLazyInitializationExcludeFilter();
    }

    @Bean
    @ConditionalOnMissingBean
    public TaskSchedulerBuilder taskSchedulerBuilder(TaskSchedulingProperties properties,
            ObjectProvider<TaskSchedulerCustomizer> taskSchedulerCustomizers) {
        TaskSchedulerBuilder builder = new TaskSchedulerBuilder();
        builder = builder.poolSize(properties.getPool().getSize());
        Shutdown shutdown = properties.getShutdown();
        builder = builder.awaitTermination(shutdown.isAwaitTermination());
        builder = builder.awaitTerminationPeriod(shutdown.getAwaitTerminationPeriod());
        builder = builder.threadNamePrefix(properties.getThreadNamePrefix());
        builder = builder.customizers(taskSchedulerCustomizers);
        return builder;
    }

}
```

任务调度自动配置类在任务执行自动配置类之后启用注册，如果当前上下文中没有以下任何一个实例：

- SchedulingConfigurer。
- TaskScheduler。
- ScheduledExecutorService。

那么它会自动注册一个默认的 ThreadPoolTaskScheduler 线程池任务调度器实例，它同样基于 JDK 并发包中的 Executor 接口实现了自己的任务调度器，除了 ScheduledExecutorService 是 JDK 并发包中的调度类，其他两个都是 Spring 调度框架中的类。

同时可以注册一个 TaskSchedulerBuilder 实例，对应的参数绑定类为 TaskSchedulingProperties，通过一系列 spring.task.scheduling.* 参数可以实现自定义的任务调度线程池配置。

8.1.2　线程池工作流程

任务执行参数绑定类的源码如下：

```
@ConfigurationProperties("spring.task.execution")
public class TaskExecutionProperties {

    private final Pool pool = new Pool();

    private final Shutdown shutdown = new Shutdown();

    public static class Pool {

        /**
         * Queue capacity. An unbounded capacity does not increase the pool and therefore
         * ignores the "max-size" property.
         */
        private int queueCapacity = Integer.MAX_VALUE;

        /**
         * Core number of threads.
         */
        private int coreSize = 8;

        /**
         * Maximum allowed number of threads. If tasks are filling up the queue, the pool
```

```
     * can expand up to that size to accommodate the load. Ignored if the queue is
     * unbounded.
     */
    private int maxSize = Integer.MAX_VALUE;

    /**
     * Whether core threads are allowed to time out. This enables dynamic growing and
     * shrinking of the pool.
     */
    private boolean allowCoreThreadTimeout = true;

    /**
     * Time limit for which threads may remain idle before being terminated.
     */
    private Duration keepAlive = Duration.ofSeconds(60);

    // ...

}

...
```

各线程池参数的说明如下表所示。

参数名	说明	默认值
queueCapacity	队列大小	Integer 最大值
coreSize	核心线程数	8
maxSize	最大线程数	Integer 最大值
allowCoreThreadTimeout	是否允许核心线程超时	true
keepAlive	线程保活时长	60 秒

除了 allowCoreThreadTimeout 参数，其他参数都是 JDK 线程池提供的，线程池工作流程及各参数的使用如下图所示。

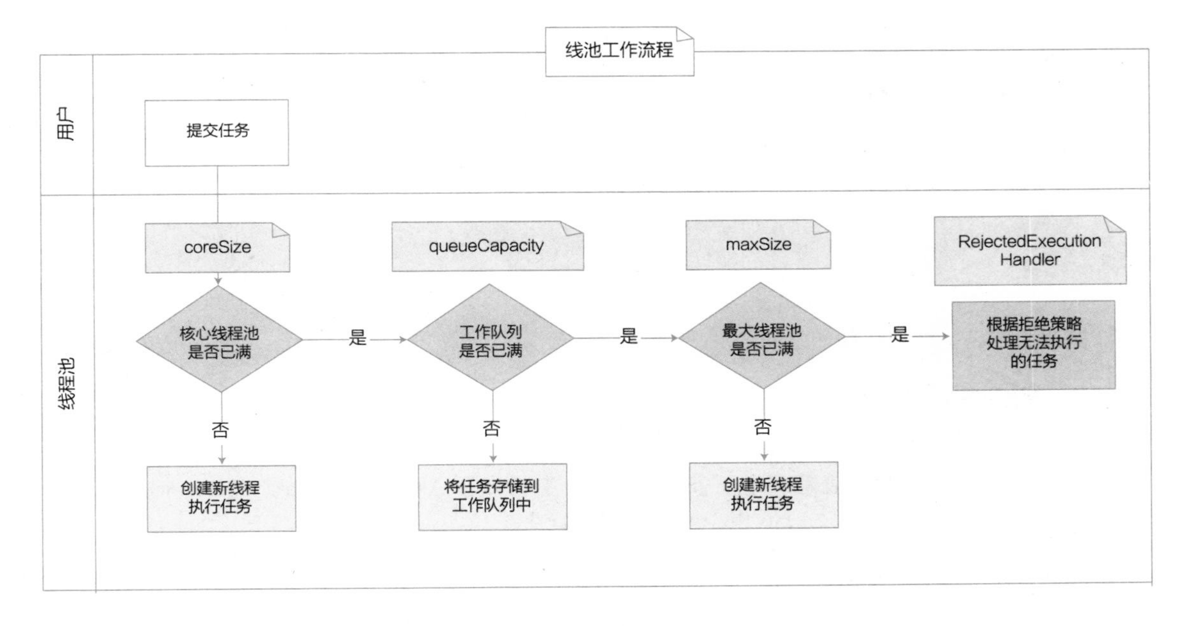

线程池大概的工作流程如下：

- 如果线程数 < 核心线程数 coreSize，就会一直创建新线程来处理任务。
- 如果线程数 = 核心线程数 coreSize，且任务队列 queueCapacity 没有满，就会先将任务存放到队列中，空闲线程会从队列中取出任务并执行。
- 如果线程数 = 核心线程数 coreSize，任务队列 queueCapacity 也满了，并且线程数 < 最大线程数 maxSize，就会一直创建新线程来处理任务。
- 如果核心线程数 coreSize、任务队列 queueCapacity、最大线程数 maxSize 三个都满了，则执行任务拒绝策略，默认为抛弃策略（见 ExecutorConfigurationSupport 类的定义）。
- 如果线程空闲时间达到 keepAlive 时间，则销毁线程，默认只销毁核心线程之外的线程，allowCoreThreadTimeout 用于控制是否销毁核心线程。

8.1.3　实现计划任务

Spring 中的计划任务十分轻量级，使用起来也比 Quartz 要简单得多。在使用 Spring Boot 计划任务之前，需要先开启计划任务，一般建议在应用主类上启用，如以下示例所示。

```
@EnableScheduling
```

```
@EnableAsync
@SpringBootApplication
public class Application {

    public static void main(String[] args) {
        SpringApplication.run(Application.class);
    }

}
```

其中最关键的两个注解如下所示。

- @EnableScheduling：表示启用任务调度，它导入了 SchedulingConfiguration 调度配置类，是通过注册一个 ScheduledAnnotationBeanPostProcessor 实例实现的。
- @EnableAsync：表示启用任务异步执行，它导入了 AsyncConfigurationSelector 配置选择器类，其根据不同的 AOP 模式选择不同的配置类。

这两个注解都是 spring-context 包中的注解，不需要引入额外的依赖，这里实现了一个简单的计划任务，每 3 秒打印一次日志，如下面的示例代码所示。

```
@Slf4j
@Component
public class SimpleTask {

    @Scheduled(cron = "*/3 * * * * *")
    public void printTask() {
        log.info(" 这是一个简单的计划任务。");
    }

}
```

在要实现计划任务的方法上使用 @Scheduled 注解启用任务调度，使用它需要先开启对应的 @EnableScheduling 注解，它使用的是 ThreadPoolTaskScheduler 线程池任务调度器。

启动应用后会看到应用每 3 秒打印一次日志，如下面的日志所示。

```
--- [TaskScheduler-1] c...SimpleTask        : 这是一个简单的计划任务。
--- [TaskScheduler-1] c...SimpleTask        : 这是一个简单的计划任务。
--- [TaskScheduler-1] c...SimpleTask        : 这是一个简单的计划任务。
--- [TaskScheduler-1] c...SimpleTask        : 这是一个简单的计划任务。
--- [TaskScheduler-1] c...SimpleTask        : 这是一个简单的计划任务。
--- [TaskScheduler-1] c...SimpleTask        : 这是一个简单的计划任务。
--- [TaskScheduler-1] c...SimpleTask        : 这是一个简单的计划任务。
```

```
--- [TaskScheduler-1] c...SimpleTask         : 这是一个简单的计划任务。
--- [TaskScheduler-1] c...SimpleTask         : 这是一个简单的计划任务。
--- [TaskScheduler-1] c...SimpleTask         : 这是一个简单的计划任务。
```

任务正常执行，可以注意到每次都是 TaskScheduler-1 这个任务调度线程在执行，而没有使用 Spring Boot 自动配置的 ThreadPoolTaskExecutor 线程池，需要使用 @Async 注解开启方法异步执行才会使用该线程池，即对应之前开启的 @EnableAsync 来开启异步执行注解。

现在加上 @Async 注解后再开启任务异步执行，如以下示例所示。

```
@Slf4j
@Component
public class SimpleTask {

    private int count;

    @Async
    @Scheduled(cron = "*/3 * * * * *")
    public void printTask() {
        log.info(" 这是一个简单的计划任务。");
    }

}
```

修改代码后，再来观察一下任务执行日志，如以下日志所示。

```
--- [     task1-1] c...SimpleTask         : 这是一个简单的计划任务。
--- [     task1-2] c...SimpleTask         : 这是一个简单的计划任务。
--- [     task1-3] c...SimpleTask         : 这是一个简单的计划任务。
--- [     task1-4] c...SimpleTask         : 这是一个简单的计划任务。
--- [     task1-5] c...SimpleTask         : 这是一个简单的计划任务。
--- [     task1-6] c...SimpleTask         : 这是一个简单的计划任务。
--- [     task1-7] c...SimpleTask         : 这是一个简单的计划任务。
--- [     task1-8] c...SimpleTask         : 这是一个简单的计划任务。
--- [     task1-1] c...SimpleTask         : 这是一个简单的计划任务。
--- [     task1-2] c...SimpleTask         : 这是一个简单的计划任务。
--- [     task1-3] c...SimpleTask         : 这是一个简单的计划任务。
--- [     task1-4] c...SimpleTask         : 这是一个简单的计划任务。
--- [     task1-5] c...SimpleTask         : 这是一个简单的计划任务。
```

现在可以看到，每次任务执行的线程都不一样了，因为自动配置的任务执行线程池默认是 8 个核心线程，系统又只有一个任务在运行，所以每次都会创建一个新的线程去运行，直到达到 8 个核心线程数，当核心线程数满了之后就会按顺序取线程池中已有的线程执行。

8.1.4 Cron 表达式

Spring Task 使用了类似于 Cron 的表达式，并扩展了常规的 UN*X 中的定义，和常规的 Cron 表达式略有不同，位数也不同，也不是完全兼容常规的 Cron 表达式，这一点需要注意。

@Scheduled 注解的 Cron 表达式参数的源码如下：

```
/**
 * A cron-like expression, extending the usual UN*X definition to include triggers
 * on the second, minute, hour, day of month, month, and day of week.
 * <p>For example, {@code "0 * * * * MON-FRI"} means once per minute on weekdays
 * (at the top of the minute - the 0th second).
 * <p>The fields read from left to right are interpreted as follows.
 * <ul>
 * <li>second</li>
 * <li>minute</li>
 * <li>hour</li>
 * <li>day of month</li>
 * <li>month</li>
 * <li>day of week</li>
 * </ul>
 * <p>The special value {@link #CRON_DISABLED "-"} indicates a disabled cron
 * trigger, primarily meant for externally specified values resolved by a
 * <code>${...}</code> placeholder.
 * @return an expression that can be parsed to a cron schedule
 * @see org.springframework.scheduling.support.CronExpression#parse(String)
 */
String cron() default "";
```

可以看到，Spring 计划任务中的 Cron 表达式只能由 6 个字段组成，再继续查看 CronExpression # parse 解析方法的源码，如下面源码中的注释所示。

```
Parse the given crontab expression  string into a CronExpression. The string has six
single space-separated time and date fields:
  ┌───────────────────────── second (0-59)
  │ ┌───────────────────────── minute (0 - 59)
  │ │ ┌───────────────────────── hour (0 - 23)
  │ │ │ ┌───────────────────────── day of the month (1 - 31)
  │ │ │ │ ┌───────────────────────── month (1 - 12) (or JAN-DEC)
  │ │ │ │ │ ┌───────────────────────── day of the week (0 - 7)
  │ │ │ │ │ │              (0 or 7 is Sunday, or MON-SUN)
  │ │ │ │ │ │
  * * * * * *
```

Spring Cron 表达式中的 6 个字段分别表示：

- 秒（0 ～ 59）。
- 分（0 ～ 59）。
- 小时（0 ～ 23）。
- 日期（1 ～ 31）。
- 月份（1 ～ 12，或者 JAN ～ DEC）。
- 周几（0 ～ 7，其中 0 和 7 表示周日，或者 MON ～ SUN）。

常用的 Cron 表达式的使用说明如下表所示。

Cron 表达式	说明
*/10 * * * * *	每 10 秒执行一次
0 0 * * * *	每小时执行一次
0 0 8-10 * * *	每天 8 ～ 10 点整点执行一次
0 0 6,19 * * *	每天 6 点和 19 点执行一次
0 0/30 8-10 * * *	每天 8 ～ 10 点之间每 30 分钟执行一次
0 0 9-17 * * MON-FRI	每周一到周五 9 点到 17 点整点执行一次
0 0 0 11 11 ?	每年双 11 的 0 点执行一次

支持以下关键字代替对应的字段：

- @yearly（每年）。
- @monthly（每月）。
- @weekly（每周）。
- @daily（每日）。
- @hourly（每小时）。

比如，每年 1 月 1 号 0 点执行一次 0 0 0 1 1 *，就可以直接写成 @yearly。

更多的用法可以参考以下方法：

```
org.springframework.scheduling.support.CronExpression#parse
```

Spring Cron 表达式没有指定年份的字段，默认表示每年，那么问题来了：

> 如果想在具体的某年某日，如 2025/02/25 这一天执行任务，那么怎么配置呢？

笔者也没有找到办法，Spring Cron 表达式不能指定年份，实际工作中也很少会有跨年的计划任务，如果需要跨年配置就只能换成 Quartz 计划任务框架了，Quartz 计划任务框架的功能要更强大。

8.1.5 自定义线程池

从线程池工作流程一节中可以知道，Spring Boot 任务执行的默认队列大小为 Integer 的最大值，即 231-1，可以理解为无限大，如果存放大量的执行任务，就可能导致内存溢出，这也是为什么不建议使用默认线程池的原因。

实际工作中需要自定义任务线程池，以规避默认线程池带来的内存溢出风险，既可以通过配置更改默认的线程池参数，也可以通过完全自定义一个线程池 Bean 来实现。

1. 通过参数配置线程池

比如以下自定义线程池配置示例：

```
spring:
  task:
    execution:
      pool:
        coreSize: 5
        maxSize: 10
        queueCapacity: 50
        keepAlive: 10s
        allowCoreThreadTimeout: false
    scheduling:
      pool:
        size: 3
```

这里配置了任务执行线程池及任务调度线程池，任务执行线程池的配置如下：

- **核心线程数**：5。
- **最大线程数**：10。
- **任务队列大小**：50。
- **线程保活时长**：10 秒。
- **是否允许核心线程超时**：否。

任务调度线程池（默认为 1）的配置如下：

- **线程数**：3。

任务执行拒绝策略不能通过参数配置实现，但可以通过自定义线程池 Bean 的方式实现。

2. 自定义线程池 Bean

根据前面章节的任务自动配置过程可以知道，自定义线程池可以通过注册自定义的 TaskExecutorBuilder 或者 Executor 来实现，因为 TaskExecutorBuilder 不能自定义策略，所以这里通过注册自定义 Executor 的扩展类 ThreadPoolTaskExecutor 的方式来实现。

自定义线程池 Bean 的配置如以下示例所示。

```
@Configuration
public class TaskConfig {

    @Lazy
    @Bean
    public ThreadPoolTaskExecutor taskExecutor(TaskExecutionProperties
taskExecutionProperties) {
        ThreadPoolTaskExecutor taskExecutor = new ThreadPoolTaskExecutor();

        PropertyMapper map = PropertyMapper.get().alwaysApplyingWhenNonNull();
        TaskExecutionProperties.Pool pool = taskExecutionProperties.getPool();
        map.from(pool::getQueueCapacity).to(taskExecutor::setQueueCapacity);
        map.from(pool::getCoreSize).to(taskExecutor::setCorePoolSize);
        map.from(pool::getMaxSize).to(taskExecutor::setMaxPoolSize);
        map.from(pool::getKeepAlive).asInt(Duration::getSeconds).to(taskExecutor::
setKeepAliveSeconds);
        map.from(pool::isAllowCoreThreadTimeout).to(taskExecutor::
setAllowCoreThreadTimeOut);
        map.from("my-task-").whenHasText().to(taskExecutor::setThreadNamePrefix);

        // 默认不设置就是 AbortPolicy
        taskExecutor.setRejectedExecutionHandler(new ThreadPoolExecutor.AbortPolicy());

        return taskExecutor;
    }

}
```

只要命名为 taskExecutor，@Async 注解就会自动选用这个线程池，也可以配置多个线程池，不同的任务类型使用不同的线程池，比如这里再注册一个 taskExecutor2，如以下配置示例所示。

```
@Configuration
public class TaskConfig {

    @Lazy
```

```
    @Bean
    public ThreadPoolTaskExecutor taskExecutor1(TaskExecutionProperties taskExecution-
Properties) {
        ThreadPoolTaskExecutor taskExecutor = new ThreadPoolTaskExecutor();

        PropertyMapper map = PropertyMapper.get().alwaysApplyingWhenNonNull();
        TaskExecutionProperties.Pool pool = taskExecutionProperties.getPool();
        map.from(pool::getQueueCapacity).to(taskExecutor::setQueueCapacity);
        map.from(pool::getCoreSize).to(taskExecutor::setCorePoolSize);
        map.from(pool::getMaxSize).to(taskExecutor::setMaxPoolSize);
        map.from(pool::getKeepAlive).asInt(Duration::getSeconds).to (taskExecutor::
setKeepAliveSeconds);
        map.from(pool::isAllowCoreThreadTimeout).to(taskExecutor::
setAllowCoreThreadTimeOut);
        map.from("my-task1-").whenHasText().to(taskExecutor::setThreadNamePrefix);

        // 默认不设置就是 AbortPolicy
        taskExecutor.setRejectedExecutionHandler(new ThreadPoolExecutor.AbortPolicy());

        return taskExecutor;
    }

    @Lazy
    @Bean
    public ThreadPoolTaskExecutor taskExecutor2() {
        ThreadPoolTaskExecutor taskExecutor = new ThreadPoolTaskExecutor();

        PropertyMapper map = PropertyMapper.get().alwaysApplyingWhenNonNull();
        TaskExecutionProperties.Pool pool = taskExecutionProperties.getPool();
        map.from(10).to(taskExecutor::setQueueCapacity);
        map.from(3).to(taskExecutor::setCorePoolSize);
        map.from(5).to(taskExecutor::setMaxPoolSize);
        map.from(20).to(taskExecutor::setKeepAliveSeconds);
        map.from(true).to(taskExecutor::setAllowCoreThreadTimeOut);
        map.from("my-task2-").whenHasText().to(taskExecutor::setThreadNamePrefix);

        return taskExecutor;
    }


}
```

如果存在多个线程池 Bean，但没有一个被命名为 taskExecutor，那么使用 @Async 注解时就需要指定要使用的线程池 Bean 的名称，示例如下：

```
@Slf4j
@Component
public class SimpleTask {

    @Async("taskExecutor2")
    @Scheduled(cron = "*/3 * * * * *")
    public void printTask() {
        log.info(" 这是一个简单的计划任务。");
    }

}
```

这里指定了使用 taskExecutor2 线程池，如果不指定，也可以在线程池 Bean 上使用 @Primary 标识默认的线程池，如果 @Async 注解没有指定线程池名称就会使用默认的线程池，比如：

```
@Configuration
public class TaskConfig {

    @Lazy
    @Bean
    @Primary
    public ThreadPoolTaskExecutor taskExecutor1(TaskExecutionProperties
taskExecutionProperties) {

// ...
```

这里指定了 taskExecutor1 为默认的线程池，@Async 注解就能默认直接使用这个线程池。然后重启应用，再来观察任务日志，如以下日志所示。

```
--- [     my-task1-1] c...SimpleTask        : 这是一个简单的计划任务。
--- [     my-task1-2] c...SimpleTask        : 这是一个简单的计划任务。
--- [     my-task1-3] c...SimpleTask        : 这是一个简单的计划任务。
--- [     my-task1-4] c...SimpleTask        : 这是一个简单的计划任务。
--- [     my-task1-5] c...SimpleTask        : 这是一个简单的计划任务。
--- [     my-task1-1] c...SimpleTask        : 这是一个简单的计划任务。
--- [     my-task1-2] c...SimpleTask        : 这是一个简单的计划任务。
--- [     my-task1-3] c...SimpleTask        : 这是一个简单的计划任务。
--- [     my-task1-4] c...SimpleTask        : 这是一个简单的计划任务。
--- [     my-task1-5] c...SimpleTask        : 这是一个简单的计划任务。
```

从日志可以看到，自定义任务线程池后，执行任务的线程名称和任务执行频次都有了变化，更多的任务线程池的扩展用法可以参考本节内容。

> 自定义任务调度不再赘述。

8.2 Quartz 计划任务

8.2.1 概述

一般简单的计划任务可以使用 Spring 计划任务，更为复杂的计划任务建议使用专业的 Quartz 框架，或者其他分布式任务框架，比如 Elastic Job、xxl-job 等。据笔者了解，它们也都支持 Spring Boot Starter 一站式启动器快速集成，本节就以 Quartz 为例进行介绍，这三个框架中只有 Quartz 是 Spring Boot 官方支持的。

Quartz 是一款老牌、开源的 Java 作业调度框架，功能十分丰富，也十分成熟稳定，自笔者工作以来，就一直流行至今，专业且强大，是 Java 领域使用最广泛的框架之一。

Spring Boot 除了自动配置 Spring 自带的计划任务和调度框架，还支持对 Quartz 作业调度框架的自动配置，提供了 spring-boot-starter-quartz 一站式启动器，如以下依赖所示。

```
<dependency>
    <groupId>org.springframework.boot</groupId>
    <artifactId>spring-boot-starter-quartz</artifactId>
</dependency>
```

只要加入该启动器依赖就能实现自动配置，它会自动引入 Quartz 框架及相关必要的依赖，如下图所示。

```
∨ spring-boot-starter-quartz : 3.0.0 [compile]
    > quartz : 2.3.2 [compile]
    > spring-boot-starter : 3.0.0 [compile]
    > spring-context-support : 6.0.2 [compile]
    > spring-tx : 6.0.2 [compile]
```

自动配置类为 QuartzAutoConfiguration，通过自动注册 Spring 中的 SchedulerFactoryBean 就能完成 Scheduler 调度器的创建、配置、生命周期管理等。对应的参数绑定类为 QuartzProperties，通过一系列 spring.quartz 参数可以自定义 SchedulerFactoryBean 配置。

8.2.2　实现计划任务

和前面的章节一样，这里使用 Quartz 实现了一个简单的计划任务，每 3 秒打印一次日志，如下面的示例所示。

```
@Slf4j
public class SimpleTask extends QuartzJobBean {

    @Override
    protected void executeInternal(JobExecutionContext context) {
        log.info(" 这是一个简单的 Quartz 计划任务。");
    }

}
```

在 Spring Boot 中，只需要继承 Spring 提供的 QuartzJobBean 抽象类并实现其 executeInternal 抽象方法即可，QuartzJobBean 抽象类实现并封装了 Quartz 的 Job 作业接口。

和 Spring 任务不同的是，Quartz 需要针对每个任务提供额外的配置，以下几个类型的 Bean 也会自动关联到 Scheduler 调度器：

- JobDetail：用于创建一个具体的作业实例。
- Calendar：用于指定 / 排除特定的时间。
- Trigger：用于定义作业触发时机。

下面是 JobDetaill 和 Trigger 的使用配置示例：

```
@Configuration
public class TaskConfig {

    @Bean
    public JobDetail simpleTask() {
        return JobBuilder.newJob(SimpleTask.class)
                .withIdentity("simple-task")
                .withDescription(" 简单任务 ")
                .storeDurably()
                .build();
    }

    @Bean
    public Trigger simpleTaskTrigger() {
```

```
        CronScheduleBuilder cronScheduleBuilder = CronScheduleBuilder.cronSchedule
("0/3 * * * * ? *");
        return TriggerBuilder.newTrigger()
                .withIdentity("simple-task-trigger")
                .forJob(simpleTask())
                .withSchedule(cronScheduleBuilder)
                .build();
    }

}
```

Quartz 的 Cron 表达式的格式如下所示。

```
秒 分 时 日 月 周 [ 年 ]
```

相比 Spring 计划任务，Quartz 可以多配置一位，即最后一位年，比如“0/3 * * * * ? 2022-2030”即表示 2022 到 2030 年之间每 3 秒执行一次。表达式可以是 6 位，也可以是 7 位，最后一位年并不是必需的。

任务配置好之后，查看应用启动后的部分日志，如以下启动日志所示。

```
RAMJobStore initialized.
Scheduler meta-data: Quartz Scheduler (v2.3.2) 'quartzScheduler' with instanceId
'NON_CLUSTERED'
  Scheduler class: 'org.quartz.core.QuartzScheduler' - running locally.
  NOT STARTED.
  Currently in standby mode.
  Number of jobs executed: 0
  Using thread pool 'org.quartz.simpl.SimpleThreadPool' - with 10 threads.
  Using job-store 'org.quartz.simpl.RAMJobStore' - which does not support persistence.
and is not clustered.

Starting Quartz Scheduler now
Scheduler quartzScheduler_$_NON_CLUSTERED started.
Started Application in 1.501 seconds (process running for 1.765)
```

可以看到一些 Quartz 初始化的相关信息，包括存储类型、线程池类型及数量，以及是否集群模式等。

任务执行日志如下所示。

```
--- [eduler_Worker-1] c....SimpleTask          : 这是一个简单的 Quartz 计划任务。
--- [eduler_Worker-2] c....SimpleTask          : 这是一个简单的 Quartz 计划任务。
```

```
--- [eduler_Worker-3] c....SimpleTask          : 这是一个简单的 Quartz 计划任务。
--- [eduler_Worker-4] c....SimpleTask          : 这是一个简单的 Quartz 计划任务。
--- [eduler_Worker-5] c....SimpleTask          : 这是一个简单的 Quartz 计划任务。
--- [eduler_Worker-6] c....SimpleTask          : 这是一个简单的 Quartz 计划任务。
--- [eduler_Worker-7] c....SimpleTask          : 这是一个简单的 Quartz 计划任务。
--- [eduler_Worker-8] c....SimpleTask          : 这是一个简单的 Quartz 计划任务。
--- [eduler_Worker-9] c....SimpleTask          : 这是一个简单的 Quartz 计划任务。
--- [eduler_Worker-10] c....SimpleTask         : 这是一个简单的 Quartz 计划任务。
--- [eduler_Worker-1] c....SimpleTask          : 这是一个简单的 Quartz 计划任务。
--- [eduler_Worker-2] c....SimpleTask          : 这是一个简单的 Quartz 计划任务。
```

Quartz 默认是异步多线程执行的，可以看到默认的 10 个线程在依次执行任务。

8.2.3　自定义配置

Quartz 默认会加载 Quartz 包目录中的 /org/quartz/quartz.properties 配置文件，包括 Quartz 调度器、线程池、持久化的相关配置，该配置文件的全部内容如下所示。

```
# Default Properties file for use by StdSchedulerFactory
# to create a Quartz Scheduler Instance, if a different
# properties file is not explicitly specified.
#

org.quartz.scheduler.instanceName: DefaultQuartzScheduler
org.quartz.scheduler.rmi.export: false
org.quartz.scheduler.rmi.proxy: false
org.quartz.scheduler.wrapJobExecutionInUserTransaction: false

org.quartz.threadPool.class: org.quartz.simpl.SimpleThreadPool
org.quartz.threadPool.threadCount: 10
org.quartz.threadPool.threadPriority: 5
org.quartz.threadPool.threadsInheritContextClassLoaderOfInitializingThread: true

org.quartz.jobStore.misfireThreshold: 60000

org.quartz.jobStore.class: org.quartz.simpl.RAMJobStore
```

Spring Boot 中的 QuartzProperties 参数绑定类提供了一个 Map 类型的 properties 参数，可以自定义覆盖这些配置，如以下源码所示。

```
@ConfigurationProperties("spring.quartz")
public class QuartzProperties {
```

```
    // ...

      /**
       * Additional Quartz Scheduler properties.
       */
    private final Map<String, String> properties = new HashMap<>();

    // ...

}
```

比如要更改默认的线程数，如以下配置示例所示。

```
spring:
  quartz:
    properties:
      org:
        quartz:
          threadPool:
            threadCount: 5
```

再重启应用查看任务执行日志，如以下日志所示。

```
--- [eduler_Worker-1] c....SimpleTask          : 这是一个简单的 Quartz 计划任务。
--- [eduler_Worker-2] c....SimpleTask          : 这是一个简单的 Quartz 计划任务。
--- [eduler_Worker-3] c....SimpleTask          : 这是一个简单的 Quartz 计划任务。
--- [eduler_Worker-4] c....SimpleTask          : 这是一个简单的 Quartz 计划任务。
--- [eduler_Worker-5] c....SimpleTask          : 这是一个简单的 Quartz 计划任务。
--- [eduler_Worker-1] c....SimpleTask          : 这是一个简单的 Quartz 计划任务。
--- [eduler_Worker-2] c....SimpleTask          : 这是一个简单的 Quartz 计划任务。
```

从日志线程名可以看到，现在线程池中只有 5 个线程在执行任务。

8.2.4 持久化任务数据

首先看一下 QuartzProperties 参数类的源码：

```
@ConfigurationProperties("spring.quartz")
public class QuartzProperties {

    /**
```

```
     * Quartz job store type.
     */
    private JobStoreType jobStoreType = JobStoreType.MEMORY;

    ...

}

public enum JobStoreType {

    /**
     * Store jobs in memory.
     */
    MEMORY,

    /**
     * Store jobs in the database.
     */
    JDBC

}
```

如源码所示，Quartz 支持两种作业持久化存储方式：

- MEMORY。
- JDBC。

Quartz 存放作业的相关数据默认使用基于 MEMORY 内存的方式，基于内存意味着每次应用重启后作业数据就会丢失，如果需要持久化作业数据，那么 Spring Boot 还支持基于 JDBC 存储的自动配置，只要 DataSource 数据源已注册，就可以通过配置切换持久化方式，如以下配置示例所示。

```
spring:
  quartz:
    job-store-type: jdbc
    jdbc:
      initialize-schema: always
```

参数 initialize-schema 支持以下三个枚举值：

- ALWAYS：总是初始化数据库。
- EMBEDDED：只初始化嵌入式数据库。
- NEVER：从不初始化数据库。

参数 initialize-schema 的值为 ALWAYS，表示脚本在每次应用重启后都会删除重建表，这样持久化就没有什么意义了，可以在应用首次启动之前手动执行一次 Quartz 中自带的数据库脚本，或者通过 spring.quartz.jdbc.schema 参数指定自定义的数据库构建脚本。

Quartz 中自带的数据库脚本路径可以参考 QuartzProperties 参数类，如下面的源码所示。

```
@ConfigurationProperties("spring.quartz")
public class QuartzProperties {

    // ...

    public static class Jdbc {

        private static final String DEFAULT_SCHEMA_LOCATION = "classpath:org/quartz/impl/"
                + "jdbcjobstore/tables_@@platform@@.sql";
```

可以看到在 Quartz 对应的 org/quartz/impl/ 包目录中，Quartz 提供了多个主流数据库的构建脚本，以 MySQL 数据库为例，它提供了两个不同数据库引擎的构建脚本，如下图所示。

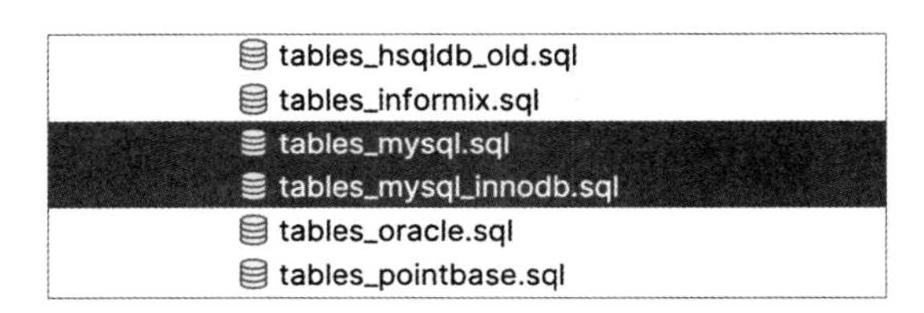

上面的一个脚本使用的是 MySQL 当前版本的默认数据库引擎，下面的脚本则使用指定的 InnoDB 数据库引擎。

在应用启动并完成数据库初始化后，可以通过 show tables 命令查看 Quartz 持久化相关的数据表，如以下参考命令所示。

```
mysql> show tables;
+------------------------------+
| Tables_in_javastack          |
+------------------------------+
| QRTZ_BLOB_TRIGGERS           |
| QRTZ_CALENDARS               |
| QRTZ_CRON_TRIGGERS           |
| QRTZ_FIRED_TRIGGERS          |
| QRTZ_JOB_DETAILS             |
| QRTZ_LOCKS                   |
| QRTZ_PAUSED_TRIGGER_GRPS     |
| QRTZ_SCHEDULER_STATE         |
| QRTZ_SIMPLE_TRIGGERS         |
```

```
| QRTZ_SIMPROP_TRIGGERS         |
| QRTZ_TRIGGERS                 |
+-------------------------------+
11 rows in set (0.00 sec)
```

默认情况下以数据库中持久化的任务配置为准，比如现在修改任务触发机制为每 5 秒执行一次，如以下示例代码所示。

```
@Bean
public Trigger simpleTaskTrigger() {
    CronScheduleBuilder cronScheduleBuilder = CronScheduleBuilder.cronSchedule
("0/5 * * * * ? *");
    return TriggerBuilder.newTrigger()
            .withIdentity("simple-task-trigger")
            .forJob(simpleTask())
            .withSchedule(cronScheduleBuilder)
            .build();
}
```

重启应用，该任务还是每 3 秒执行一次，因为应用中的新配置不会覆盖数据库中持久化的配置，如果需要覆盖持久化的任务配置，则可以通过以下参数进行配置：

```
spring:
  quartz:
    overwrite-existing-jobs: true
```

这样应用中最新的任务配置就会覆盖数据库中已持久化的配置，再次重启应用，该任务就会变成每 5 秒执行一次。

8.2.5　动态维护任务

以上是通过 Bean 的方式配置任务，如果任务过多，应用就会出现大量的配置代码，不是很优雅，也不好维护。

除了使用 Bean 配置任务的方式，还可以使用前面介绍的 SchedulerFactoryBean 来动态维护任务，比如动态添加、删除任务等，我们把上面的 Bean 配置方式改一下，如以下配置示例所示。

```
@RequiredArgsConstructor
@Configuration
public class TaskConfig {

    public static final String SIMPLE_TASK = "simple-task";
```

```
    private final SchedulerFactoryBean schedulerFactoryBean;

    /**
     * 动态添加任务
     * @throws SchedulerException
     */
    @PostConstruct
    public void init() throws SchedulerException {
        Scheduler scheduler = schedulerFactoryBean.getScheduler();
        boolean exists = scheduler.checkExists(JobKey.jobKey(SIMPLE_TASK));
        if (!exists) {
            scheduler.scheduleJob(simpleTask(), simpleTaskTrigger());
        }
    }

    // @Bean
    public JobDetail simpleTask() {
        return JobBuilder.newJob(SimpleTask.class)
                .withIdentity(SIMPLE_TASK)
                .withDescription("简单任务")
                .storeDurably()
                .build();
    }

    // @Bean
    public Trigger simpleTaskTrigger() {
        CronScheduleBuilder cronScheduleBuilder = CronScheduleBuilder.cronSchedule
("0/3 * * * * ? *");
        return TriggerBuilder.newTrigger()
                .withIdentity("simple-task-trigger")
                .forJob(simpleTask())
                .withSchedule(cronScheduleBuilder)
                .build();
    }

}
```

首先把 @Bean 注解都注释掉，然后使用注入的 SchedulerFactoryBean 实例获取 Quartz 调度器来添加并启动任务，这里的逻辑是，如果任务没有持久化才添加并启动任务，读者可以自行扩展。

实际开发中可以把创建任务的各个方法抽取出来形成公共方法，这样做的好处就是既可以省去大量的配置代码，还可以把任务配置放到配置文件中或者配置中心来统一管理维护，如果有需要，还能以对外 API 接口的形式来动态维护任务。

第 9 章 Spring Boot 缓存与消息队列

本章会介绍 Spring Boot 缓存与消息队列的集成和使用，涉及 ActiveMQ、RabbitMQ、Kafka、Redis 等，这些都是 Java 应用必不可少的中间件技术，它们都是支撑互联网高吞吐量、高并发的基石。Spring Boot 为这些技术都提供了自动配置，并简化了这些第三方技术的使用难度。

9.1 缓存

9.1.1 概述

Spring 框架提供了对缓存操作的基本抽象：

- org.springframework.cache.Cache。
- org.springframework.cache.CacheManager。

Spring 框架只提供抽象，不提供具体的缓存存储，底层需要依赖第三方存储组件，如果当前应用没有注册 CacheManager 或者 CacheResolver 实例，那么 Spring Boot 会按以下顺序来检测缓存组件：

Generic → JCache（JSR-107）（EhCache 3、Hazelcast、Infinispan 等）→ Hazelcast → Infinispan → Couchbase → Redis → Caffeine → Cache2k → Simple

Spring Boot 在新的 org.springframework.boot.autoconfigure.AutoConfiguration.imports 自动配置文件中提供了对缓存的自动配置，自动配置类为 CacheAutoConfiguration，对应的参数绑定类为 CacheProperties，通过一系列 spring.cache.* 参数可以自定义缓存配置。

在 CacheAutoConfiguration 缓存自动配置类中引入了 CacheConfigurationImportSelector 配置选择器，如下面的源码所示。

```
@AutoConfiguration(after = { CouchbaseDataAutoConfiguration.class,
HazelcastAutoConfiguration.class,
HibernateJpaAutoConfiguration.class, RedisAutoConfiguration.class })
@ConditionalOnClass(CacheManager.class)
@ConditionalOnBean(CacheAspectSupport.class)
@ConditionalOnMissingBean(value = CacheManager.class, name = "cacheResolver")
@EnableConfigurationProperties(CacheProperties.class)
@Import({ CacheConfigurationImportSelector.class, CacheManagerEntityManagerFactoryDepends-
OnPostProcessor.class })
public class CacheAutoConfiguration {

    ...

static class CacheConfigurationImportSelector implements ImportSelector {

    @Override
    public String[] selectImports(AnnotationMetadata importingClassMetadata) {
        CacheType[] types = CacheType.values();
        String[] imports = new String[types.length];
        for (int i = 0; i < types.length; i++) {
            imports[i] = CacheConfigurations.getConfigurationClass(types[i]);
        }
        return imports;
    }

}
```

由该选择器通过 CacheConfigurations 缓存配置类来选择对应的缓存组件的配置类，如下面的源码所示。

```
final class CacheConfigurations {

private static final Map<CacheType, String> MAPPINGS;

static {
Map<CacheType, String> mappings = new EnumMap<>(CacheType.class);
mappings.put(CacheType.GENERIC, GenericCacheConfiguration.class.getName());
mappings.put(CacheType.HAZELCAST, HazelcastCacheConfiguration.class.getName());
mappings.put(CacheType.INFINISPAN, InfinispanCacheConfiguration.class.getName());
mappings.put(CacheType.JCACHE, JCacheCacheConfiguration.class.getName());
```

```
mappings.put(CacheType.COUCHBASE, CouchbaseCacheConfiguration.class.getName());
mappings.put(CacheType.REDIS, RedisCacheConfiguration.class.getName());
mappings.put(CacheType.CAFFEINE, CaffeineCacheConfiguration.class.getName());
mappings.put(CacheType.CACHE2K, Cache2kCacheConfiguration.class.getName());
mappings.put(CacheType.SIMPLE, SimpleCacheConfiguration.class.getName());
mappings.put(CacheType.NONE, NoOpCacheConfiguration.class.getName());
MAPPINGS = Collections.unmodifiableMap(mappings);
}

...

}
```

mappings 集合维护了所有 Spring Boot 支持的缓存组件的映射关系，由具体的缓存类型 CacheType 映射到具体的缓存配置类，以 Redis 为例，它会选择并启用 RedisCacheConfiguration 缓存配置类。

9.1.2 开启缓存

Spring Boot 为缓存提供了一个 spring-boot-starter-cache 一站式启动器，如以下依赖配置所示。

```
<dependency>
    <groupId>org.springframework.boot</groupId>
    <artifactId>spring-boot-starter-cache</artifactId>
</dependency>
```

它主要添加了 spring-boot-starter 基础依赖和缓存所需要的 spring-context-support 依赖，如下图所示。

```
v spring-boot-starter-cache : 3.0.0 [compile]
  > spring-boot-starter : 3.0.0 [compile]
  > spring-context-support : 6.0.2 [compile]
```

加入该启动器依赖后，在应用主配置上添加 @EnableCaching 注解即可开启缓存。还可以通过参数指定要使用的缓存组件，比如，使用 Redis 作为缓存组件，如下面的配置示例所示。

```
spring:
  cache:
    type: redis
```

也可以在某种应用中或者某些场合下禁用缓存，如下面的配置示例所示。

```
spring:
  cache:
    type: none
```

缓存类型 type 的可选配置值有 COUCHBASE、GENERIC、REDIS、HAZELCAST、CACHE2K、CAFFEINE、JCACHE、INFINISPAN、NONE、SIMPLE。

读者也可以参考 CacheType 枚举类进行配置。

9.1.3 默认简单缓存

在前面的缓存概述中，笔者介绍了 Spring Boot 检测缓存组件的顺序，可以看到最后检测到的是 Simple 类型，也就是说，当检测到没有使用任何缓存组件时，Spring Boot 默认使用的是基于内存的简单缓存，它使用基于线程安全的 ConcurrentHashMap 集合进行存储。

从 CacheConfigurations 缓存配置类映射集合中可以找到 Simple 类型对应的配置类是 SimpleCacheConfiguration 类，该类的源码如下所示。

```
@Configuration(proxyBeanMethods = false)
@ConditionalOnMissingBean(CacheManager.class)
@Conditional(CacheCondition.class)
class SimpleCacheConfiguration {

@Bean
ConcurrentMapCacheManager cacheManager(CacheProperties cacheProperties,
CacheManagerCustomizers cacheManagerCustomizers) {
ConcurrentMapCacheManager cacheManager = new ConcurrentMapCacheManager();
List<String> cacheNames = cacheProperties.getCacheNames();
if (!cacheNames.isEmpty()) {
cacheManager.setCacheNames(cacheNames);
}
return cacheManagerCustomizers.customize(cacheManager);
}

}
```

它注册了一个 ConcurrentMapCacheManager 缓存管理器实例，该类的源码如下所示。

```
public class ConcurrentMapCacheManager implements CacheManager, BeanClassLoaderAware {

    private final ConcurrentMap<String, Cache> cacheMap = new ConcurrentHashMap(16);
```

```
...

}
```

可以看到它是使用 ConcurrentHashMap 集合进行内存式存储的，这意味着一旦应用重启就会丢失所有缓存中的数据，所以建议仅在学习和测试中使用这种类型，不建议在线上及访问量较大的应用中使用，否则数据丢失会造成缓存雪崩效应，导致大量请求到达后端数据库，从而导致数据库崩溃，甚至整个系统崩溃。

下面在不使用任何缓存组件的情况下来应用简单缓存，首先创建一个两数相乘的缓存服务类，如下面的示例代码所示。

```
@Slf4j
@Service
public class CacheService {

    @Cacheable("calc")
    public int multiply(int a, int b) {
        int c = a * b;
        log.info("{} * {} = {}", a, b, c);
        return c;
    }

}
```

在服务类入口方法上使用 @Cacheable 注解标识该方法使用缓存，然后创建一个缓存测试接口，注入并调用缓存服务类来获取两个数的乘积结果，如下面的示例代码所示。

```
@RequiredArgsConstructor
@RestController
public class CacheController {

    private final CacheService cacheService;

    @RequestMapping("/multiply")
    public int multiply(@RequestParam("a") int a,
                        @RequestParam("b") int b) {
        return cacheService.multiply(a, b);
    }

}
```

再访问该接口，测试效果如下图所示。

首次调用该接口时会进入设置了缓存的 CacheService#multiply 方法，然后输出日志，并把返回数据存储到缓存中，后续调用时，只要全部参数相同就不会进入该缓存方法，而是直接从缓存中读取数据并返回。如果任意一个参数不同，则会当作一个新缓存来处理。

这里的 @Cacheable 注解便是 Spring AOP 的切面编程的应用，背后的原理便是动态代理，通过在程序运行时创建代理对象来代理原始对象中的方法，然后在调用原始对象方法的前后植入逻辑以实现缓存操作的目的。

9.1.4 Redis 缓存

本节还以 Redis 数据库为例，介绍 Spring Boot 如何集成 Redis 来使用缓存，Redis 的安装、集成在第 7 章介绍过了，这里不再赘述。

首先添加 spring-boot-starter-data-redis 一站式启动器依赖，如下面的配置所示。

```
<dependency>
    <groupId>org.springframework.boot</groupId>
    <artifactId>spring-boot-starter-data-redis</artifactId>
</dependency>
```

只需要一个依赖，Spring Boot 就会自动配置 Redis 缓存管理器，在 RedisCacheConfiguration 配置类中注册了一个 RedisCacheManager 缓存管理器，也就是 CacheManager 接口的 Redis 实现。

这里使用默认的 Redis 本地连接配置和缓存配置，如有需要请根据 RedisProperties 参数类进行配置，缓存相关的配置则可以根据 CacheProperties 参数类进行定制，Redis 缓存配置的参数形式为 spring.cache.redis.*。

重启应用，再访问上一节简单缓存中添加的接口，测试结果如下图所示。

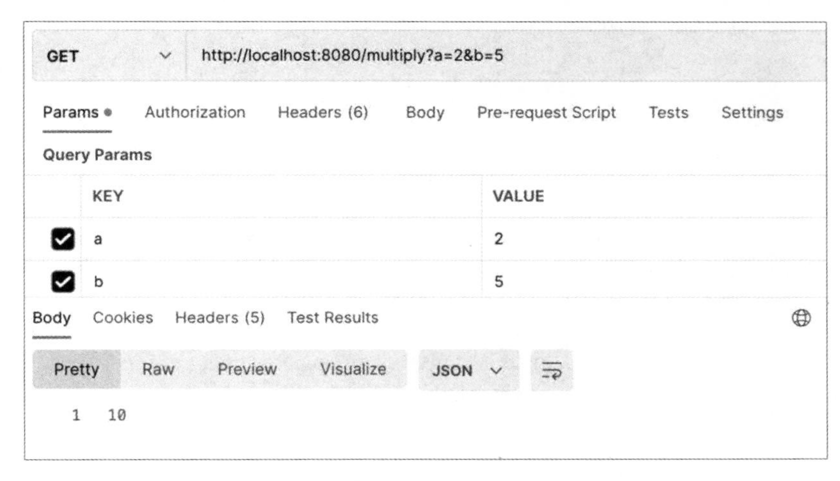

在使用了几次不同参数调用之后，查询方法是否在 Redis 中缓存了结果：

```
127.0.0.1:6379> keys *
1) "calc::SimpleKey [2,3]"
2) "calc::SimpleKey [2,5]"
3) "calc::SimpleKey [2,51]"
```

可以看到 Redis 缓存了方法结果，格式为：

> *方法名* ::KEY [*方法参数数组*]

Spring Boot 支持在应用启动时创建缓存名称，还能对 Redis 中的缓存设置过期时间，如下面的配置示例所示。

```
spring:
  cache:
    cache-names: "calc,test"
    redis:
      time-to-live: "10s"
```

这里创建了 calc 和 test 两个缓存名称，并且都只有 10 秒的有效期，也就是说，10 秒之后缓存就会失效，调用相同参数的接口会重新进入方法并重新缓存。

application 配置文件只能对所有缓存进行统一设置，如果需要对缓存进行更多配置控制，比如对多个不同的缓存设置不同的过期时间，则可以注册一个 =RedisCacheManagerBuilderCustomizer 类型的实例，参考示例如下所示。

```
@Configuration
public class CacheConfiguration {

    /**
     * 优先配置文件中的配置
     * @return
     */
    @Bean
    public RedisCacheManagerBuilderCustomizer myRedisCacheManagerBuilderCustomizer() {
        return (builder) -> builder
                .withCacheConfiguration("calc", RedisCacheConfiguration
                        .defaultCacheConfig().entryTtl(Duration.ofSeconds(5)))
                .withCacheConfiguration("test", RedisCacheConfiguration
                        .defaultCacheConfig().entryTtl(Duration.ofMinutes(10)));

    }

}
```

这里设置了 calc 类型的缓存过期时间为 5 秒，test 类型的缓存过期时间为 10 秒，通过 RedisCacheManagerBuilderCustomizer 自定义的缓存配置要优先于 application 配置文件中的缓存配置，也就是说，如果同时设置了缓存配置，则以 RedisCacheManagerBuilderCustomizer 中的为准。

还可以添加自定义的 RedisCacheConfiguration 配置类，通过覆盖缓存相关 Bean 的方式来完全控制默认的组件配置，这在想自定义默认序列化策略时会很有用。

> 其他缓存组件都是类似的用法，可以参考其自动配置类和参数配置类。

9.2 消息系统

9.2.1 概述

Java 程序员在面试时，经常会被面试官问到的一个问题：

> 有了多线程，为什么还要消息队列？

笔者曾经就被问到过这样的问题，这两个显然不能放在一起比较，它们是两个不同维度的东西，所应用的场景和发挥的能力也不一样，一般是结合使用的。所以，如果一定要比较，那么可以从下面几个方面来分析。

1. 业务解耦

使用多线程，应用的所有业务代码都彼此关联在一起，耦合性太强。消息队列则不存在业务关联关系，各个业务系统彼此独立，各个系统之间通过发送消息来实现异步消费，可以很方便地进行解耦。

2. 削峰填谷

使用多线程，资源消耗还是在本机器上，如果要处理突如其来的高并发流量，则不能很快地进行动态扩展。而消息队列独立存储消息，充分利用了各个应用服务器资源，可以将多条消息按照一定的订阅规则分发到多个消费者的应用服务器上，可以很好地应对高并发流量并进行削峰填谷。

3. 故障处理

使用多线程，一是很难跟踪每个线程的执行情况，二是出现问题时很难对问题线程进行恢复处理。而消息队列则拥有标准的消息管理机制，可以可视化地看到每条消息的执行情况，出现问题时可以设置重试策略，或者手动重新发送消息等。

4. 平台限制

使用多线程会受限于具体的编程语言，它解决的是编程层面的问题。而消息队列则不限定具体的编程语言，消息的存储和实现都在消息队列服务器端，它解决的是架构层面的问题。

消息队列的优势远不止上面几点，虽然相较于多线程有很多优势，但需要引进第三方消息组件，所以它同时会给系统带来可用性、复杂性、一致性等诸多问题，比如要考虑消息中间件的稳定性问题，以及消息重复消费、消息丢失、顺序消费等问题。

Spring Boot 支持以下几种常用的消息协议 / 系统：

- JMS，全称为 Java Message Service，即 Java 消息服务。Spring 框架提供了 JmsTemplate 模板，它简化了 JMS API 的使用，Spring Boot 还提供了该模板的自动配置。
- AMQP，全称为 Advanced Message Queuing Protocol，即高级消息队列协议。Spring 框架也为 AMQP 的主要实现 RabbitMQ 提供了类似的模板 RabbitTemplate，Spring Boot 还为 RabbitTemplate 和 RabbitMQ 提供了自动配置。
- Apache Kafka，Spring 框架同样提供了对 KafkaTemplate 模板的支持，Spring Boot 则提供了类似的自动配置。

- STOMP，全称为 Simple Text-Orientated Messaging Protocol，即面向消息的简单文本协议，Spring Boot 同样提供了类似的自动配置。

可以说 Spring Boot 几乎支持所有主流的消息系统，它为“开箱即用”的 API 提供了自动配置，Spring Boot 在新的 org.springframework.boot.autoconfigure.AutoConfiguration.imports 自动配置文件中提供了这几种消息系统的自动配置，如下所示。

- JmsAutoConfiguration。
- ArtemisAutoConfiguration。
- RabbitAutoConfiguration。
- KafkaAutoConfiguration。

这几种消息协议 / 系统的 Spring 依赖如下表所示。

消息协议 / 系统	Spring 依赖
JMS（ActiveMQ Artemis）	spring-boot-starter-artemis
AMQP（RabbitMQ）	spring-boot-starter-amqp
Kafka	spring-kafka

需要注意的是，Spring Boot 提供了对 Kafka 消息系统的自动配置类，不过并没有提供 Starter 启动器依赖，它是通过 spring-kafka 这个项目提供支持的。另外，目前 RocketMQ 在国内的应用也比较多，并已成为 Apache 顶级项目。虽然 Spring Boot 没有提供官方的对 RocketMQ 的支持，但是可以在 RocketMQ/Spring Cloud Alibaba 中找到相关的 Starter 启动器依赖。

本章将以目前最主流的、也是官方支持的消息系统 ActiveMQ、RabbitMQ 和 Kafka 为例，介绍 Spring Boot 如何集成并简单使用这些消息系统。

9.2.2 ActiveMQ

1. 概述

JMS 是 Java 开发平台上面向消息通信的技术标准和规范，可用于在多个应用之间或分布式系统中发送消息进行异步通信。它是一个与具体平台无关的消息通信 API，绝大多数提供商都支持 JMS 规范，JMS 最主流的开源技术有 ActiveMQ、RocketMQ 等。

直白点理解，JMS 和 JDBC 的概念一样，JMS 只是一种消息规范，提供了一套消息通信 API，ActiveMQ、RocketMQ 等都是 JMS 规范的实现，这两个都是 Apache 旗下的顶级开源项目，也都支持 JMS 规范。

JMS 规范目前有以下两个版本：

- JMS 1.1。
- JMS 2.0。

JMS 2.0 主要在 JMS 1.1 的基础上进行了改进，简化了消息通信 API，同时向后兼容 JMS 1.1 的 API，JMS 2.0 不仅在易用性方面得到了大大提升，也大大减少了开发代码量。

ActiveMQ 是 JMS 规范最早、最流行、最典型的代表，由 Java 语言开发，是 Apache 开源的最主流的消息系统之一，ActiveMQ 主要有以下两个版本：

- ActiveMQ Classic：它指的是 ActiveMQ 最原始、最经典的版本，基于 JMS 1.1 规范。
- ActiveMQ Artemis：它指的是下一代的 ActiveMQ，用来代替 ActiveMQ Classic。ActiveMQ Artemis 不但支持最新的 JMS 2.0 规范，还支持 AMQP、STOMP、MQTT、OpenWire 等多种消息协议，底层还使用了基于 Netty 的异步 I/O，大大提升了性能。

ActiveMQ 5.x 及之前的版本叫 ActiveMQ Classic，现在的 ActiveMQ 6.x+ 叫 ActiveMQ Artemis，功能全面增强，所以现在推荐基于 ActiveMQ Artemis 集成消息系统。

2. ActiveMQ Artemis 环境搭建

1）安装 ActiveMQ Artemis

去官网下载最新的二进制包，地址为链接 10。

笔者这里下载了最新的版本 apache-artemis-2.26.0-bin.tar.gz，下载之后进行解压，并设置 ARTEMIS_HOME 和 PATH 环境变量，参考命令如下所示。

```
$ tar -zvxf apache-artemis-2.26.0-bin.tar.gz
$ vi ~/.bash_profile
$ source ~/.bash_profile
```

2）创建一个 Broker 实例

Broker 实例是一个包含所有配置和运行数据的目录，所以建议不要在 ARTEMIS_HOME 主目录下创建 Broker 实例，避免下次升级 ActiveMQ Artemis 时造成 Broker 实例数据丢失。

创建一个 Broker 目录及实例，参考命令如下：

```
$ mkdir /data/ActiveMQ
$ cd /data/ActiveMQ
$ artemis create mybroker
```

这里的 artemis 命令也支持 Windows 版本（artemis.cmd），在创建 Broker 实例时，控制台会提示创建默认用户名等初始化信息，如下面的信息所示。

```
Creating ActiveMQ Artemis instance at: /data/ActiveMQ/mybroker

--user:
Please provide the default username:
admin

--password: is mandatory with this configuration:
Please provide the default password:


--allow-anonymous | --require-login:
Allow anonymous access?, valid values are Y,N,True,False
n

Auto tuning journal ...
done! Your system can make 0.05 writes per millisecond, your journal-buffer-timeout
will be 19964000

You can now start the broker by executing:

   "/data/ActiveMQ/mybroker/bin/artemis" run

Or you can run the broker in the background using:

   "/data/ActiveMQ/mybroker/bin/artemis-service" start
```

笔者这里创建了一个简单的用户 admin/123456，且不允许匿名访问。

3）启动 Broker 实例

因为 ActiveMQ 是由 Java 语言开发的，所以运行 ActiveMQ Artemis 最新版本需要 JDK 11+ 环境。

在上一节创建 Broker 实例的控制台可以看到以下启动命令：

```
$ /data/ActiveMQ/mybroker/bin/artemis
```

如果想在后台启动，则参考命令如下：

```
$ /data/ActiveMQ/mybroker/bin/artemis-service
```

笔者这里使用后者，在后台启动 Broker 实例的参考命令如下：

```
$ /data/ActiveMQ/mybroker/bin/artemis-service start
Starting artemis-service
artemis-service is now running (17397)
```

除了使用 start 参数，还支持使用 stop、restart、force-stop、status 等命令参数，启动之后可以访问 ActiveMQ Artemis 控制台了。

ActiveMQ Artemis 控制台的登录地址如下：

```
http://localhost:8161/
```

输入刚才创建 Broker 实例时的用户名和密码即可登录，如下图所示。

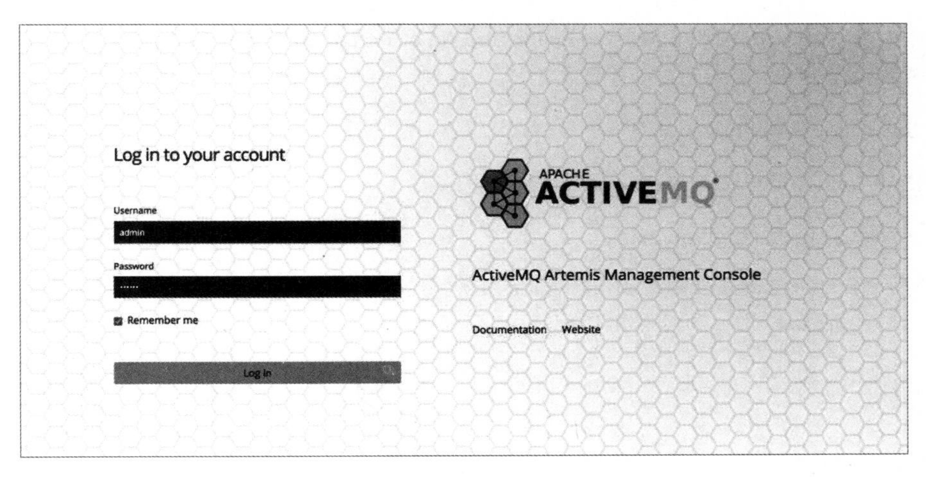

登录后的结果如下图所示。

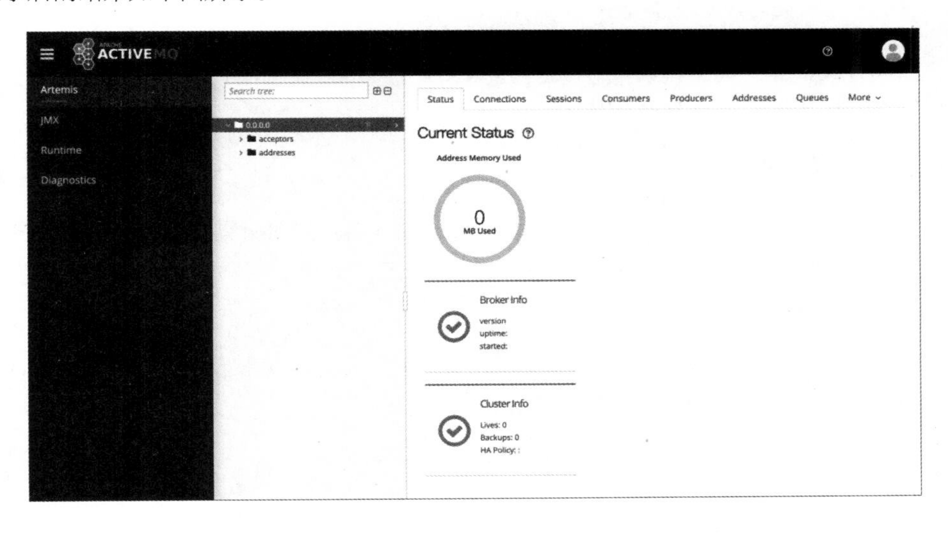

在 ActiveMQ Artemis 控制台可以查看一些实例运行的基础信息，以及连接、会话、生产者、消费者、队列等相关信息的搜索和操作菜单。

3. Spring Boot 集成 ActiveMQ Artemis

Spring Boot 提供了 spring-boot-starter-artemis 一站式启动器，如下面的依赖配置所示。

```
<dependency>
    <groupId>org.springframework.boot</groupId>
    <artifactId>spring-boot-starter-artemis</artifactId>
</dependency>
```

在新的 org.springframework.boot.autoconfigure.AutoConfiguration.imports 自动配置文件中注册了以下 JMS 和 ActiveMQ 相关的自动配置类：

- JmsAutoConfiguration。
- JndiConnectionFactoryAutoConfiguration。
- ArtemisAutoConfiguration。

只需配置一系列 spring.jms.* 和 spring.artemis.* 参数就可以完成自动配置，如下面的配置示例所示。

```
spring:
  jms:
    cache:
      enabled: true # 默认 true
      session-cache-size: 5 # 默认 1
  artemis:
    mode: native
    broker-url: tcp://localhost:61616
    user: admin
    password: 123456
```

在 ArtemisAutoConfiguration 自动配置类中，spring.jms.cache.enabled 默认为缓存开启状态（true），默认注册使用的是可缓存的 CachingConnectionFactory 工厂创建连接池，否则（false）使用的是 ActiveMQConnectionFactory。

如果使用了 spring.artemis.pool.enabled 连接池开启参数，则此时使用的是原生的 JmsPool-ConnectionFactory 工厂创建连接池，如以下配置所示。

```
spring:
```

```
  jms:
    cache:
      enabled: true # 默认 true
      session-cache-size: 5 # 默认 1
  artemis:
    mode: native
    broker-url: tcp://localhost:61616
    user: admin
    password: 123456
    pool:
      enabled: true # 默认 false
      max-connections: 50 # 默认 1
      idle-timeout: 5s # 默认 30s
      max-sessions-per-connection: 100 # 默认 500
```

此外还需要加上 pooled-jms 连接池的依赖，如以下依赖配置所示。

```
<dependency>
    <groupId>org.messaginghub</groupId>
    <artifactId>pooled-jms</artifactId>
</dependency>
```

Spring Boot 已经对该依赖纳入了管理，所以无须额外的版本号，更多的配置参数可以查看 JMS 和 ActiveMQ 对应的参数绑定类 JmsProperties 和 ArtemisProperties。

这里创建一个发送消息接口，以及接收消息的队列监听方法。在发送消息时可以直接注入 JmsTemplate 模板并使用，然后通过 @JmsListener 注解接收消息，如以下示例所示。

```
@RequiredArgsConstructor
@RestController
@Slf4j
public class MsgController {

    private final JmsTemplate jmsTemplate;

    /**
     * 发送消息
     * @param msg
     * @return
     */
    @RequestMapping("/send")
    public String sendMsg(@RequestParam("msg") String msg) {
        jmsTemplate.convertAndSend("test-queue", msg);
```

```
        return " 已发送 ";
    }

    /**
     * 接收消息
     * @param msg
     */
    @JmsListener(destination = "test-queue")
    public void receiveMsg(String msg) {
        log.info(" 收到 ActiveMQ 消息: {}", msg);
    }

}
```

调用发送消息接口进行测试，如下图所示。

结果如下图所示。

```
o.s.b.w.embedded.tomcat.TomcatWebServer  : Tomcat started on port(s): 8080 (http)
c.j.springboot.activemq.Application      : Started Application in 2.244 seconds (p
o.a.c.c.C.[Tomcat].[localhost].[/]       : Initializing Spring DispatcherServlet '
o.s.web.servlet.DispatcherServlet        : Initializing Servlet 'dispatcherServlet
o.s.web.servlet.DispatcherServlet        : Completed initialization in 0 ms
c.j.springboot.activemq.MsgController    : 收到 ActiveMQ 消息: 测试消息
```

从日志可以看到消息成功发送并被消费了，再回到 ActiveMQ 控制台 Queues 队列操作页面，可以看到队列也自动创建了，如下图所示。

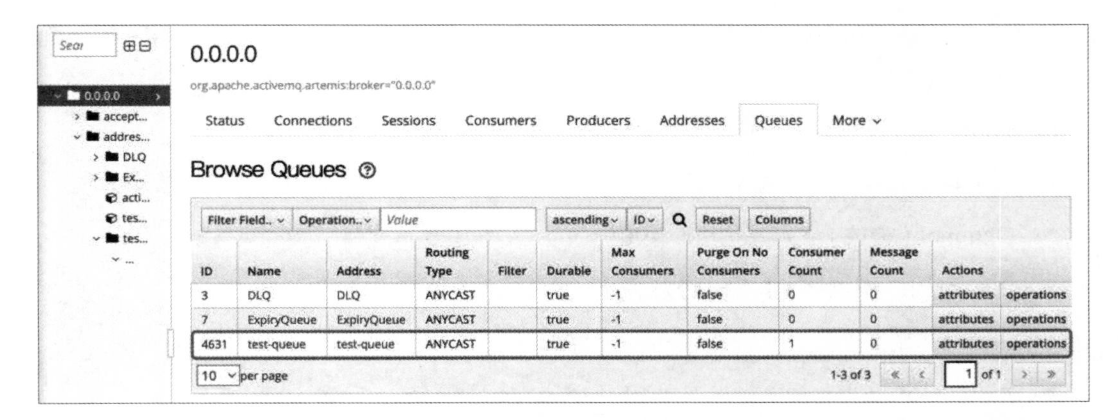

单击具体的队列可以对该队列进行查看等操作，比如手动发送消息到该队列，如下图所示。

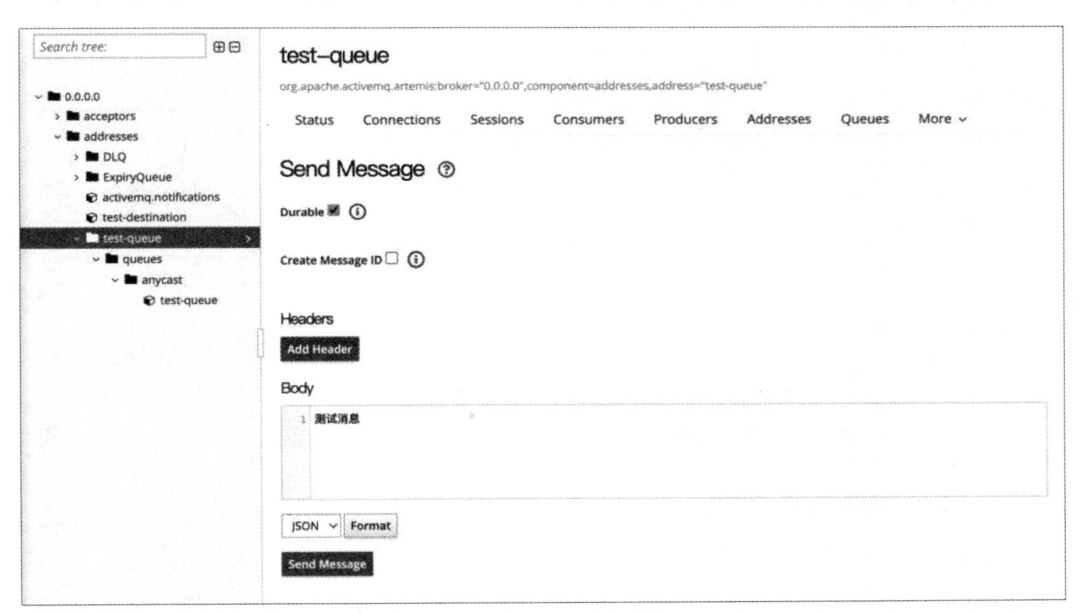

以上只是介绍了如何使用 Spring Boot 集成 JMS&ActiveMQ，以及进行简单的消息发送和接收操作，更多的应用读者可以参考其自动配置类并进行扩展。

9.2.3 RabbitMQ

1. 概述

RabbitMQ 是基于 AMQP 的轻量级、可靠的、可扩展、可移植的消息系统，基于 Erlang 语言编写，是目前应用最广泛的 MQ 系统之一，在使用 Spring Boot 集成 RabbitMQ 之前需要了解以下核心知识点。

RabbitMQ 包含的重要组件如下：

- ConnectionFactory（**连接管理器**）：用于应用与 RabbitMQ 之间建立连接的管理器。
- Channel（**信道**）：消息推送时使用的通道。
- Exchange（**交换器**）：用于分发消息。
- Queue（**队列**）：用于存储生产者写入的消息。
- RoutingKey（**路由键**）：用于交换器把生成者的消息写入不同的队列。
- BindingKey（**绑定键**）：用于将交换器绑定到不同的队列。

RabbitMQ 交换器（Exchange）支持的消息分发模式如下：

- direct（**默认**）：路由模式，发送消息时指定不同的 Routing Key，交换器会根据不同的 Routing Key 将消息写入不同的队列。
- fanout：广播模式，交换器会将消息发送到它绑定的所有队列中，也是性能最好、使用最多的模式。
- headers：交换器使用消息内容中的 headers 属性值与队列定义的参数进行匹配，性能较差，几乎不使用。
- topic：匹配模式，和路由模式差不多，但允许使用通配符匹配来分发消息。

RabbitMQ 的使用非常广泛，几乎是 Java 应用领域消息队列事实上的标准。

2. RabbitMQ 环境搭建

1）安装 RabbitMQ

RabbitMQ 几乎支持所有主流的操作系统，包括 Docker 镜像，这里以 CentOS 为例进行安装演示。

RabbitMQ 是基于 Erlang 语言编写的，所以需要安装 Erlang 语言环境，和 Java 应用一样，不同的 RabbitMQ 版本也需要依赖兼容的 Erlang 语言版本，下图是 RabbitMQ 官方提供的兼容表。

RabbitMQ and Erlang/OTP Compatibility Matrix

The table below provides an Erlang compatibility matrix of currently supported RabbitMQ release series. For RabbitMQ releases that have reached end of life, see Unsupported Series Compatibility Matrix.

RabbitMQ version	Minimum required Erlang/OTP	Maximum supported Erlang/OTP	Notes
3.11.2 3.11.1 3.11.0	25.0	25.1	• As of Erlang 25.1, OpenSSL 3.0 support in Erlang is considered to be mature enough for production. • Erlang 25 before 25.0.2 is affected by CVE-2022-37026, a CVE with critical severity (CVSS 3.x Base Store: 9.8)
3.10.10 3.10.9 3.10.8	24.2	25.1	• As of Erlang 25.1, OpenSSL 3.0 support in Erlang is considered to be mature enough to consider for production. • Erlang 25 before 25.0.2 and 24 before 24.3.4.2 are affected by CVE-2022-37026, a CVE with critical severity (CVSS 3.x Base Store: 9.8)
3.10.7 3.10.6 3.10.5	23.2	25.1	• Erlang 25 is the recommended series. • Erlang 25 before 25.0.2 and 24 before 24.3.4.2 are affected by CVE-2022-37026, a CVE with critical severity (CVSS 3.x Base Store: 9.8) • Erlang 23 support was discontinued on July 31st, 2022.

市面上有多个可用于 Erlang/OTP 的 RPM 安装包，但推荐使用来自 RabbitMQ 团队的零依赖 Erlang RPM 安装包，它紧跟最新的 Erlang/OTP 补丁发布，下载地址为链接 11。

可以通过命令一并下载安装，相关参考命令如下：

```
$sudo yum update -y
$ sudo yum -q makecache -y --disablerepo=’*’ --enablerepo=’rabbitmq_erlang’
--enablerepo=’rabbitmq_server’
$ sudo yum install socat logrotate -y
$ sudo yum install --repo rabbitmq_erlang --repo rabbitmq_server erlang rabbitmq-server -y
```

2）启动 RabbitMQ

安装 RabbitMQ 捆绑包时，RabbitMQ 服务器默认不作为守护进程启动，在系统启动时默认启动守护进程，需要以管理员身份运行命令，参考命令如下：

```
$ sudo chkconfig rabbitmq-server on
```

启动、状态查询、停止的相关参考命令如下：

```
$ service rabbitmq-server start
$ service rabbitmq-server status
```

```
$ service rabbitmq-server stop
```

3）启用Web控制台

安装好RabbitMQ之后，还需要安装一个Web控制台，Web控制台提供了各种可视化的操作管理，有助于运维和排查问题，这需要用到RabbitMQ管理插件，它默认包含在RabbitMQ发行版中，但必须先启用才能使用，参考命令如下：

```
$ rabbitmq-plugins enable rabbitmq_management
```

启用Web控制台之后，访问http://xxxx:15672/控制台页面，如下图所示。

输入默认的游客账号和密码（guest/guest）即可登录，登录后进入控制台首页，如下图所示。

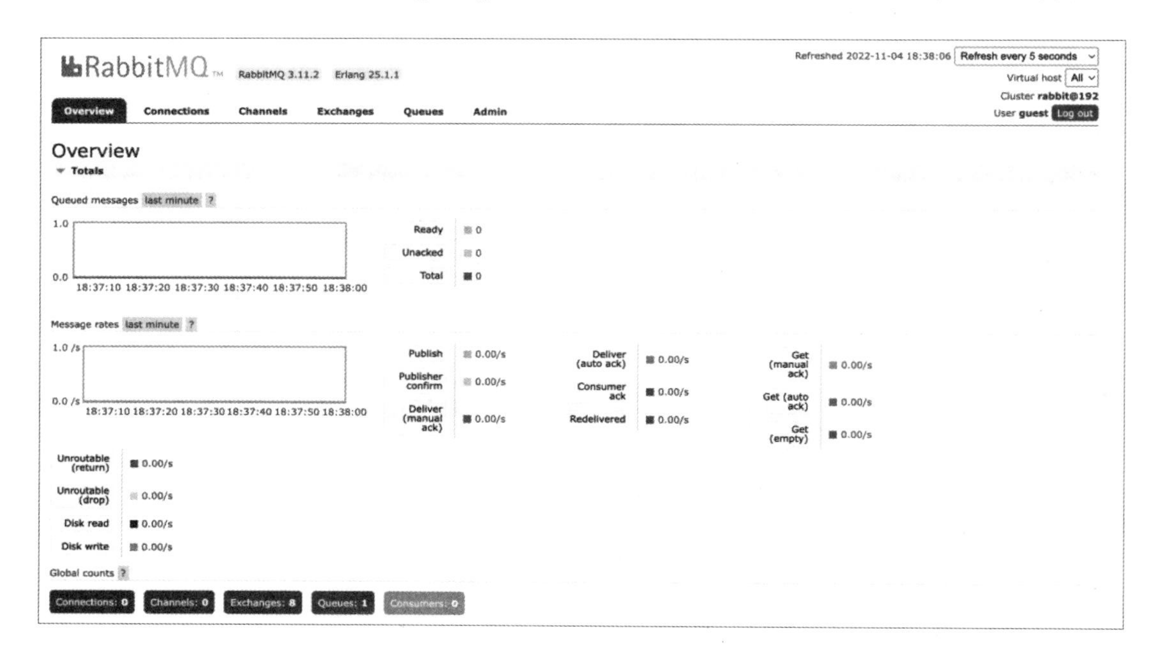

4）添加远程用户

默认的 guest 用户仅限于 localhost 本机访问，后续与 Spring Boot 集成就不行了，所以需要添加一个用户，这里先添加一个超级管理员。

添加账号，参考命令如下：

```
$ ./rabbitmqctl add_user admin 123456
```

添加访问权限，参考命令如下：

```
$ ./rabbitmqctl set_permissions -p "/" admin ".*" ".*" ".*"
```

设置超级权限，参考命令如下：

```
$ ./rabbitmqctl set_user_tags admin administrator
```

设置之后就可以通过 admin 用户登录控制台和远程连接了。

3. Spring Boot 集成 RabbitMQ

Spring Boot 提供了 spring-boot-starter-amqp 一站式启动器，如下面的依赖配置所示。

```
<dependency>
    <groupId>org.springframework.boot</groupId>
    <artifactId>spring-boot-starter-amqp</artifactId>
</dependency>
```

在新的 org.springframework.boot.autoconfigure.AutoConfiguration.imports 自动配置文件中注册了 RabbitMQ 的自动配置类 RabbitAutoConfiguration，对应的参数绑定类为 RabbitProperties，只需配置一系列 spring.rabbitmq.* 参数就可以完成自动配置。

连接配置示例如下：

```
spring:
  rabbitmq:
    host: localhost
    port: 5672
    username: admin
    password: 123456
```

还可以像数据库数据源一样通过一个地址参数直连，如以下配置示例所示。

```
spring:
  rabbitmq:
    addresses: "amqp://admin:123456@localhost"
```

更多的配置参数可以查看 RabbitMQ 对应的参数绑定类 RabbitProperties。

这里创建一个发送消息接口，以及接收消息的队列监听方法，在发送消息时可以直接注入 RabbitTemplate 模板并使用，然后通过 @RabbitListener 注解接收消息，如以下示例所示。

```
@RequiredArgsConstructor
@RestController
@Slf4j
public class MsgController {

    private final RabbitTemplate rabbitTemplate;

    /**
     * 发送 Direct 模式消息
     * @param msg
     * @return
     */
    @RequestMapping("/send")
    public String sendMsg(@RequestParam("msg") String msg) {
        rabbitTemplate.convertAndSend("test-direct-exchange",
                "test-direct-routing-key", msg);
        return "已发送";
    }

    /**
     * 接收 Direct 模式消息
     * @param msg
     */
    @RabbitListener(queues = "test-direct-queue")
    public void receiveMsg(String msg) {
        log.info("收到 RabbitMQ 消息：{}", msg);
    }

}
```

RabbitMQ 不能自动创建对应的队列、交换器，以及队列与交换器的绑定，它们可以在 RabbitMQ 控制台页面中创建，也可以通过代码的方式创建，比如以下配置示例：

```
@Configuration
public class RabbitMQConfig {

    /**
     * 创建 Direct 模式队列
```

```
     * @return
     */
    @Bean
    public Queue testDirectQueue() {
        return new Queue("test-direct-queue");
    }

    /**
     * 创建 Direct 模式交换器
     * @return
     */
    @Bean
    public DirectExchange TestDirectExchange() {
        return new DirectExchange("test-direct-exchange");
    }

    /**
     * 创建 Direct 队列与交换器绑定
     * @param testDirectQueue
     * @return
     */
    @Bean
    public Binding testDirectBinding(Queue testDirectQueue) {
        return BindingBuilder.bind(testDirectQueue)
                .to(TestDirectExchange()).with("test-direct-routing-key");
    }

}
```

上面创建的是 Direct 直连类型的交换器，即根据消息携带的路由键（routing key）将消息投递到与之关联的队列中。

调用发送消息接口进行测试，如下图所示。

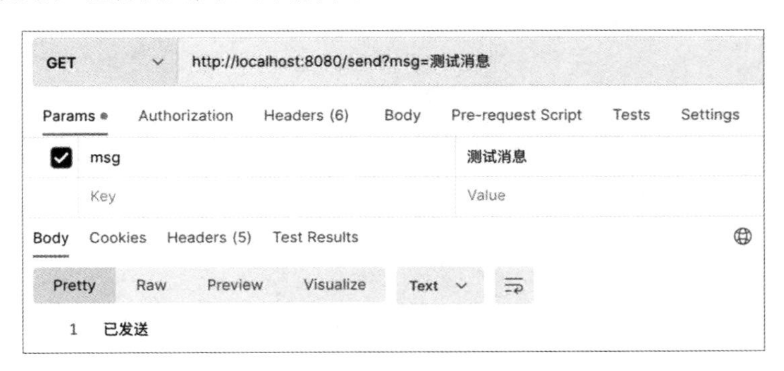

结果如下图所示。

```
o.s.b.w.embedded.tomcat.TomcatWebServer  : Tomcat started on port(s): 8080 (http) wi
o.s.a.r.c.CachingConnectionFactory       : Attempting to connect to: [localhost:5672
o.s.a.r.c.CachingConnectionFactory       : Created new connection: rabbitConnectionF
c.j.springboot.rabbitmq.Application      : Started Application in 2.139 seconds (pro
o.a.c.c.C.[Tomcat].[localhost].[/]       : Initializing Spring DispatcherServlet 'di
o.s.web.servlet.DispatcherServlet        : Initializing Servlet 'dispatcherServlet'
o.s.web.servlet.DispatcherServlet        : Completed initialization in 2 ms
c.j.springboot.rabbitmq.MsgController    : 收到 RabbitMQ 消息：测试消息
```

从日志可以看到消息成功发送并被消费了，再回到RabbitMQ控制台页面，可以看到创建的队列、交换器等信息，如下面的三个图所示。

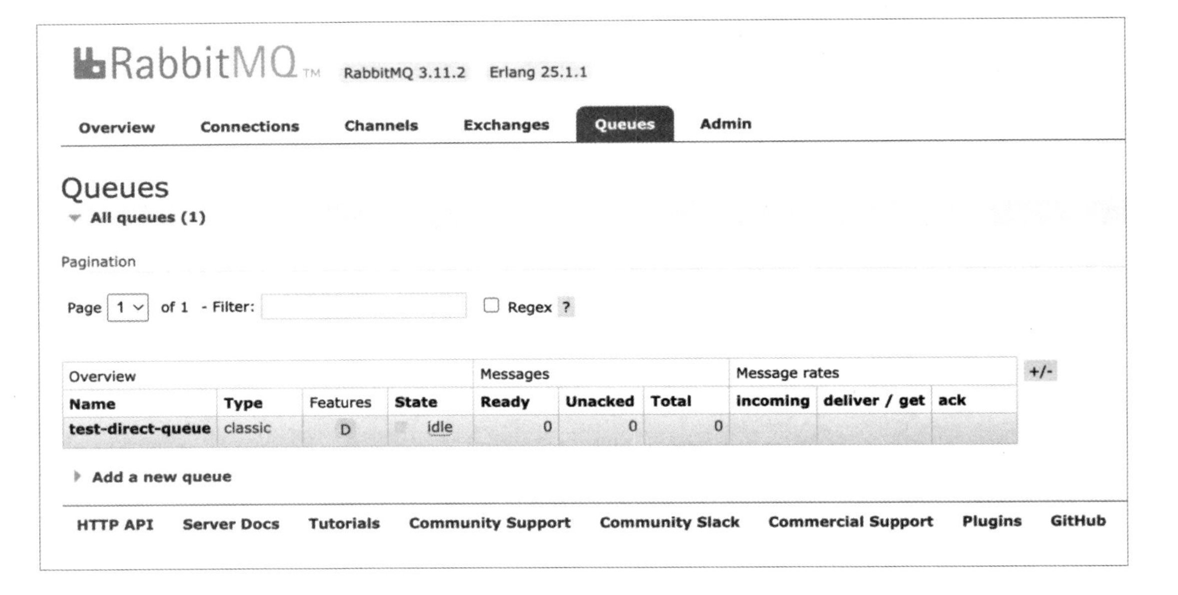

在 Queues菜单页上单击 test-direct-queue 项进入该队列，也能进行交换器和路由键的绑定、手动发送消息等操作，如下图所示。

以上只是介绍了如何使用 Spring Boot 集成 RabbitMQ，以及简单的使用方式，更多的高级用法可以参考其自动配置类进行扩展。

9.2.4 Kafka

1. 概述

Kafka 是一个高吞吐量、分布式的开源消息系统，它更是一个开源的分布式事件流平台，它可以进行及时、海量的流数据处理，现在已经被广泛用于高性能的数据管道、数据流分析、数据集成等负责核心任务的应用。

常见的 Kafka 的应用场景如下：

- 消息队列（MQ）。
- 流处理。
- 日志收集。
- 用户活动跟踪。
- 指标收集。

Kafka 最初由 LinkedIn 公司开发，后来捐献给了 Apache 软件基金会并成为顶级项目。Kafka 是由 Scala 和 Java 语言编写的，它们都是 JVM 系语言，所以 Kafka 编译后的源代码就是 .class 文件。

Kafka 2.8.0 中已经有了对 ZooKeeper 的早期替换方案，如下图的官方版本说明所示。

2.8.0

- Released April 19, 2021
- Release Notes
- Source download: kafka-2.8.0-src.tgz (asc, sha512)
- Binary downloads:
  - Scala 2.12 - kafka_2.12-2.8.0.tgz (asc, sha512)
  - Scala 2.13 - kafka_2.13-2.8.0.tgz (asc, sha512)

  We build for multiple versions of Scala. This only matters if you are using Scala and you want a version built for the same Scala version you use. Otherwise any version should work (2.13 is recommended).

Kafka 2.8.0 includes a number of significant new features. Here is a summary of some notable changes:

- Early access of replace ZooKeeper with a self-managed quorum
- Add Describe Cluster API
- Support mutual TLS authentication on SASL_SSL listeners

但目前最新的 Kafka 3.3.1 中还没有彻底移除 ZooKeeper，还可以使用 ZooKeeper 来管理 Kafka 集群，同时还可以使用最新的 KRaft 来管理 Kafka 集群，并且新增了一个 @metadata 的内部 Topic 主题来存储元数据信息。在后续的版本中，Kafka 会彻底移除对 ZooKeeper 的依赖。

移除 ZooKeeper 的原因可能有以下两点：

（1）Kafka 需要维护另外一个中间件，增加了额外的运维成本，完全没有必要。

（2）依赖 ZooKeeper 使得 Kafka 显得比较笨重，移除 ZooKeeper 依赖可以促使自己变得更轻量化，带来更优的性能、

2. Kafka 环境搭建

1）安装 Kafka

首先在官网下载 Kafka，下载地址为链接 12。

Kafka 几乎支持所有主流的操作系统，包括 Docker 镜像，这里以 CentOS 为例进行安装演示。

下载之后解压即可，参考命令如下：

```
$ tar -xzf kafka_2.13-3.3.1.tgz
$ cd kafka_2.13-3.3.1
```

2）启动 Kafka

> Kafka 启动运行需要依赖 Java 8+。

虽然 Apache Kafka 准备移除 ZooKeeper，但目前（Kafka 3.3.1）还是没有正式移除，现在还支持使用 ZooKeeper 或 KRaft 的方式启动 Kafka，二者选其一。

以 ZooKeeper 方式启动 Kafka

先启动 ZooKeeper 服务，参考命令如下：

```
$ bin/zookeeper-server-start.sh config/zookeeper.properties &
```

再启动 Kafka 服务，参考命令如下：

```
$ bin/kafka-server-start.sh config/server.properties &
```

以 KRaft 方式启动 Kafka

生成一个 Cluster UUID，参考命令如下：

```
$ KAFKA_CLUSTER_ID=" $(bin/kafka-storage.sh random-uuid)"
```

格式化日志目录，参考命令如下：

```
$ bin/kafka-storage.sh format -t $KAFKA_CLUSTER_ID -c config/kraft/server.properties
```

启动 Kafka 服务，参考命令如下：

```
$ bin/kafka-server-start.sh config/kraft/server.properties &
```

3）发送 / 消费 Kafka 消息

成功启动 Kafka 服务后，就有了一个基本的 Kafka 运行环境并可以使用了，下面使用 Kafka 命令行工具来进行简单的发送 / 消费消息的测试。

首先创建一个 spring-boot-test 测试主题，参考命令如下：

```
$ bin/kafka-topics.sh --create --topic spring-boot-test --bootstrap-server localhost:9092
```

然后启动一个生产者客户端发送消息，参考命令如下：

```
$ bin/kafka-console-producer.sh --topic spring-boot-test --bootstrap-server localhost:9092
>测试消息 1
>测试消息 2
>结束
>^C%
```

每一行输入结束后，再按回车键，即表示发送多条独立的消息。

然后新开一个命令行窗口，启动一个消费者客户端消费消息，参考命令如下：

```
$ bin/kafka-console-consumer.sh --topic spring-boot-test --from-beginning --bootstrap-server
localhost:9092
测试消息 1
测试消息 2
结束
```

按 Ctrl+C 键可以结束运行生产者和消费者客户端，只要生产者和消费者客户端没有被关闭，就可以一直进行发送消息和消费消息的测试。

消息一旦生产就被持久化了，消费者可以一直进行消费。Kafka 中有一个偏移量的概念，每个消费者会保存消费的偏移量值，用来标识它消费的位置，当消费消息的时候，偏移量会线性增加。偏移量由消费者进行控制，消费者可以将偏移量重置到之前的位置以重新消费消息，也可以跳过一些消息消费。

3. Spring Boot 集成 Kafka

前面介绍了 Kafka 的自动配置是由 spring-kafka 项目提供的，Spring Boot 并没有再提供专用的 Starter 启动器，Spring Boot 集成 Kafka 时只需要引入 spring-kafka 项目依赖，如以下依赖配置所示。

```
<dependency>
    <groupId>org.springframework.kafka</groupId>
    <artifactId>spring-kafka</artifactId>
</dependency>
```

spring-kafka 项目依赖主要包含了 Spring 消息、事务、重试模块，以及 Kafka 客户端等，如下图所示。

```
∨ spring-kafka : 3.0.0 [compile]
      jsr305 : 3.0.2 [runtime]
    > kafka-clients : 3.3.1 [compile]
    > micrometer-observation : 1.10.2 [compile]
    > spring-context : 6.0.2 [compile]
    > spring-messaging : 6.0.2 [compile]
      spring-retry : 2.0.0 [compile]
    > spring-tx : 6.0.2 [compile]
```

在新的 org.springframework.boot.autoconfigure.AutoConfiguration.imports 自动配置文件中注册了 Kafka 的自动配置类 KafkaAutoConfiguration，对应的参数绑定类为 KafkaProperties，只需配置一系列 spring.kafka.* 参数就可以完成自动配置。

连接配置示例如下所示。

```
spring:
  kafka:
    bootstrap-servers: localhost:9092
    consumer:
      group-id: testGroup
```

更多的配置参数可以查看 Kafka 对应的参数绑定类 KafkaProperties。

在集成 Kafka 并发送消息之前需要先创建一个主题，既可以通过之前介绍的 Kafka 命令行创建主题，也可以通过代码的方式创建主题，比如以下配置示例：

```java
@Configuration
public class KafkaConfig {

    public static final String SPRING_BOOT_TEST_TOPIC = "spring-boot-test-topic";

    @Bean
    public NewTopic testTopic() {
        return new NewTopic(SPRING_BOOT_TEST_TOPIC, 4, (short) 1);
    }

}
```

Spring Boot 启动时会加载该配置类并创建一个区分数为 4、副本数为 1 的测试主题，如果该主题已经创建过，则会忽略。

Kafka 中的分区是指对 Topic（主题）进行分区，即把一个主题拆分成多个分区，如下图所示。

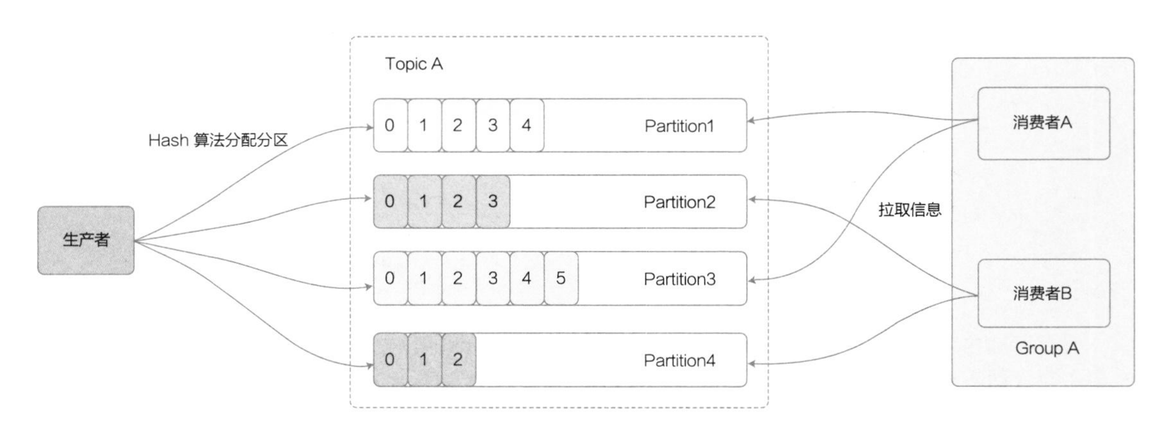

对 Topic 分区的好处是，每个 Topic 可以不受单台服务器的限制，扩展性非常好，可以实现负载均衡，支持消费者并行消费，提高吞吐量以处理更多的消息。

如上图所示，该主题被分成了 4 个分区，两个不同的生产者客户端向该主题写入消息，如果消息指定了 KEY，则使用 KEY 策略，KEY 相同的消息写入同一个分区，如果消息没有指定 KEY，则默认使用轮询策略，另外还可以使用随机策略，每条消息随机写入不同的分区。

消息只能被写入其中一个分区，每个分区都是有序且顺序不可变的记录集，并且可以持续追加，分区中的每条消息记录都会有唯一的 offset（偏移量）进行区分。一个消费者可以消费多个分区的消息，但一个分区只能被一个消费者组中的一个消费者消费。另外，如果一个 Topic 只有一个分区，那么就只能有一个消费者进行消费。

每个 Topic 可以建立多个分区，而副本是对分区的复制，每个分区可以建立多个副本，可以有多个追随者副本（Follower Replica），但只能有一个领导者副本（Leader Replica）。

> **需要注意的是：**
> Kafka 不支持减少 Topic 分区数，不然会导致数据丢失，但可以通过删除原来的 Topic，然后创建新的 Topic 的方式重新调整分区数。

这里创建一个发送消息接口，以及消费指定主题的监听方法，在应用中发送 Kafka 消息时可以直接注入 KafkaTemplate 模板并使用，然后通过 @KafkaListener 注解监听对应的主题即可消费消息，如以下示例所示。

```
@RequiredArgsConstructor
@RestController
@Slf4j
public class MsgController {

    private final KafkaTemplate kafkaTemplate;

    /**
     * 发送消息
     * @param msg
     * @return
     */
    @RequestMapping("/send")
    public String sendMsg(@RequestParam("msg") String msg) {
        kafkaTemplate.send(KafkaConfig.SPRING_BOOT_TEST_TOPIC, "test-key", msg);
        return " 已发送 ";
    }

    /**
     * 接收消息
     * @param msg
     */
```

```
    @KafkaListener(topics = KafkaConfig.SPRING_BOOT_TEST_TOPIC)
    public void receiveMsg(String msg) {
        log.info("收到 Kafka 消息: {}", msg);
    }

}
```

调用发送消息接口进行测试，如下图所示。

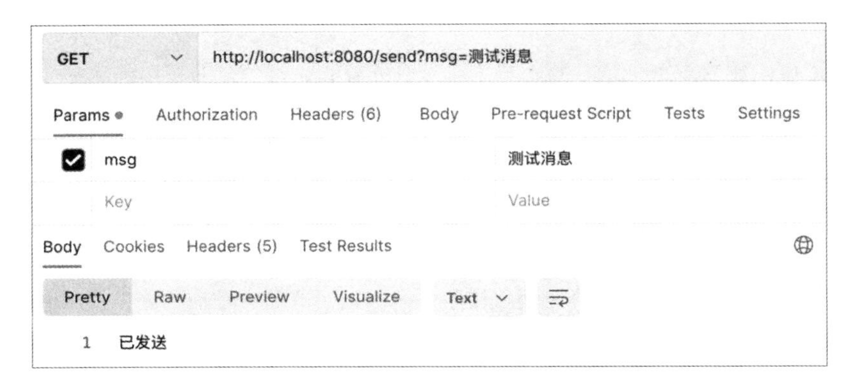

结果如下图所示。

```
o.a.kafka.common.utils.AppInfoParser     : Kafka startTimeMs: 1672714954659
org.apache.kafka.clients.Metadata        : [Producer clientId=producer-1] Res
org.apache.kafka.clients.Metadata        : [Producer clientId=producer-1] Res
org.apache.kafka.clients.Metadata        : [Producer clientId=producer-1] Res
org.apache.kafka.clients.Metadata        : [Producer clientId=producer-1] Res
org.apache.kafka.clients.Metadata        : [Producer clientId=producer-1] Clu
o.a.k.c.p.internals.TransactionManager   : [Producer clientId=producer-1] Pro
c.j.springboot.kafka.MsgController       : 收到 Kafka 消息: 测试消息
```

从日志可以看到消息成功发送并被消费了。

以上只是介绍了如何使用 Spring Boot 集成 Kafka，以及简单的使用方式，更多的高级用法可以参考其自动配置类并进行扩展。

第 10 章 Spring Boot 调试与单元测试

任何项目在上线之前都必须经过本地开发、调试、测试阶段，本章将介绍如何在 Spring Boot 项目中进行断点跟踪调试，这和传统的项目略有区别。本章还将介绍 Spring Boot 特有的开发者工具，它可以帮助开发者在开发阶段调试应用，帮助开发者大大提升开发效率，最后介绍 Spring Boot 如何正确地进行单元测试。

10.1 断点调试

10.1.1 使用 main 方法启动调试

根据 Spring Boot 项目不同的启动方式，断点调试的方式也不一样。

如果直接使用 main 方法启动应用，那么选择 Debug 模式启动即可，如下图所示。

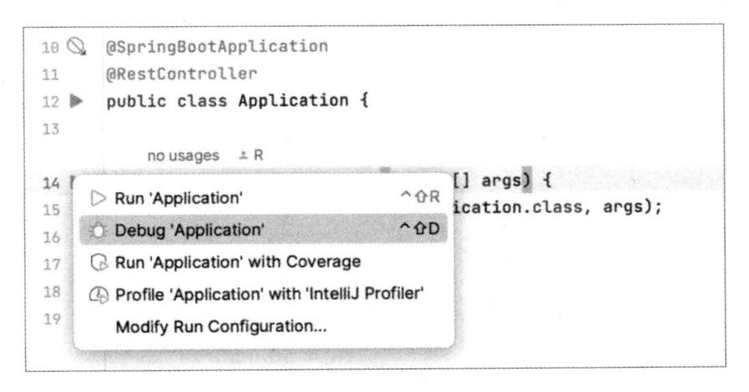

当代码运行到断点处的时候就可以进行调试了，如下图所示。

```
        R
19      @RequestMapping("/hello")
20      public String helloWorld() {
            return "hello world.";
22      }
23
```

10.1.2 使用 Maven 插件启动调试

使用 main 方法启动调试有一定的缺陷，因为启动应用时不能使用 Maven 的相关插件，比如不能使用 Maven 资源插件过滤配置资源，导致无法过滤参数占位符而应用启动失败，所以实际项目中使用 Spring Boot Maven 插件启动应用的方式较多。

使用 Spring Boot Maven 插件启动应用，它会“fork”一个子进程来运行应用，所以它不能像 main 方法那样直接“debug”启动调试，而需要像调试远程应用一样调试，断点调试起来略显麻烦。首先需要在Spring Boot Maven插件中配置jvmArguments调试参数启动应用，然后新建一个远程调试。

1. 启用调试模式

启用调试模式有以下两种方式：

（1）在插件配置中指定调试模式的 jvmArguments 参数，如以下配置示例所示。

```
<project>
    <build>
        <plugins>
            <plugin>
                <groupId>org.springframework.boot</groupId>
                <artifactId>spring-boot-maven-plugin</artifactId>
                <configuration>
                    <jvmArguments>
                        -Xdebug -Xrunjdwp:transport=dt_socket,server=y,suspend=y,address=5005
                    </jvmArguments>
                </configuration>
            </plugin>
        </plugins>
    </build>
</project>
```

（2）使用 mvn 命令启动时，指定调试模式的 jvmArguments 参数，完整运行命令如下所示。

```
$ mvn spring-boot:run -Dspring-boot.run.jvmArguments="-Xdebug -Xrunjdwp:transport=dt_socket, ser
ver=y,suspend=y,address=5005"
```

主流的开发工具都集成了 Maven 插件，无须 mvn 关键字，否则可能运行出错，如下图所示。

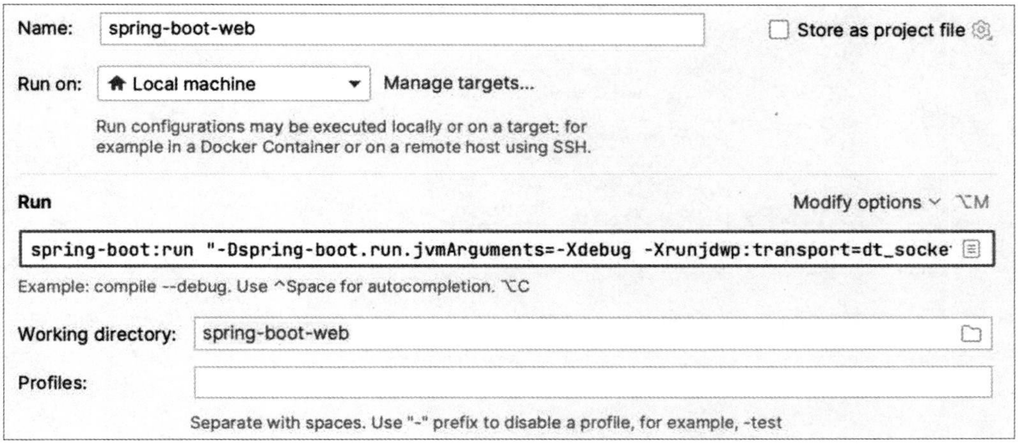

Name: spring-boot-web　　Store as project file

Run on: Local machine　Manage targets...

Run configurations may be executed locally or on a target: for example in a Docker Container or on a remote host using SSH.

Run　　Modify options ⌥M

spring-boot:run "-Dspring-boot.run.jvmArguments=-Xdebug -Xrunjdwp:transport=dt_socke

Example: compile --debug. Use ^Space for autocompletion. ⌥C

Working directory: spring-boot-web

Profiles:

Separate with spaces. Use "-" prefix to disable a profile, for example, -test

以上创建了一个 spring-boot-web 启动项，然后使用正常运行模式运行即可，应用会在指定的 5005 端口上监听，如下图所示。

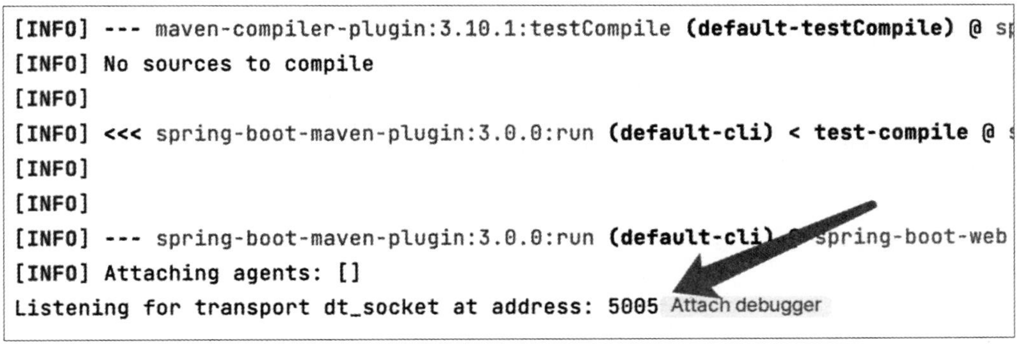

```
[INFO] --- maven-compiler-plugin:3.10.1:testCompile (default-testCompile) @ sp
[INFO] No sources to compile
[INFO]
[INFO] <<< spring-boot-maven-plugin:3.0.0:run (default-cli) < test-compile @ s
[INFO]
[INFO]
[INFO] --- spring-boot-maven-plugin:3.0.0:run (default-cli) @ spring-boot-web
[INFO] Attaching agents: []
Listening for transport dt_socket at address: 5005 Attach debugger
```

一般建议在命令行中指定参数进入调试模式，只在有需要的时候启用调试。

2. 远程调试

在 IDE中新建一个 Remote 调试任务，主流的 IDE 开发工具都是支持的，如下图所示。

为了与上一步的 spring-boot-web 启动项进行区分，这里将对应的远程调试命名为 spring-boot-web-debug，一般只需要确定 Host 和 Port 参数，然后使用 Debug 模式启动该 Remote 调试任务，Debug 控制台会显示已经连接成功的日志，如下图所示。

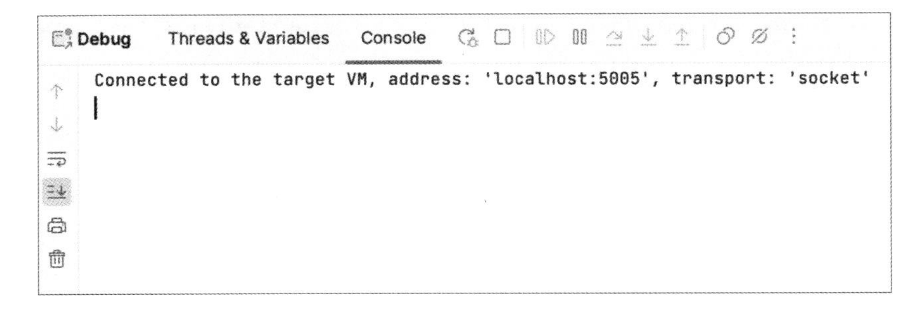

一旦连接成功，上一步的 spring-boot-web 启动项才会继续完成启动，之后就可以打断点调试了，如下图所示。

所以，Spring Boot Maven 插件启动的应用，断点调试都需要经过这两次启动操作才行，略显麻烦。

10.2 开发者工具

10.2.1 概述

Spring Boot 提供了一系列开发者工具，主要用于应用开发阶段，作用包括自动禁用缓存、支持应用自动重启、实时重载等，使用开发者工具可以大大提高开发效率。

只需要引入开发者工具依赖即可启用开发者工具，如下所示。

```
<dependency>
    <groupId>org.springframework.boot</groupId>
    <artifactId>spring-boot-devtools</artifactId>
    <optional>true</optional>
</dependency>
```

其中，Optional 表示该依赖只能在本项目及其子项目中传递，而不会传递到引用该项目的项目中，其他项目需要主动引用该依赖才行。

开发者工具是在开发阶段使用的，所以，打包应用时，Spring Boot 默认不会把开发者工具打包进去，如果有需要，也可以通过配置把开发者工具打进包中，如以下配置所示。

```
<build>
    <plugins>
        <plugin>
            <groupId>org.springframework.boot</groupId>
```

```
            <artifactId>spring-boot-maven-plugin</artifactId>
            <configuration>
                <excludeDevtools>false</excludeDevtools>
            </configuration>
        </plugin>
    </plugins>
</build>
```

其中，excludeDevtools 参数设置为 false 表示不排除开发者工具依赖。另外，因为开发者工具是开发阶段才使用的工具，所以使用 java -jar 命令运行整包时，开发者工具会被自动禁用，虽然可以通过 -Dspring.devtools.restart.enabled=true 参数来启用生产环境中的开发者工具，但不建议这样做，有安全风险，建议不要在生产环境中包含开发者工具。

10.2.2 默认值

下表是 Spring Boot 开发者工具使用的参数默认值。

参数名称	默认值
server.error.include-binding-errors	always
server.error.include-message	always
server.error.include-stacktrace	always
server.servlet.jsp.init-parameters.development	true
server.servlet.session.persistent	true
spring.freemarker.cache	false
spring.graphql.graphiql.enabled	true
spring.groovy.template.cache	false
spring.h2.console.enabled	true
spring.mustache.servlet.cache	false
spring.mvc.log-resolved-exception	true
spring.reactor.debug	true
spring.template.provider.cache	false
spring.thymeleaf.cache	false
spring.web.resources.cache.period	0
spring.web.resources.chain.cache	false

这里以缓存项为例说明为什么需要默认值。Spring Boot 的某些功能组件可能会提供缓存，这在生产环境中肯定是能提升性能的，但在开发阶段可能会因为没有及时生效而排查困难，所以 Spring Boot 开发者工具默认是禁用所有缓存项的。

如果不想让应用使用这些开发者工具带来的默认参数值，则可以通过指定参数来禁用默认值，如以下配置示例所示。

```
spring:
  devtools:
    add-properties: false
```

10.2.3 自动重启

1. 概述

Spring Boot 开发者工具会监控类路径下的文件，一旦有文件发生变更，应用就能自动快速重启，而不需要手动重启应用。自动重启技术使用了两个类加载器：

- **基础类加载器**：用于加载不会变更的文件，比如第三方的 jar 包。
- **重启类加载器**：用于加载应用类文件。

这样做的好处是每次重启只需要创建一个重启类加载器即可，基础类加载器已加载好的不用动，虽然不能和 JRebel 热部署方式相比，但和冷启动相比，自动重启技术拥有更高的启动效率。

需要注意的是：

> 使用 AspectJ 切面编程时不支持自动重启。

2. 触发机制

自动重启触发机制如下：

- Eclipse：修改保存就能自动重启。
- IntelliJ IDEA：需要重新构建（Build）项目才能自动重启。
- **构建工具**：在 IDE 构建插件中使用 mvn compile/gradle build 命令也能触发重启。

IntelliJ IDEA 默认是手动构建应用的，当然也可以开启自动构建应用，如下面两个图所示（IntelliJ IDEA 的版本为 2022.3）。

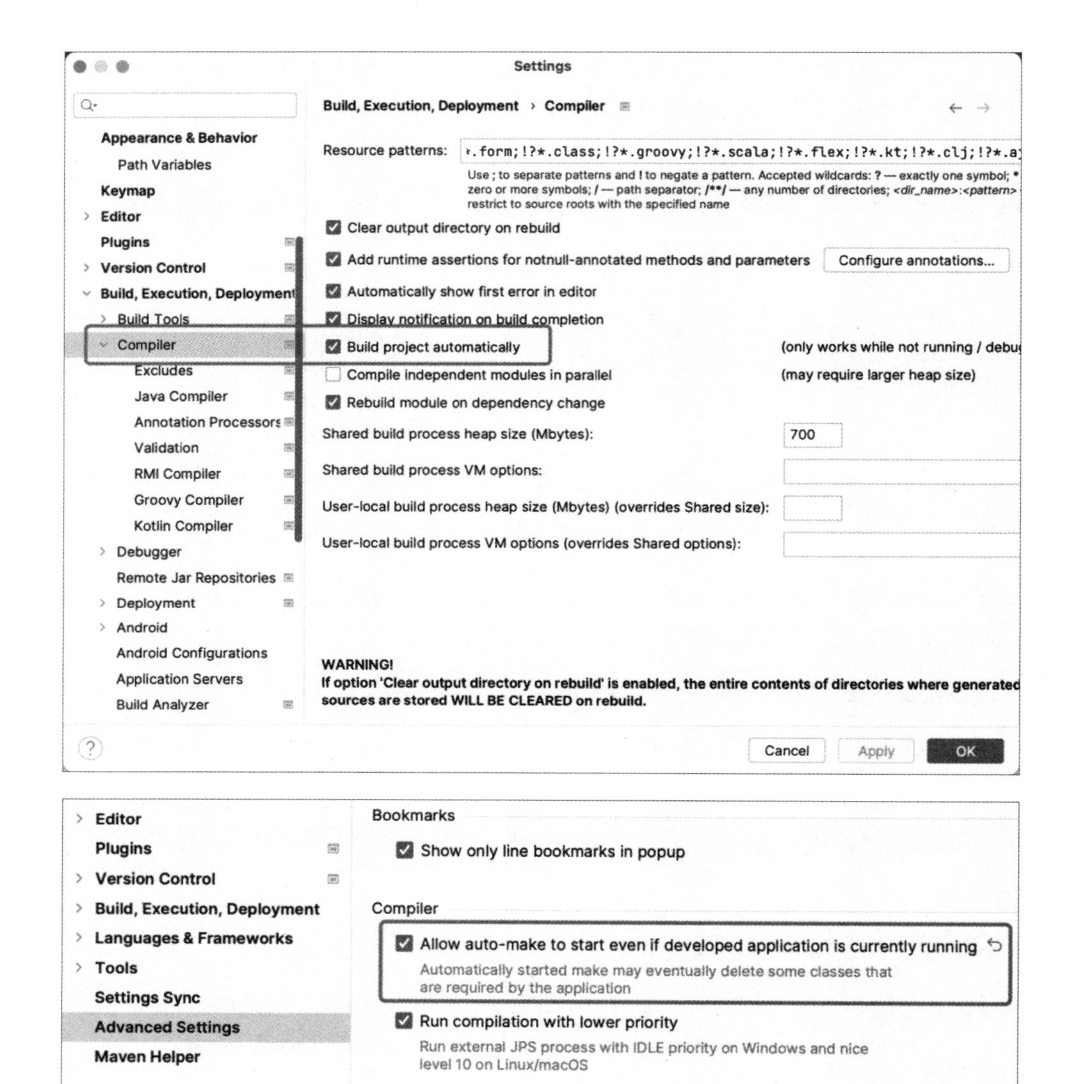

IDE 开发工具每次修改文件之后都会自动编译并重启，一方面会频繁消耗资源，另一方面可能不符合开发者的要求，比如，开发者想在修改了一定量文件之后再决定是否自动重启，这时可以把 IDE 修改为手动编译模式，或者指定一个触发文件，只有该触发文件被修改之后应用才自动重启。

配置示例如下所示。

```
spring:
  devtools:
```

```
restart:
  trigger-file: .reloadtrigger
```

首先在配置文件中指定自动重启的触发文件，然后在资源目录下创建配置中指定的.reloadtrigger触发文件，这时，只有该触发文件被修改（更新）后应用才会自动重启。

另外，并不一定要手动更新这个触发文件，在各种主流的IDE开发工具中都可以设置一个触发动作，比如在IntelliJ IDEA中可以设置运行更新时的动作，设置只更新触发文件，如下图配置所示。

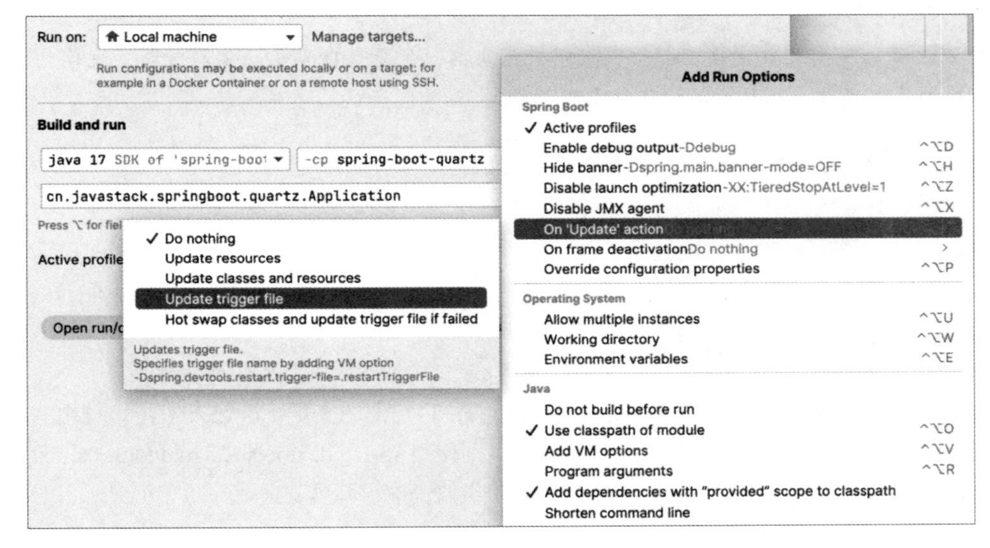

然后在Services面板上会有一个更新的按钮，如下图所示。

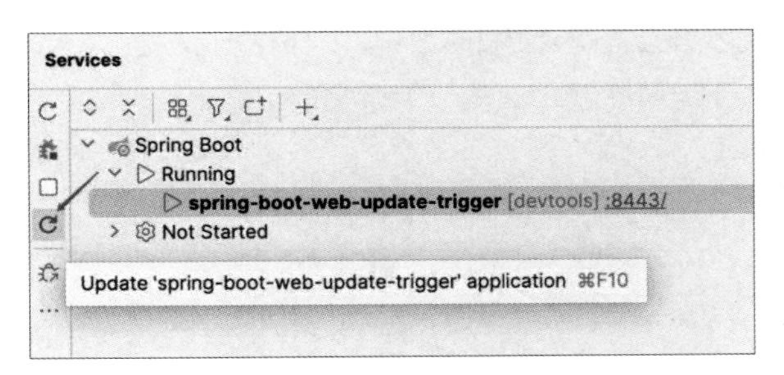

不过这个设置只限于使用main方法启动的Spring Boot应用，不支持使用Spring Boot Maven插件方式启动的应用，笔者没有找到相关设置和按钮，需要手动更新触发文件。

3. 排除资源

某些资源是没有必要触发应用自动重启的，默认情况下，以下目录下的资源不会触发自动重启：

- /META-INF/maven。
- /META-INF/resources。
- /resources。
- /static。
- /public。
- /templates。

> 虽然以上目录下的文件变更不会触发应用自动重启，但是会触发应用实时重载，也能达到快速刷新的效果，具体介绍见后续的实时重载一节。

如果不想使用以上默认值，则可以通过 spring.devtools.restart.exclude 参数自定义排除项，如以下配置示例所示。

```
spring:
  devtools:
    restart:
      exclude: static/**,public/**
```

这里只排除 static 和 public 两个目录下的所有文件，它们的变更不会触发应用自动重启。如果要保留所有的默认目录，再添加其他的额外目录，则使用 spring.devtools.restart.additional-exclude 参数即可，这个和配置管理一章中的 Profile 配置用法类似，这里不再赘述。

4. 禁用日志报告

应用自动重启时，每次都会打印变更日志报告，如下图所示。

```
==========================
CONDITION EVALUATION DELTA
==========================


Positive matches:
-----------------

   None


Negative matches:
-----------------

   JacksonHttpMessageConvertersConfiguration.MappingJackson2HttpMessageConverterConfiguration#mappingJackson2H
      Did not match:
         - @ConditionalOnMissingBean (types: org.springframework.http.converter.json.MappingJackson2HttpMessag
```

如果不想应用自动重启时显示报告日志，可以通过配置禁用，如以下配置所示。

```
spring:
  devtools:
    restart:
      log-condition-evaluation-delta: false
```

5. 禁用自动重启

如果需要保留开发者工具的所有功能，但又不想启用应用自动重启功能，则可以通过配置单独禁用自动重启功能，如以下配置所示。

```
spring:
  devtools:
    restart:
      enabled: false
```

这个配置虽然会禁用自动重启及文件监测功能，但还是会初始化自动重启类加载器，自动重启类加载器的存在可能导致一些反序列化方面的问题，比如可能发生以下两个问题：

- 和 Java原生 ObjectInputStream 反序列化类不兼容，可以考虑使用 Spring 中的 ConfigurableObjectInputStream 类并结合 Thread.currentThread().getContextClassLoader() 类加载器。
- 有一些第三方的序列化类库没有考虑上下文类加载器，也可能存在反序列化问题。

所以，如果自动重启类加载器和其他类库存在兼容性问题，则可以在应用启动之前通过系统参数彻底禁用自动重启功能，如以下代码所示。

```
@SpringBootApplication
public class MyApplication {

    public static void main(String[] args) {
        System.setProperty("spring.devtools.restart.enabled", "false");
        SpringApplication.run(MyApplication.class, args);
    }

}
```

10.2.4 实时重载

Spring Boot 开发者工具支持应用文件变更后网页实时重载内容，即不需要手动刷新网页，当类路径下的资源发生变更时可以触发浏览器自动刷新，自动刷新的前提是需要浏览器安装 LiveReload 扩展插件，并开启应用开发者工具中的自动重启功能。

Spring Boot 开发者工具内嵌了一个 LiveReload 服务器，这个可以从应用启动日志中发现，如下图所示。

```
Will secure any request with [org.springframework.security.web.session.DisableEncod
LiveReload server is running on port 35729
Tomcat started on port(s): 8443 (https) 8080 (http) with context path ''
Started Application in 2.785 seconds (process running for 3.162)
Initializing Spring DispatcherServlet 'dispatcherServlet'
Initializing Servlet 'dispatcherServlet'
Completed initialization in 1 ms
```

LiveReload 服务器运行在 35729 端口，用来与浏览器插件进行实时交互。

需要注意的是：

同时只能运行一个 LiveReload 服务器，如果在 IDE 中启动了多个应用，那么只有第一个启动的应用才支持实时重载。

LiveReload 的浏览器扩展插件的下载地址为链接 13。

也可以直接去应用商店快速安装，比如在 Chrome 应用商店中搜索“livereload”，然后选择“添加至 Chrome”即可，如下图所示。

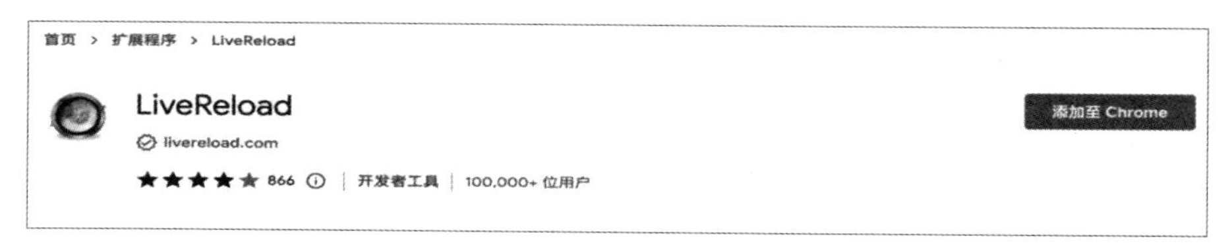

安装完插件之后，在需要自动刷新的网页上面单击 LiveReload 插件图标，让图标从“圆心”变成“实心”状态，如下图所示。

圆心图标表示该网页未开启自动刷新功能，实心图标则表示该网页开启了自动刷新功能，开启自动刷新功能后会发现网络面板出现了 WebSocket 长连接请求，如下面两个图所示。

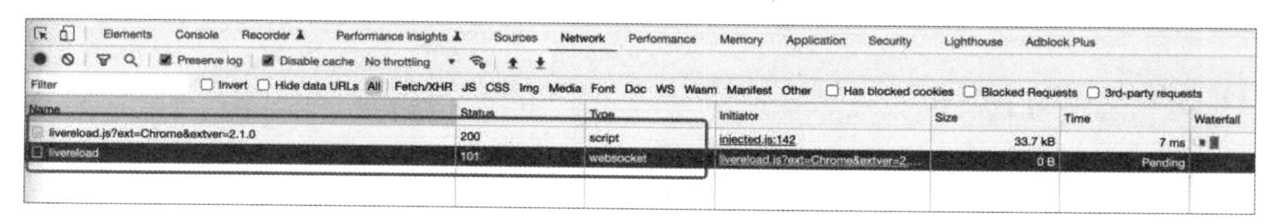

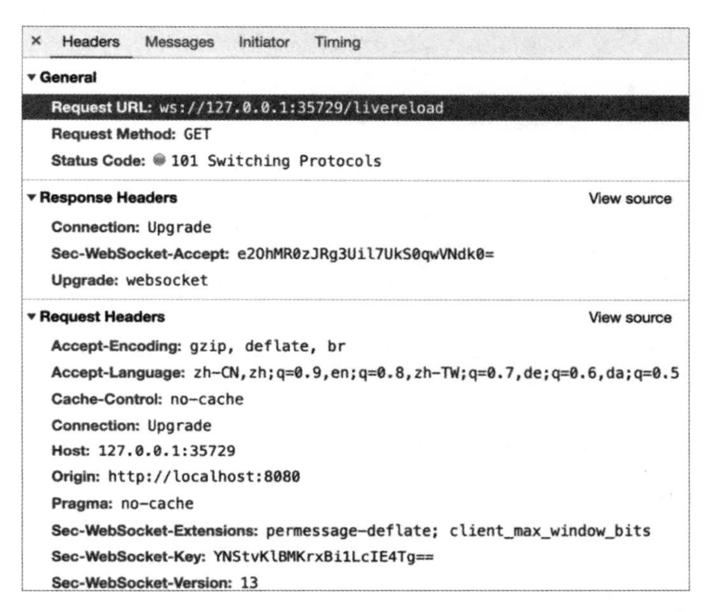

URL请求地址中的35729便是LiveReload服务器的端口，和LiveReload服务器建立长连接后将实时动态通知浏览器插件，这便是浏览器能自动刷新的原因。随便修改一处代码，自动重启之后网页就会自动刷新。

再回到自动重启一节中提到的以下资源目录：

- /META-INF/maven。
- /META-INF/resources。
- /resources
- /static。
- /public。
- /templates。

前面提到，修改这些目录下的资源虽然默认不会触发应用自动重启，但是会触发应用实时重载，因为开发者工具默认会禁用各种缓存，所以不需要自动重启服务器，这些目录下的资源被修改后，网页就可以实时重载最新的内容。

如果不想开启实时重载，则可以通过配置禁用，如以下配置所示。

```
spring:
  devtools:
```

```
    livereload:
      enabled: false
```

10.2.5 全局配置

在实际开发工作中，同时开发的项目可能会非常多，现在都是微服务多模块开发，一个服务就是一个模块，即一个项目，如果逐个地配置开发者工具会比较麻烦，也不好维护管理，所以开发者工具也支持全局配置，这样所有 Spring Boot 项目都能应用生效。

全局配置文件位于 $HOME/.config/spring-boot 目录下，配置文件名为 spring-boot-devtools，支持 .properties 和 .yaml 两种配置文件类型，可以是以下几种文件后缀：

- spring-boot-devtools.properties。
- spring-boot-devtools.yaml。
- spring-boot-devtools.yml。

$HOME 默认指的是用户的主目录，如果要自定义这个目录，则可以通过以下两种方式设置：

- 设置 SPRING_DEVTOOLS_HOME 环境变量。
- 设置 spring.devtools.home 系统参数。

笔者这里添加了一个 spring-boot-devtools.yml 全局配置文件，禁用自动重启日志报告，以及指定了一个触发文件，如以下配置示例所示。

```
spring:
 devtools:
   restart:
     log-condition-evaluation-delta: false
     trigger-file: ".reloadtrigger"
```

建议使用全局配置文件，这样就不用在每个项目中单独配置开发者工具了。

需要注意的是：

（1）spring-boot-devtools 配置文件不支持任何基于 Profile 的配置。

（2）如果项目同时涉及 Spring Boot 新老版本开发，则可以使用 $HOME/.spring-boot-devtools.properties 配置文件，新老版本可以兼容这种配置文件。

.spring-boot-devtools.properties 配置文件示例如下：

```
spring.devtools.restart.log-condition-evaluation-delta=false
spring.devtools.restart.trigger-file=.reloadtrigger
```

注意：.properties 配置不能带双引号。

10.3　单元测试

10.3.1　概述

Spring Boot 提供了丰富的测试功能，主要由以下两个模块组成：

- spring-boot-test：提供测试核心功能。
- spring-boot-test-autoconfigure：提供对测试的自动配置。

Spring Boot 提供了一个 spring-boot-starter-test 一站式启动器，如以下依赖配置所示。

```
<dependency>
    <groupId>org.springframework.boot</groupId>
    <artifactId>spring-boot-starter-test</artifactId>
    <scope>test</scope>
</dependency>
```

测试启动器依赖不仅包含以上两个 Spring Boot 模块，还包含 Spring Test 测试模块，以及其他第三方测试类库，如下所示。

- JUnit 5：Java 最主流的单元测试框架。
- AssertJ：一款快速断言库。
- Hamcrest：一款单元测试匹配库。
- Mockito：一款 Mock 测试框架。
- JSONassert：一款 JSON 断言库。
- JsonPath：一款 JSON XPath 库。

更多测试相关的依赖可见具体的依赖关系树，如下图所示。

```
∨ spring-boot-starter-test : 3.0.0 [test]
    > assertj-core : 3.23.1 [test]
      hamcrest : 2.2 [test]
    > jakarta.xml.bind-api : 4.0.0 [test]
    > json-path : 2.7.0 [test]
    > jsonassert : 1.5.1 [test]
    > junit-jupiter : 5.9.1 [test]
    > mockito-core : 4.8.1 [test]
    > mockito-junit-jupiter : 4.8.1 [test]
    > spring-boot-starter : 3.0.0 [compile]
    > spring-boot-test : 3.0.0 [test]
    > spring-boot-test-autoconfigure : 3.0.0 [test]
    > spring-core : 6.0.2 [compile]
    > spring-test : 6.0.2 [test]
    > xmlunit-core : 2.9.0 [test]
```

以上这些都是 Spring Boot 提供的常用的测试类库，如果上面的测试类库还不能满足你的需要，也可以任意添加以上没有的类库。

现在基本上使用的是 JUnit 5，如果应用还在使用 JUnit 4 写的单元测试用例，那么也可以使用 JUnit 5 的 Vintage 引擎来运行，如下面的依赖配置所示。

```
<dependency>
    <groupId>org.junit.vintage</groupId>
    <artifactId>junit-vintage-engine</artifactId>
    <scope>test</scope>
    <exclusions>
        <exclusion>
            <groupId>org.hamcrest</groupId>
            <artifactId>hamcrest-core</artifactId>
        </exclusion>
    </exclusions>
</dependency>
```

需要排除 hamcrest-core 依赖，因为该依赖已经改坐标了，并且默认内置在 Spring Boot 依赖管理中，如上面的依赖关系树所示，最新的 Hamcrest 依赖已经是 org.hamcrest:hamcrest 坐标了。

Spring Boot提供了一个 @SpringBootTest 注解，用在单元测试类上以启用支持 Spring Boot 特性的单元测试，如果使用的是 JUnit 4，那么测试类上还需要额外的 @RunWith(SpringRunner. class) 注解，然后在测试类方法上添加 @Test 注解即可，每一个 @Test 注解修饰的方法就是一个单元测试方法。

@SpringBootTest 注解有一个最重要的 webEnvironment 环境参数，支持以下几种环境设置：

- **MOCK（默认）：** 加载一个 Web ApplicationContext 并提供一个 Mock Web Environment，但不会启动内嵌的 Web 服务器，并可以结合 @AutoConfigureMockMvcor 和 @AutoConfigure-WebTestClient 注解一起使用进行 Mock 测试。

- RANDOM_PORT：加载一个 WebServerApplicationContext，以及提供一个真实的 Web Environment，并以随机端口启动内嵌服务器。
- DEFINED_PORT：和 RANDOM_PORT 一样，不同的是 DEFINED_PORT 是以应用指定的端口运行的，默认端口为 8080。
- NONE：加载一个 ApplicationContext，但不会提供任何 Web Environment。

如果使用的 @SpringBootTest 注解不带任何参数，则默认为 Mock 环境。

10.3.2 真实环境测试

在 @SpringBootTest 注解中指定基于随机端口的真实 Web 环境，然后在类成员变量或者方法参数上注入 TestRestTemplate 实例，就可以完成对 Spring MVC 接口的真实环境测试。

下面是一个基于随机端口的真实环境的测试用例：

```
@SpringBootTest(webEnvironment = SpringBootTest.WebEnvironment.RANDOM_PORT)
public class MvcTest {

    @Test
    public void getUserTest(@Autowired TestRestTemplate testRestTemplate) {
        Map<String, String> multiValueMap = new HashMap<>();
        multiValueMap.put("username", "Java 技术栈 ");
        Result result = testRestTemplate.getForObject("/user/get?username={username}",
                Result.class, multiValueMap);
        assertThat(result.getCode()).isEqualTo(0);
        assertThat(result.getMsg()).isEqualTo("ok");
    }

}
```

测试当前应用下的 /user/get 接口，传入对应的用户名参数，最后检查接口返回结果是否和预期一致，测试结果如下图所示。

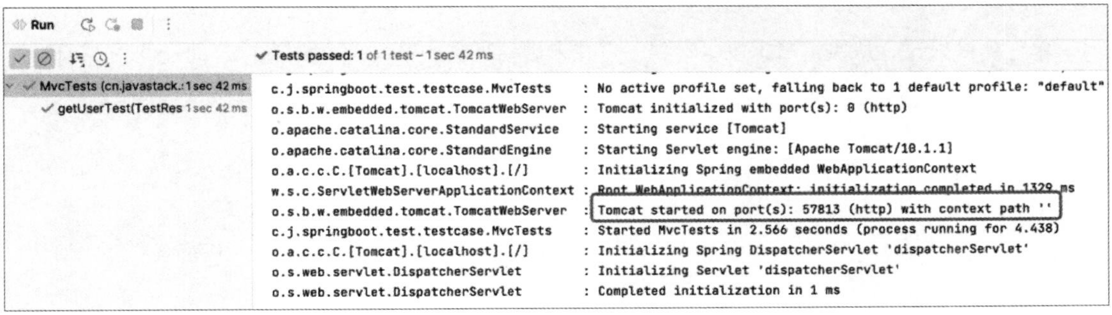

```
Run
Tests passed: 1 of 1 test – 1 sec 42 ms
MvcTests (cn.javastack.: 1 sec 42 ms
  getUserTest(TestRes 1 sec 42 ms
c.j.springboot.test.testcase.MvcTests      : No active profile set, falling back to 1 default profile: "default"
o.s.b.w.embedded.tomcat.TomcatWebServer    : Tomcat initialized with port(s): 0 (http)
o.apache.catalina.core.StandardService     : Starting service [Tomcat]
o.apache.catalina.core.StandardEngine      : Starting Servlet engine: [Apache Tomcat/10.1.1]
o.a.c.c.C.[Tomcat].[localhost].[/]         : Initializing Spring embedded WebApplicationContext
w.s.c.ServletWebServerApplicationContext   : Root WebApplicationContext: initialization completed in 1329 ms
o.s.b.w.embedded.tomcat.TomcatWebServer    : Tomcat started on port(s): 57813 (http) with context path ''
c.j.springboot.test.testcase.MvcTests      : Started MvcTests in 2.566 seconds (process running for 4.438)
o.a.c.c.C.[Tomcat].[localhost].[/]         : Initializing Spring DispatcherServlet 'dispatcherServlet'
o.s.web.servlet.DispatcherServlet          : Initializing Servlet 'dispatcherServlet'
o.s.web.servlet.DispatcherServlet          : Completed initialization in 1 ms
```

单元测试通过，从执行日志可以看到，它启动了一个嵌入式的 Tomcat 容器来测试真实的 Web 应用环境。

10.3.3 Mock 环境测试

通过在类上面使用 @AutoConfigureMockMvc 注解，然后在类成员变量或者方法参数上注入 MockMvc 实例，就可以完成对 Spring MVC 接口的 Mock 测试。

下面是一个基于默认 Mock 环境的测试用例：

```
@SpringBootTest
@AutoConfigureMockMvc
class MockMvcTests {

    @Test
    public void getUserTest(@Autowired MockMvc mvc) throws Exception {
        mvc.perform(MockMvcRequestBuilders.get("/user/get?username={username}", "test"))
                .andExpect(status().isOk())
                .andExpect(content().string("{\"code\":0,\"msg\":\"ok\",\"
data\":\"test\"}"));
    }

}
```

测试当前应用下的 /user/get 接口，传入对应的用户名参数，最后检查请求状态是否 OK（200），响应的内容是否和预期一致，测试结果如下图所示。

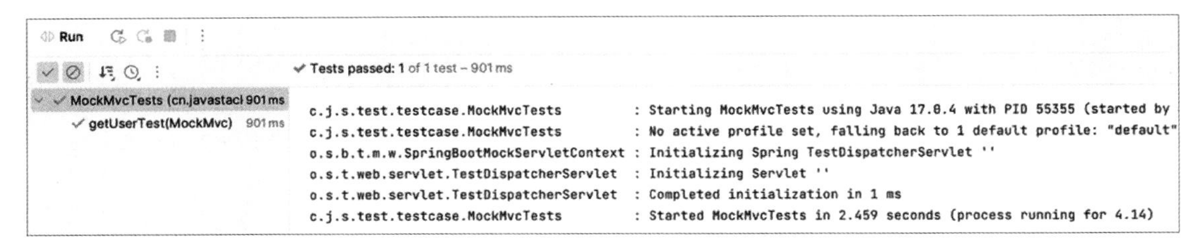

单元测试通过，从执行日志可以看到，它并未启动真实的 Web 环境来测试，而是使用 Mock 环境测试的。

10.3.4 Mock 组件测试

某些时候可能还需要模拟一些组件，比如某些服务只有上线之后才能调用，在开发阶段不可用，这时就需要 Mock 模拟测试了，提供各种模拟组件以完成测试。

Spring Boot 提供了一个 @MockBean 注解，可为 Spring 中的 Bean 组件定义基于 Mockito 的 Mock 测试，它可以创建一个新 Bean 以覆盖 Spring 环境中已有的 Bean，它可以用在测试类、成员变量上，或者 @Configuration 配置类、成员变量上，被模拟的 Bean 在每次测试结束后自动重置。

假现现在有一个远程的服务 userService，本地不能调用，现在进行 Mock 测试，如以下使用示例所示。

```
@SpringBootTest
class MockBeanTests {

//    @Autowired
//    private UserService userService;

    @MockBean
    private UserService userService;

    @Test
    public void countAllUsers() {
        BDDMockito.given(this.userService.countAllUsers()).willReturn(88);
        assertThat(this.userService.countAllUsers()).isEqualTo(88);
    }

}
```

这里的 @MockBean 注解使用在 UserService 变量上，表明这个 userService 实例在当前测试用例中是被 Mock 覆盖的，如果要模拟的 Bean 有多个，则可以使用 @Qualifier 注解指定，然后通过 Mockito 提供的代理工具类方法创建模拟返回数据，运行该服务的测试方法，当模拟数据和预期结果一致时才会测试通过。

这里通过 BDDMockito 工具类模拟 userService#countAllUsers 方法并让它返回统计的用户总数（88），最后检查该方法的返回值是否和预期一致，测试结果如下图所示。

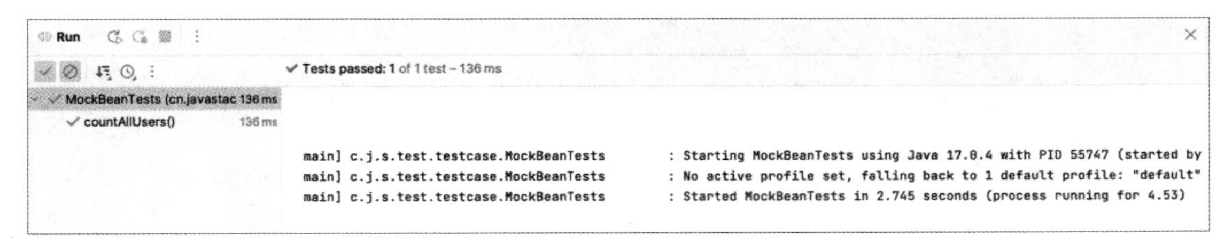

单元测试通过，也可以使用 @SpyBean 注解代替 @MockBean 注解，两者的区别是：

- @SpyBean—如果没有提供 Mockito 代理方法，则会调用真实的 Bean 来获取数据。

- @MockBean—不管有没有提供 Mockito 代理方法，都会调用 Mock 的 Bean 来获取数据。

@MockBean、@SpyBean 注解既可作用于 Mock 环境，也可作用于真实环境，它只是用来模拟、替换环境中指定的 Bean 而已，但不能用于模拟在应用上下文刷新期间 Bean 的行为，因为在执行测试用例时应用上下文已经刷新完成了，所以不可能再去模拟了，这种情况下建议使用 @Bean 方法来创建模拟配置。

10.3.5 技术框架测试

除了支持 Spring Boot 环境的单元测试，Spring Boot 还提供了各种技术组件的单元测试配置，比如 JSON、Spring MVC、WebFlux、Spring Data JAP、Spring Data JDBC、Spring Data Redis 等。

比如想测试 JSON 序列化 / 反序列化功能，其自动配置类为 JsonTestersAutoConfiguration，它提供了对各个 JSON 框架的测试支持，只需要在单元测试类上添加一个 @JsonTest 注解，注入对应的 JacksonTester、GsonTester、JsonbTester 实例就可以直接测试。

首先在资源目录下创建 jack.json 文件，JSON 文件的内容如下所示。

```
{"id":10001, "name":"Jack", "birthday": "2000-10-08 21:00:00"}
```

以下是使用 Jackson 序列化 / 反序列化的测试示例：

```
@JsonTest
class JsonTests {

    @Autowired
    private JacksonTester<User> json;

    @Test
    void serialize() throws Exception {
        User user = new User(10001L, "Jack",
                LocalDateTime.of(2000, 10, 8, 21, 0, 0));
        System.out.println(this.json.write(user));
        assertThat(this.json.write(user)).isEqualToJson("/jack.json");
        assertThat(this.json.write(user)).hasJsonPathStringValue("@.name");
        assertThat(this.json.write(user)).
                extractingJsonPathStringValue("@.name").isEqualTo("Jack");
    }

    @Test
    void deserialize() throws Exception {
```

```
        String content = "{\"id\":10002, \"name\":\"Petty\", \"birthday\":
\"2021-01-21T02:32:00\"}";
        assertThat(this.json.parse(content))
                .isEqualTo(new User(10002L, "Petty",
                        LocalDateTime.of(2021, 1, 21, 2, 32, 0)));
        assertThat(this.json.parseObject(content).getName()).isEqualTo("Petty");
    }

}
```

这里提供了两个测试方法：

- **serialize**：测试 JSON 序列化功能，判断序列化后的内容和资源目录下的 jack.json 文件的内容是否一致，是否有 name 这个字段，并且该字段值是否为 Jack。
- **deserialize**：测试 JSON 反序列化功能，判断反序列化后的用户对象和新创建的用户对象是否一致，并且反序列化后的用户对象的 name 值是否为 Petty。

测试结果如下面的两个图所示。

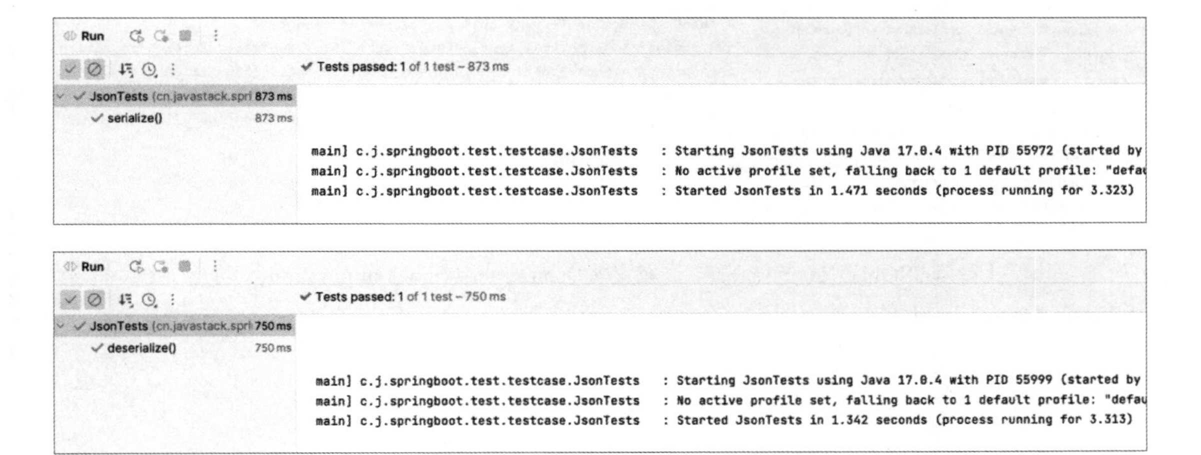

更多技术框架的自动配置可以参考 spring-boot-test-autoconfigure 测试自动配置包中的自动配置文件，如下图所示。

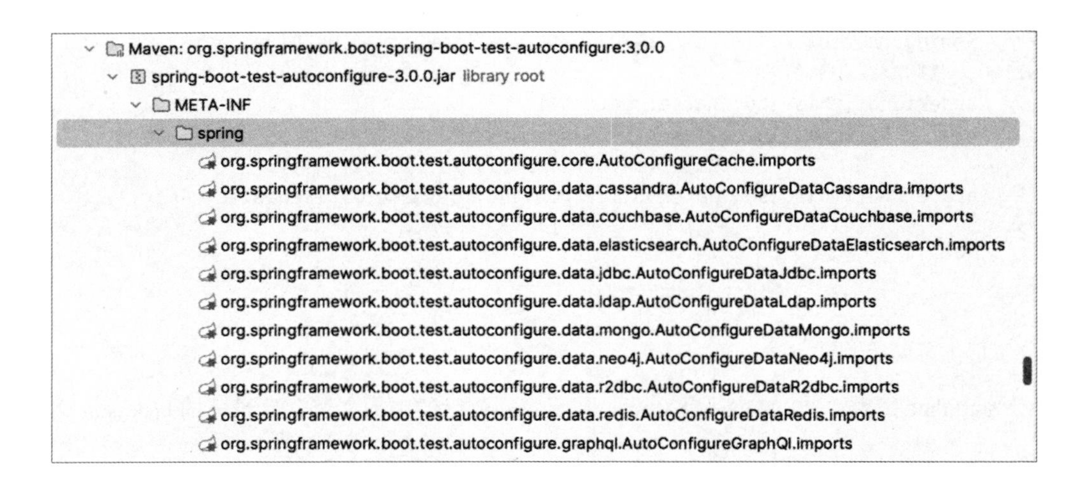

比如上面测试的 JSON 序列化和反序列化功能，它注册了 Spring Boot 所支持的几种 JSON 工具类的自动配置类，如下图所示。

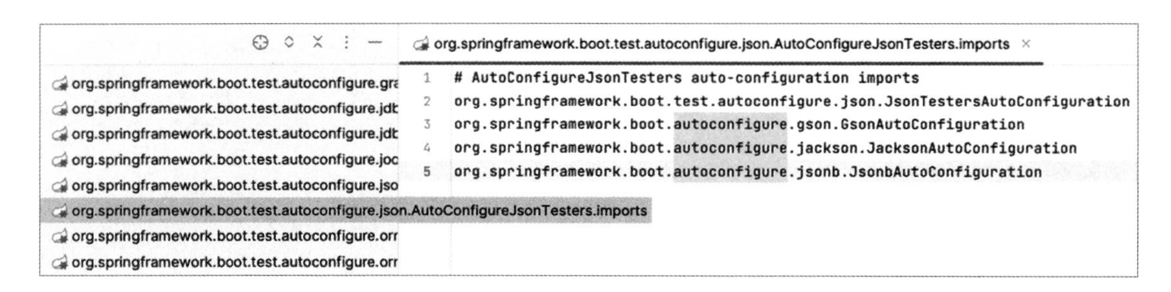

这里可以看到 JSON 测试支持的自动配置类 JsonTestersAutoConfiguration，以及 JSON 框架 Jackson、Gson、Jsonb 的自动配置类都在此处进行了注册。

第 11 章 Spring Boot 打包与部署

Spring Boot 应用部署到生产环境中的方式多种多样，既支持嵌入式容器直接运行应用，也支持将应用部署到外置的 Servlet 容器中。当然，重点是嵌入式容器直接运行应用的方式，这也是 Spring Boot 的核心特性，这种方式可以直接使用 java 命令运行应用，也可以免去 java 命令直接像运行系统脚本一样运行应用，或者直接以操作系统服务运行应用，既可以整包运行应用，也可以拆包运行应用，还可以将应用生成 Docker 镜像部署到 Docker 容器中，或者将应用部署到云平台中，Spring Boot 3.0 还支持以 GraalVM 原生镜像运行应用，可以说几乎满足所有的部署场景的需要。本章从应用打包开始，讲解上述涉及的所有内容。

11.1 应用打包（jar）

11.1.1 概述

传统的 Java 应用都需要打一个 war 包，然后把它部署到 tomcat/webapps 目录下运行，虽然 Java 支持把应用打成 jar 包，但它并没有提供一个标准的方式把工程代码打包成可执行的 Java Web 应用，因为它没有提供内置的 Servlet 容器。

有了 Spring Boot 框架之后，这一切都变得简单，开发者可以很方便地将 Spring Boot 应用打成一个可执行 jar 包部署，同时支持打成 war 包部署，但 Spring Boot 首推以 jar 的形式打包和发布应用，因为使用 jar 包部署比较灵活，方便应用快速启动和扩容。

11.1.2 快速打包

基于 Maven 的应用，如果不指定 packaging 属性，那么默认就是 jar 打包方式，如以下配置所示。

```
<packaging>jar</packaging>
```

packaging 默认就是 jar 类型，可以省略不用指定。另外还需要引入 spring-boot-maven-plugin 打包插件，如以下配置所示。

```
<build>
    <plugins>
        <plugin>
            <groupId>org.springframework.boot</groupId>
            <artifactId>spring-boot-maven-plugin</artifactId>
        </plugin>
    </plugins>
</build>
```

然后使用 mvn package 或者 mvn install 命令打包，如以下参考命令所示。

```
$ mvn package
$ mvn install
```

两个命令二选一即可，前者 package 是只打包到应用的 target 目录下，后者 install 是同时打包到应用的 target 目录下并安装到本地 Maven 仓库中，如果是打包为可执行的应用，则建议使用 package 命令。

在开发工具中可以省略 mvn 命令的前缀，如下图所示。

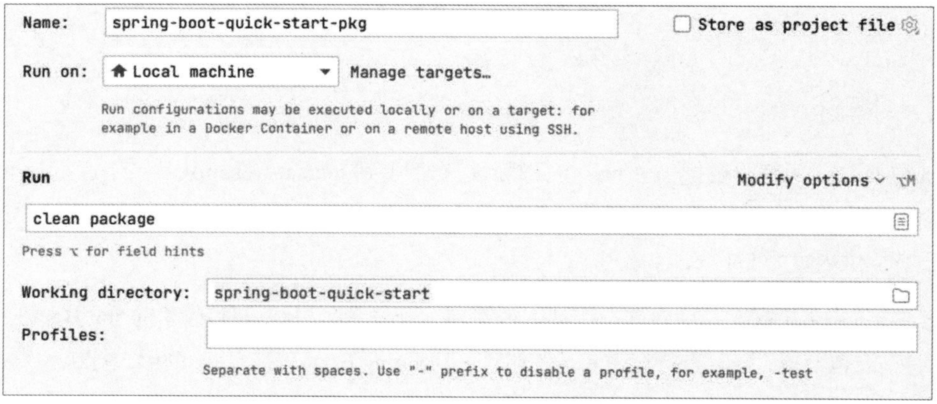

打包命令一般可以结合 clean 使用，即先清理之前的包再打包新的包，执行打包命令后会在当

前应用 target 目录下生成两个 jar 包，如下图所示。

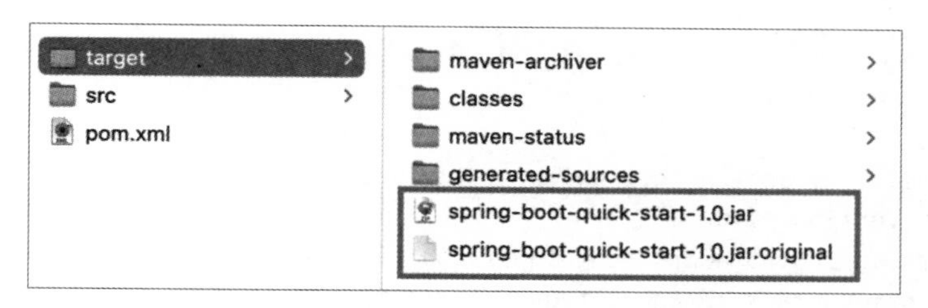

应用先打包为 spring-boot-quick-start-1.0.jar 原始 jar 包，包含所有依赖项，它是不能直接运行的，再通过 spring-boot-maven-plugin 插件中的 repackage 目标重新打包为可执行的 jar 包，然后重命名原始 jar 包为 *.jar.original。

第一个 jar 包（spring-boot-quick-start-1.0.jar）是可以直接运行的 jar 包，可以通过 java 命令运行，如以下运行命令所示。

```
$ java -jar spring-boot-quick-start-1.0.jar
```

这个 jar 包包括了所有编译完的 class 类文件，还内置了 Servlet 容器，以及所有需要依赖的其他 jar 包等，这个 jar 包的目录结构如下：

```
.
├── BOOT-INF
│   ├── classes
│   │   └── cn
│   │       └── javastack
│   │           └── springboot
│   │               └── quckstart
│   └── lib
├── META-INF
│   └── maven
│       └── cn.javastack
│           └── spring-boot-quick-start
└── org
    └── springframework
        └── boot
            └── loader
                ├── archive
                ├── data
                ├── jar
                ├── jarmode
                └── util
```

在 META-INF 目录下可以找到 MANIFEST.MF 描述文件，内容如下所示。

```
Manifest-Version: 1.0
Created-By: Maven JAR Plugin 3.3.0
Build-Jdk-Spec: 17
Implementation-Title: spring-boot-quick-start
Implementation-Version: 1.0
Main-Class: org.springframework.boot.loader.JarLauncher
Start-Class: cn.javastack.springboot.quckstart.Application
Spring-Boot-Version: 3.0.0
Spring-Boot-Classes: BOOT-INF/classes/
Spring-Boot-Lib: BOOT-INF/lib/
Spring-Boot-Classpath-Index: BOOT-INF/classpath.idx
Spring-Boot-Layers-Index: BOOT-INF/layers.idx
```

它包含了一个 jar 包的相关信息，从中就可以找到应用的引导类和启动类，这个在第 4 章 Spring Boot 启动过程分析中已有详细介绍，这里不再赘述。

第二个 jar 包是包含了所有代码和资源文件的普通 jar 包，这个 jar 包的目录结构如下所示。

```
.
├── cn
│   └── javastack
│       └── springboot
│           └── quckstart
└── META-INF
    └── maven
        └── cn.javastack
            └── spring-boot-quick-start
```

再来看其 META-INF 目录下的 MANIFEST.MF 描述文件内容：

```
Manifest-Version: 1.0
Created-By: Maven JAR Plugin 3.3.0
Build-Jdk-Spec: 17
Implementation-Title: spring-boot-quick-start
Implementation-Version: 1.0
```

它没有包含应用启动入口类，只是作为一个普通的 jar 包存在，不能使用 java 命令运行，一般作为依赖被传递到其他应用中。不是所有 Spring Boot 应用都需要部署运行，比如一些通用模块的应用，它们只是作为普通 jar 包集成到可执行的 jar 包中使用而已。

11.1.3　自定义打包

如果应用使用的是继承 spring-boot-starter-parent 父依赖的方式，那么它提供了 spring-boot-maven-plugin 打包插件的基本配置，如以下配置所示。

```
<plugin>
  <groupId>org.springframework.boot</groupId>
  <artifactId>spring-boot-maven-plugin</artifactId>
  <executions>
    <execution>
      <id>repackage</id>
      <goals>
        <goal>repackage</goal>
      </goals>
    </execution>
  </executions>
  <configuration>
    <mainClass>${start-class}</mainClass>
  </configuration>
</plugin>
```

它默认提供了 repackage 打包目标、配置应用启动类，所以应用可以直接继承父项目提供的 spring-boot-maven-plugin 插件，不用做任何配置，只需要坐标即可，如下面的配置示例所示。

```
<build>
    <plugins>
        <plugin>
            <groupId>org.springframework.boot</groupId>
            <artifactId>spring-boot-maven-plugin</artifactId>
        </plugin>
    </plugins>
</build>
```

如果有需要，还可以定制配置，常用的定制配置参数如下：

（1）layout：指定打包的项目类型，根据应用原有类型默认打包为 jar、war 包，具体可以设置以下几种。

- JAR：可执行的 jar 包。
- WAR：可执行的 war 包。
- ZIP：和 jar 包相似。

- NONE：打包所有依赖项和项目资源，但不绑定任何启动加载器。

（2）mainClass：指定 Spring Boot 的启动类，如果启动类在根目录下则不用指定。

（3）classifier：指定可执行 jar 包的扩展标识符，比如下面的示例，对 jar 包名称扩展标识符进行定制。

```
<build>
    <plugins>
        <plugin>
            <groupId>org.springframework.boot</groupId>
            <artifactId>spring-boot-maven-plugin</artifactId>
            <configuration>
                <classifier>exec</classifier>
            </configuration>
        </plugin>
    </plugins>
</build>
```

重新打包后，再看一下 jar 包的名称，如下图所示。

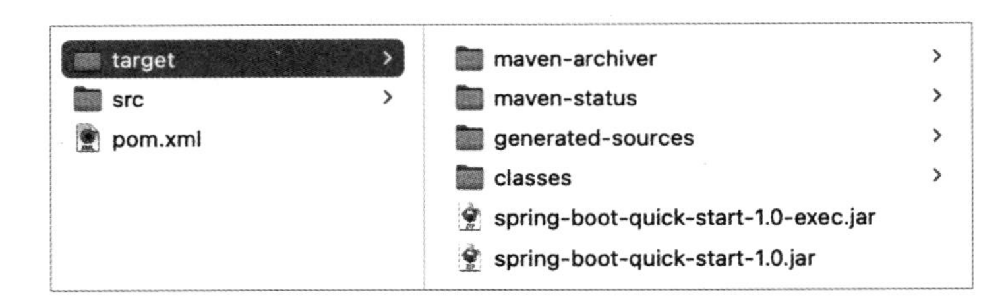

如上图所示，可执行 jar 包带上了插件配置的 exec 标识符，有了这个标识符，就可以用来区分是可执行的 jar 包，还是普通的 jar 包。

这就是继承 spring-boot-starter-parent 父依赖的好处，如果使用导入 spring-boot-dependencies 依赖的集成方式，则需要在应用中提供插件的一整套配置，如以下配置示例所示。

```
<build>
    <plugins>
        <plugin>
            <groupId>org.springframework.boot</groupId>
            <artifactId>spring-boot-maven-plugin</artifactId>
            <executions>
              <execution>
                <id>repackage</id>
                <goals>
```

```
                <goal>repackage</goal>
            </goals>
        </execution>
    </executions>
    <configuration>
        <mainClass>${start-class}</mainClass>
        <classifier>exec</classifier>
    </configuration>
</plugin>
</plugins>
</build>
```

所以，应用选用哪种 Spring Boot 集成方式很重要，如果打包失败，则需要检查应用使用的是哪种集成方式。

11.2 应用打包（war）

11.2.1 概述

Spring Boot 也支持将应用打包成 war 包，这样应用就不会内置 Servlet 容器，也不能直接运行，需要部署到外部的 Servlet 容器中。另外，基于 WebFlux 的响应式 Web 应用不会严格依赖 Servlet API，并且应用默认嵌入在 Reactor Netty 服务器中部署，所以 WebFlux 应用不支持 war 包部署。

应用打包成 war 包的好处是，可以让运维人员用熟悉的方式来统一管理和维护外置的 Servlet 容器，从而降低了容器管理的复杂性，如果是微服务，那么建议打包为 jar 包，方便快速动态扩容。

11.2.2 配置 war 包

1. 继承 SpringBootServletInitializer 抽象类

配置 war 包的第一步是继承 SpringBootServletInitializer 抽象类，并重写其 configure 配置方法，如以下示例所示。

```
@SpringBootApplication
public class Application extends SpringBootServletInitializer {

    @Override
    protected SpringApplicationBuilder configure(SpringApplicationBuilder application) {
```

```
        return application.sources(Application.class);
    }

}
```

2. 修改打包类型

在 Maven pom.xml 配置文件中，修改打包类型为 war，如以下配置所示。

```
<packaging>war</packaging>
```

3. 添加插件

如果是继承 pring-boot-starter-parent 父依赖的集成方式，那么 Spring Boot 也提供了一个 maven-war-plugin 插件，用于打包 war 应用，如以下配置所示。

```
<plugin>
  <groupId>org.apache.maven.plugins</groupId>
  <artifactId>maven-war-plugin</artifactId>
  <configuration>
    <archive>
      <manifest>
        <mainClass>${start-class}</mainClass>
        <addDefaultImplementationEntries>true</addDefaultImplementationEntries>
      </manifest>
    </archive>
  </configuration>
</plugin>
```

同理，继承的插件都提供了默认配置，在应用中只需要引入坐标即可，如以下配置所示。

```
<build>
    <plugins>
        <plugin>
            <groupId>org.apache.maven.plugins</groupId>
            <artifactId>maven-war-plugin</artifactId>
        </plugin>
    </plugins>
</build>
```

所以，如果使用导入 spring-boot-dependencies 依赖的集成方式，则需要在应用中提供插件的一整套配置。

4. 排除嵌入式容器

Spring Boot 默认使用的是 Tomcat 嵌入式容器，打包成 war 包需要排除嵌入式容器，以免应用部署到外部 Tomcat 时发生冲突，排除配置如下所示。

```
<dependencies>
    <!-- 打 war 包后不携带嵌入式 Tomcat 容器 -->
    <dependency>
        <groupId>org.springframework.boot</groupId>
        <artifactId>spring-boot-starter-tomcat</artifactId>
        <scope>provided</scope>
    </dependency>
</dependencies>
```

因为在编译启动类时，继承的 SpringBootServletInitializer 类还是会用到 Servlet 相关的依赖，所以把嵌入式 Tomcat 依赖的 scope 范围设置为 provided，表示该依赖只在编译和测试时才有效，但不会出现在外部 Servlet 的 war 包中，因为外部 Tomcat 自身会提供相关类库。

5. 兼容嵌入式容器（可选）

如果既要支持将 war 包部署到外部 Servlet 容器中，又要支持直接运行 war 包，则可以保留应用启动类中的 main 方法，并把公共启动方法抽出来，然后两边同时调用公共方法，如下面的示例所示。

```
@SpringBootApplication
@RestController
public class Application extends SpringBootServletInitializer {

    @Override
    protected SpringApplicationBuilder configure(SpringApplicationBuilder builder) {
        return startBuilder(builder);
    }

    public static void main(String[] args) {
        startBuilder(new SpringApplicationBuilder()).run(args);
    }

    private static SpringApplicationBuilder startBuilder(SpringApplicationBuilder builder) {
        return builder.sources(Application.class);
    }

}
```

同时添加 Spring Boot 的 spring-boot-maven-plugin 打包插件，这个插件是专门用来打 jar 包的，虽然上面把嵌入式容器排除了，但只是在外部 Servlet 的 war 包中排除了，此插件会再生成一个可以直接运行的 war 包，排除的嵌入式容器会以 lib-provided 目录出现在可以直接运行的 war 包中，这样就可以同时兼容嵌入式容器和外部容器。

11.2.3 开始打包

在开发工具中使用 clean packge 命令将应用打包后，前往应用的 target 目录，如下图所示。

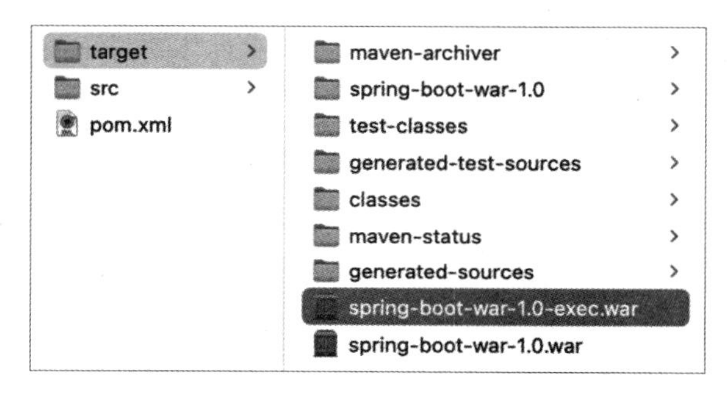

这里出现了两个 war 包，第一个 war 包（spring-boot-war-1.0-exec.war）是可以同时部署 jar 和 war 二合一的包，如果不配置兼容嵌入式容器，则第一个 war 包就不会生成。

这个 war 包的目录结构如下：

```
.
├── META-INF
│   └── maven
│       └── cn.javastack
│           └── spring-boot-war
├── org
│   └── springframework
│       └── boot
│           └── loader
│               ├── archive
│               ├── data
│               ├── jar
│               ├── jarmode
│               └── util
└── WEB-INF
    ├── classes
```

```
│   └── cn
│       └── javastack
├── lib
└── lib-provided
```

provided 范围的依赖会放到 lib-provided 目录下，如下图所示。

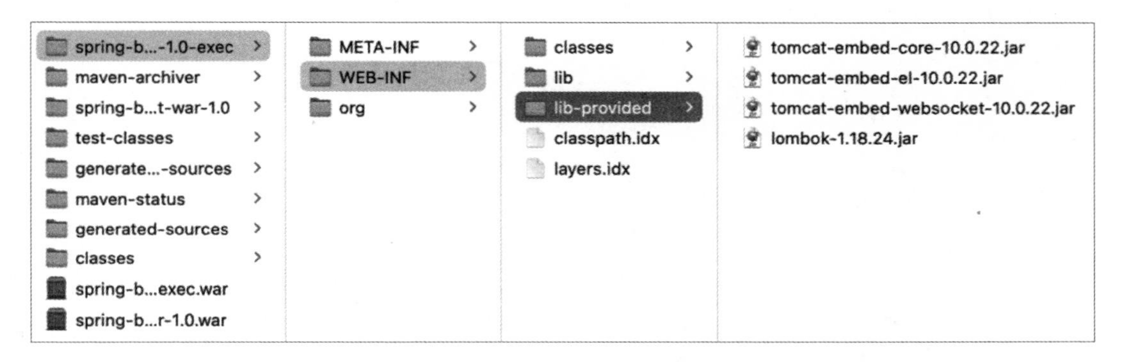

然后通过后面两个 idx 索引文件在 MANIFEST.MF 描述文件中指定相关文件：

- classpath.idx：引入了一系列 jar 包的 classpath 索引文件。
- layers.idx：分层索引文件。

这两个索引文件只有在打可直接运行的包时才会出现，这个 war 包既包含了直接运行 Spring Boot 应用的所有文件，包括嵌入式容器相关包（lib-provided），又包含了部署到外部 Servlet 容器中的所有文件，所以既可以像之前运行 jar 包一样直接运行可执行的 war 包，又可以作为 war 包部署到其他 Servlet 容器中。

比如直接运行这个 war 包，参考命令如下：

```
$ java -jar spring-boot-war-1.0-exec.war
```

这个 war包也可以直接放到外部 Servlet 容器中运行，比如放在 Tomcat 中的 webapps 目录下，启动 Tomcat 就可以自动解压缩并部署应用了，如下图所示。

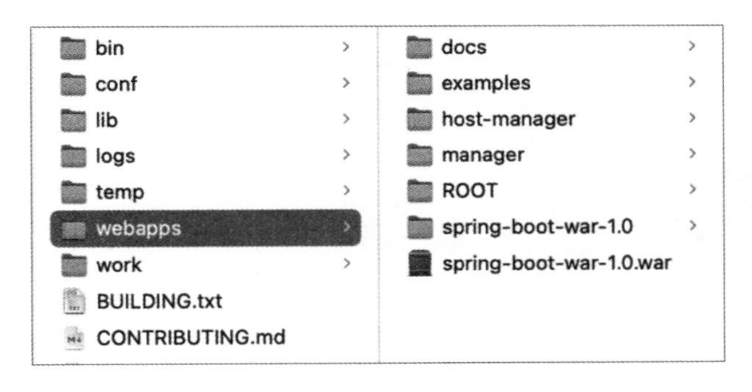

第二个 war包（spring-boot-war-1.0.war）便是仅部署到外部 Servlet 容器中的 war 包，如果不配置兼容嵌入式容器，那么只会生成这个 war 包。

这个 war 包的目录结构如下：

```
.
├── META-INF
│   └── maven
│       └── cn.javastack
│           └── spring-boot-war
└── WEB-INF
    ├── classes
    │   └── cn
    │       └── javastack
    └── lib
```

它仅包含了部署到外部 Servlet 容器中的类库，和第一个 war 包一样直接放到外部 Servlet 容器中运行，但是不能使用 java -jar 命令直接运行。

如果把 war包部署到外部 Servlet 容器中，那么 Spring Boot 应用配置文件中的 server.* 参数就不会生效了，相关参数需要在外部容器中配置。

11.3 应用运行（嵌入式容器）

11.3.1 使用 java 命令运行

前面的章节中介绍过，Spring Boot 嵌入式容器的 jar/war 包一般是通过 java -jar 命令运行的，如以下参考命令所示。

```
$ java -jar spring-boot-quick-start-1.0-exec.jar
```

关于java运行命令的更多用法，比如怎么设置 JVM 参数、传递应用参数等，可以参考相关文档，这里不再赘述。

11.3.2 直接运行

Spring Boot 也支持像运行操作系统脚本一样直接运行嵌入式容器的 jar/war 包，如下面的运行命令所示。

```
$ ./spring-boot-quick-start-1.0-exec.jar
```

这种直接运行方式需要在 spring-boot-maven-plugin 打包插件中指定 executable 配置参数为 true，如下面的配置所示。

```
<plugin>
    <groupId>org.springframework.boot</groupId>
    <artifactId>spring-boot-maven-plugin</artifactId>
    <configuration>
        <executable>true</executable>
    </configuration>
</plugin>
```

这样打出来的包既可以使用 java -jar 命令运行，也可以直接运行，它实际上就是一种 zip 格式的 jar 包，打包的时候在 zip 文件的最前面打入一个额外的脚本，这样 Linux 操作系统就会把它当作一个脚本来运行，推荐使用这种打包方式，像运行系统脚本一样运行应用包，更简单、更方便。

打包时默认嵌入的脚本支持大部分 Linux 操作系统，比如 CentOS、Ubuntu，笔者在 macOS Monterey 操作系统上测试了也是可行的，其他像 macOS、FreeBSD 等不支持的操作系统需要在 spring-boot-maven-plugin 插件中配置自定义的 embeddedLaunchScript 脚本，配置格式如下所示。

```
<plugin>
    <groupId>org.springframework.boot</groupId>
    <artifactId>spring-boot-maven-plugin</artifactId>
    <configuration>
        <classifier>exec</classifier>
        <executable>true</executable>
        <embeddedLaunchScript>...</embeddedLaunchScript>
    </configuration>
</plugin>
```

11.3.3 系统服务运行

Spring Boot 可执行的应用包可以使用 init.d 或 systemd 轻松启动并作为 UNIX/Linux 中的服务，即以 Linux 中的服务启动应用，如以下参考命令所示。

```
$ service mybatis start
$ systemctl start mybatis
```

1.init.d

init.d服务在 /etc/init.d 目录下，只需要在该目录下创建一个 Spring Boot 应用的软链接即可启动

init.d 服务，参考命令如下所示。

```
$ sudo ln -s ~/spring-boot-mybatis-1.0-exec.jar /etc/init.d/mybatis
```

软链接创建成功后就可以运行 init.d 服务了，参考命令如下所示。

```
$ sudo service mybatis start
```

2. systemd

因为 init.d 服务是串行启动的，启动时间较长，所以 systemd 是用来取代 init.d 服务的，systemd 的功能也比 init.d 更强大，systemd 现在已经成为各大 Linux 系统的标配。

systemd 服务在 /etc/systemd/system 目录下，只需要在该目录下创建一个 Spring Boot 应用的脚本文件即可启动 systemd 服务，文件格式为 xx.service，比如下面创建一个 mybatis.service 服务的配置文件，参考内容如下所示。

```
[Unit]
Description=mybatis
After=syslog.target

[Service]
User=test
ExecStart=/home/test/spring-boot-mybatis-1.0-exec.jar
SuccessExitStatus=143

[Install]
WantedBy=multi-user.target
```

然后启动服务，参考命令如下：

```
$ sudo systemctl start mybatis
```

将 systemd 服务设置为开机自动启动，参考命令如下：

```
$ sudo systemctl enable mybatis
```

更多的服务操作指令可以参考下表。

服务操作	init.d	systemd
启动服务	service xxx start	systemctl start xxx
停止服务	service xxx stop	systemctl stop xxx
重新加载配置	service xxx force-reload	systemctl daemon-reload
重启服务	service xxx restart	systemctl restart xxx

续表

服务开机启动	chkconfig --level 5 xxx on	systemctl enable xxx
服务开机启动（取消）	chkconfig --level 5 xxx off	systemctl disable xxx
查询所有开机启动服务	chkconfig --list	systemctl list-unit-files | grep enabled
查询服务是否开机启动	chkconfig --list | grep xxx	systemctl is-enabled xxx
查询所有启动失败的服务		systemctl --failed
检查服务状态	service xxx status	systemctl status xxx
删除服务	chkconfig --del xxx	删除对应的配置文件

通过 Linux 中的 init.d 和 systemd 服务可以实现优雅地部署和管理 Spring Boot 应用服务，并且不用担心系统重启而导致应用长时间终止运行。

11.3.4　拆包运行

Spring Boot 除了支持整包运行，还支持拆包运行，即先解压压缩包文件，再执行解压出来的启动类。

首先解压可执行应用包，参考命令如下：

```
$ jar -xf myapp.jar
```

然后调用 java 命令运行 Spring Boot 提供的 JarLauncher 类，参考命令如下：

```
$ java org.springframework.boot.loader.JarLauncher
```

也可以直接运行应用的原生启动类，参考命令如下：

```
$ java -cp BOOT-INF/classes:BOOT-INF/lib/* cn.javastack.mybatis.Application
```

前者可以自动检测应用的启动类，后者则是直接运行启动类，前者更实用，后者的启动时间可能要更短一点。但无论如何，拆包运行方式都要比整包运行方式快，快多少取决于整包的大小，其他方面并没什么差异，不会影响应用的正常运行。

11.4　部署 Docker 容器

11.4.1　概述

Docker 是现在最主流的开源容器，遵循 Apache 2.0 协议，使用 Go 语言开发，始于 2013 年，现已加入 Linux 基金会。Docker 的核心思想是：Build once, Run anywhere，即一次构建，处处运行一可以是一个操作系统、一个 Web 应用，或者一个数据库服务等。

Docker 环境主要包含镜像和容器，镜像只是一个只读的静态文件，不能运行，容器才是镜像的运行实例，所以需要新增一个容器。我们可以将 Docker 理解为一个轻量级的沙盒，Docker 容器完全使用沙盒机制，每个容器内都有其独立运行的环境和组件，容器之间是相互隔离互不影响的，也可以进行容器间的通信。

Docker 支持所有主流的操作系统，如 Linux/macOS/Windows，所有主流的云服务也支持 Docker，所以，我们可以将应用及依赖、环境信息等打包到一个 Docker 容器中，可以很轻松地实现应用的发布、迁移等。

Spring Boot 应用既可以基于一个 Dockerfile 文件构建 Docker 镜像，也可以使用 Maven/Gradle 插件基于 Cloud Native Buildpacks 构建 Docker 镜像。

Spring Boot 既支持基于整个应用 jar 包构建 Docker 镜像，也支持构建分层的 Docker 镜像。不建议在 Docker 镜像中构建整包，由上面的章节知道，整包运行要比拆包运行慢，这在容器环境中就更加明显。另外，将应用的所有代码和依赖库都放到 Docker 镜像的一层中不是最优的，因为平时开发过程中经常变动的可能只有 class 文件代码，其他依赖项很少变动，所以推荐构建分层的 Docker 镜像，这时 Docker 只需要替换变动的层，其他未变动的层可以从缓存中提取，还能以拆包的方式运行应用。

11.4.2 Docker 环境搭建

1. 安装 Docker

Docker 有社区版和企业版，社区版是开源免费的，企业版是收费的，当然企业版也会有更高级的特性和服务。Docker 的安装十分简单，笔者这里以 Mac 和 Docker 社区版为例进行演示。

打开 Docker 开始页，见链接 14。

根据不同的操作系统直接下载并安装 Docker Desktop 桌面管理工具，它包含了一个 Docker 可视化开发工具和一个本地的 Kubernates 环境。

> Linux 系统只支持 Ubuntu、Debian、Fedora 三个平台，其他没有界面的、不支持安装 Docker Desktop 的 Linux 系统可以独立安装 Docker 引擎，安装链接为链接 15。

安装完后使用 docker -v 命令可以查看 Docker 版本：

```
$ docker -v
Docker version 20.10.20, build 9fdeb9c
```

使用 docker version 命令还可以看到完整的版本信息：

```
$ docker version
Cannot connect to the Docker daemon at unix:///var/run/docker.sock. Is the docker daemon
running?
Client:
 Cloud integration: v1.0.29
 Version:           20.10.20
 API version:       1.41
 Go version:        go1.18.7
 Git commit:        9fdeb9c
 Built:             Tue Oct 18 18:20:35 2022
 OS/Arch:           darwin/amd64
 Context:           default
 Experimental:      true
```

看到 Docker 版本信息表示 Docker 安装成功了。

2. 启动 Docker

Docker Desktop 的安装和启动都非常简单，系统安装了 Docker Desktop 工具后，只要启动该工具就能自动启动 Docker 了，还可以在该工具中设置开机自动启动，如下图所示。

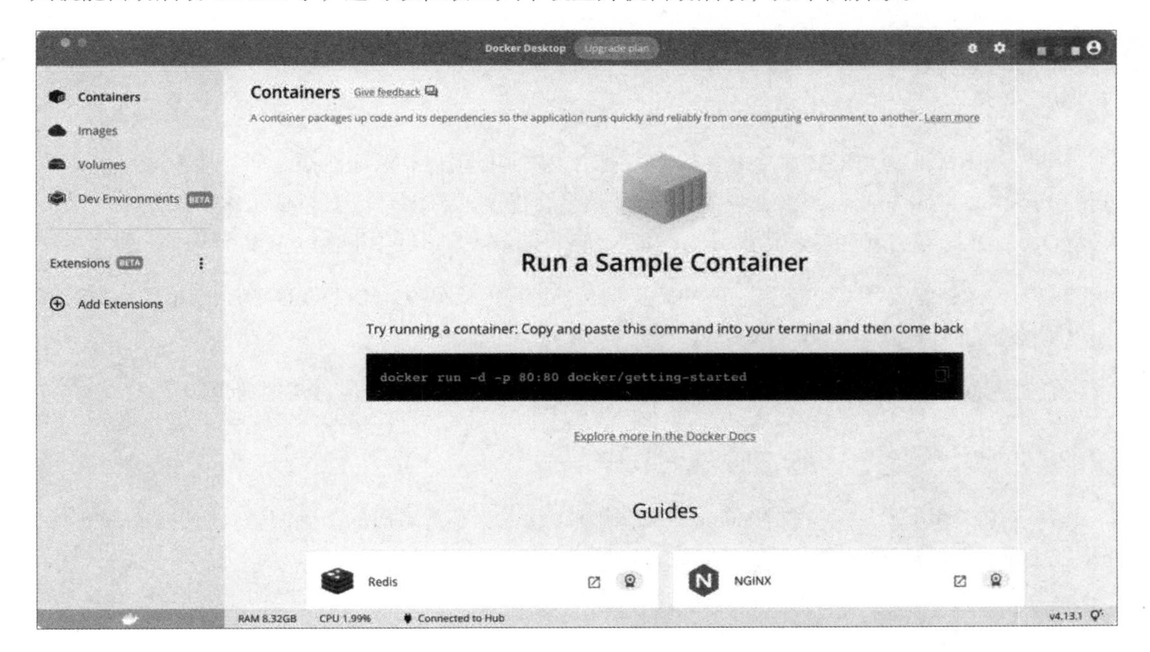

通过 Docker Desktop 工具可以进行 Docker 容器、镜像、数据卷的管理和环境设置等，可以简化 Docker 命令行的使用难度，对不是很熟悉运维的开发人员来说还是挺友好的，可以显著提升生产力。

如果是 CentOS 等不支持 Docker Desktop 的操作系统，则要先安装好 Docker 引擎，然后以系统服务运行 Docker，比如以下参考命令：

```
$ sudo systemctl start docker
```

11.4.3 基于 Dockerfile 构建镜像

1. 整包构建镜像

Dockerfile 是创建 Docker 镜像的基本配置文件，其本身是一个文本文件，由各种 Dockerfile 指令组成。

在应用的根目录下添加一个 Dockerfile 文件，参考内容如下：

```
# 整包构建
FROM eclipse-temurin:17-jre
ARG JAR_FILE=target/*.jar
COPY ${JAR_FILE} application.jar
ENTRYPOINT ["java","-Djava.security.egd=file:/dev/./urandom","-jar","
/application.jar"]
EXPOSE 8080
```

Dockerfile 文件从 FROM 指令开始，表示要引入的基础镜像，后面的指令都会基于该镜像执行，这里引入了 eclipse-temurin 镜像，它是 Eclipse 基金会提供的 JDK，基于 OpenJDK 构建，其前身是 AdoptOpenJDK，是开源免费使用的，这也是 Spring Boot 官方示例中提供的 JDK 镜像。

其中，ARG 定义了构建参数，COPY 复制了 jar 包到镜像中，ENTRYPOINT 指定了启动镜像应用的入口，EXPOSE 指定了镜像的端口。

配置好 Dockerfile 文件之后，再回到应用根目录下构建 Docker 镜像，构建命令如下：

```
$ docker build -t javastack/docker-all:1.0 .
```

其中，-t 表示 tags，即镜像的 name:tags，构建镜像时会下载对应的基础镜像，构建操作完成后，可以在 Docker Desktop 工具中看到该镜像，如下图所示。

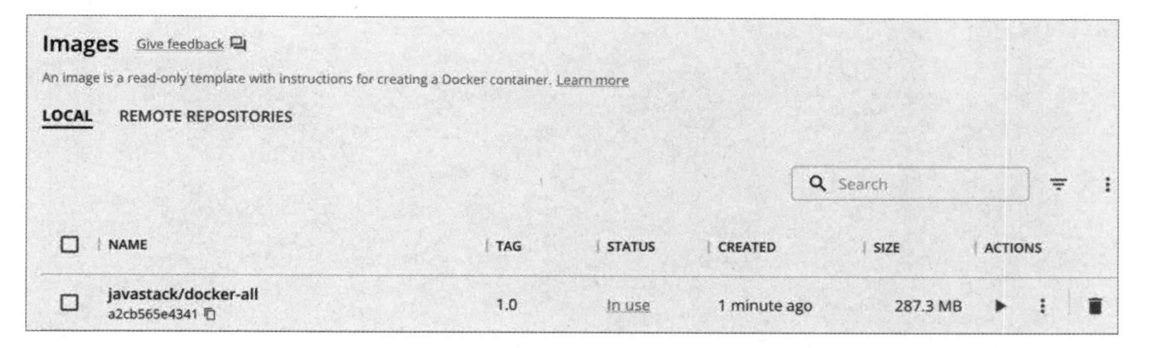

也可以通过以下 docker images 命令查询所有镜像：

```
$ docker images
REPOSITORY             TAG      IMAGE ID       CREATED         SIZE
javastack/docker-all   1.0      a2cb565e4341   2 minutes ago   287MB
```

镜像只是一个只读的静态文件，不能运行，容器才是镜像的运行实例，所以需要新增一个容器，可以在 Docker Desktop 镜像列表中单击对应的运行按钮新建一个容器，如下图所示。

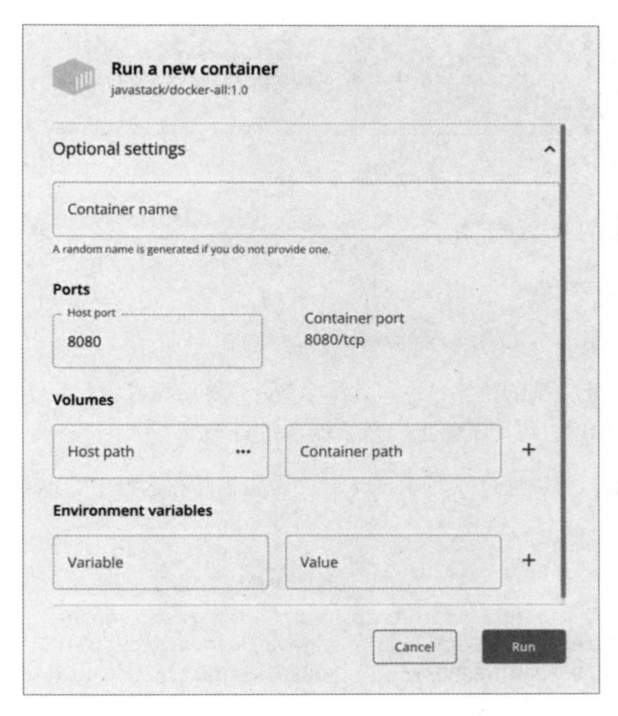

以上都为选填参数，包括容器名称，笔者这里指定了宿主机端口为 8080，然后单击 Run 按钮

运行容器，如下图所示。

Containers Give feedback

A container packages up code and its dependencies so the application runs quickly and reliably from one computing environment to another. Learn more

Only show running containers　　Search

	NAME	IMAGE	STATUS	PORT(S)	STARTED	ACTIONS
☐	naughty_khayyam b426cd5e4106	javastack/docker-all:1.0	Running	8080:8080	2 minutes agc	

也可以使用以下命令来运行容器：

```
$ docker run -dp 8080:8080 javastack/docker-all:1.0
d8f4f27963006aced0679d1238308466c46ae11e9e0e14a004b7f62e9e1dc574
```

这样就等同于基于镜像新建容器、再运行该容器，各命令参数的意义如下图所示。

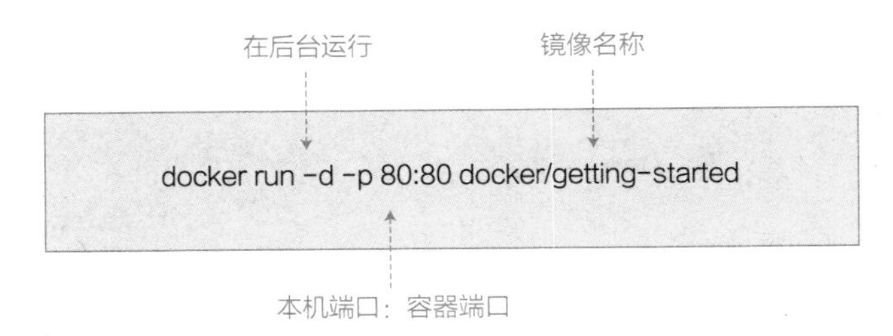

其中几个运行参数的含义如下所示。

- -d：表示在后台运行。
- -p：表示要绑定的本机端口：容器端口。

启动容器时需要指定完整的镜像名称和具体标签，容器启动成功后会显示当前容器的完整 ID，容器 ID 是容器的唯一标识，可用于停止、删除容器等操作。

容器启动完成之后，可以通过以下 docker ps 命令查询所有运行中的容器：

```
$ docker ps
CONTAINER ID   IMAGE                             COMMAND                  CREATED
STATUS         PORTS                                         NAMES
d8f4f2796300   javastack/docker-all:1.0          "java -Djava.securit…"   4 seconds ago
Up 2 seconds   0.0.0.0:8080->8080/tcp, :::8080->8080/tcp   jovial_williams
```

使用 docker ps -a 命令可以查询所有状态的容器。

再访问应用中的 /docker/hello 接口，如下图所示。

接口返回成功，说明容器正常运行了。

2. 分层构建镜像

Spring Boot 默认支持应用代码分层，在之前的章节中知道定义分层的是 layers.idx 文件，该分层也可以在插件中自定义，以下是 layers.idx 文件默认的分层：

```
- "dependencies":
  - "BOOT-INF/lib/"
- "spring-boot-loader":
  - "org/"
- "snapshot-dependencies":
- "application":
  - "BOOT-INF/classes/"
  - "BOOT-INF/classpath.idx"
  - "BOOT-INF/layers.idx"
  - "META-INF/"
```

一般变动的只有 class 类文件，引入的类库等文件是很少变动的，这样分层后，如果类库层没有变化，那么在构建 Docker 镜像时就会重用缓存中的类库层，而不会重新构建类库层，这样有利于提升构建速度。

在应用的根目录下添加一个分层构建的 Dockerfile 文件，参考内容如下：

```
FROM eclipse-temurin:17-jre as builder
WORKDIR application
ARG JAR_FILE=target/*-exec.jar
COPY ${JAR_FILE} application.jar
RUN java -Djarmode=layertools -jar application.jar extract

FROM eclipse-temurin:17-jre
WORKDIR application
COPY --from=builder application/dependencies/ ./
COPY --from=builder application/spring-boot-loader/ ./
COPY --from=builder application/snapshot-dependencies/ ./
COPY --from=builder application/application/ ./
ENTRYPOINT ["java", "org.springframework.boot.loader.JarLauncher"]
```

```
EXPOSE 8081
```

需要关注的是 RUN 指令中的 -Djarmode=layertools 参数，在应用分层包中会有一个 spring-boot-jarmode-layertools.jar 包，它可以让应用以特殊的模式启动，比如在 tartget 目录下以该模式启动，并配合 extract 命令可以把层以目录形式提取出来，参考命令如下：

```
$ java -Djarmode=layertools -jar spring-boot-docker-1.0-exec.jar extract
```

提取后的目录如下图所示。

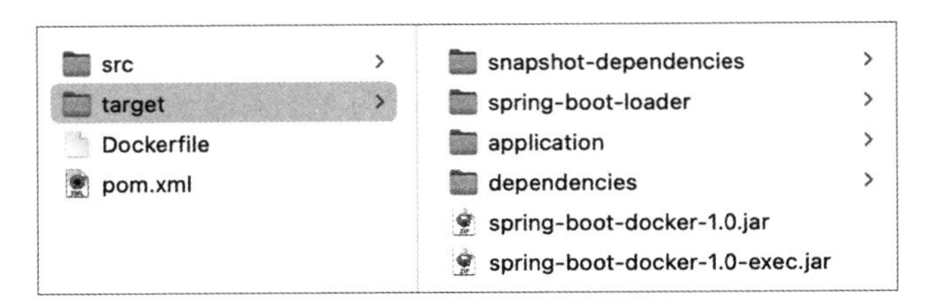

layertools 还支持使用一个 list 参数输出当前包中可以提取出来的层，参考命令如下：

```
$ java -Djarmode=layertools -jar spring-boot-docker-1.0-exec.jar list
dependencies
spring-boot-loader
snapshot-dependencies
application
```

层提取出来之后，Dockerfile 文件中的 COPY 指令就会将它们复制到镜像中，layertools 提供了一种提取包分层的方法，可以让不同层的文件存放在对应的层中，这样就可以做到镜像的分层构建。

配置好 Dockerfile 文件之后，再回到应用根目录下构建 Docker 镜像，构建命令如下：

```
$ docker build -t javastack/docker-layer:1.0 .
```

构建操作完成之后，在 Docker Desktop 中可以看到该镜像了，如下图所示。

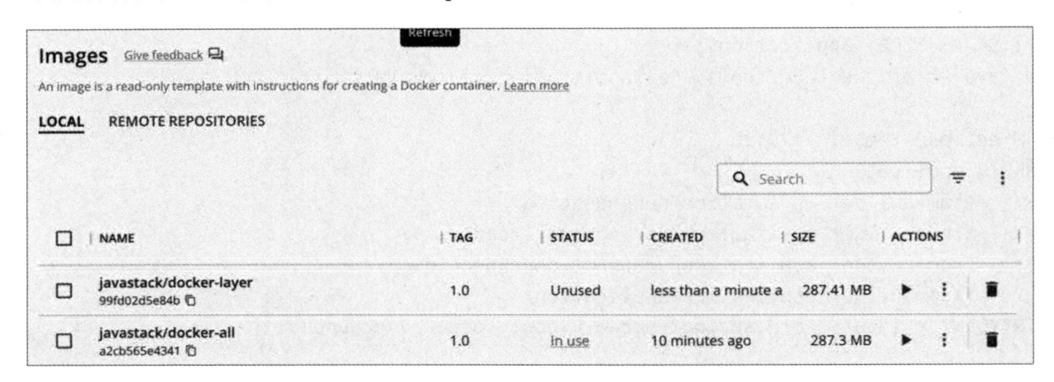

也可以通过以下 docker images 命令查询所有镜像：

```
$ docker images
REPOSITORY                 TAG        IMAGE ID       CREATED              SIZE
javastack/docker-layer     1.0        99fd02d5e84b   About a minute ago   287MB
javastack/docker-all       1.0        a2cb565e4341   11 minutes ago       287MB
```

这里可以看到整包构建和分层构建的两个镜像，大小都是一样的。

注意：分层构建不支持可直接执行的包，不然会出现以下兼容性错误。

```
Caused by: java.lang.IllegalStateException: java.lang.IllegalStateException: File
'/application/application.jar' is not compatible with layertools; ensure jar file is
valid and launch script is not enabled
```

所以，在构建分层 Docker 镜像时需要把插件中的 executable 设置为 false，如下面的配置所示。

```
<build>
  <plugins>
    <plugin>
      <groupId>org.springframework.boot</groupId>
      <artifactId>spring-boot-maven-plugin</artifactId>
      <configuration>
        <executable>false</executable>
      </configuration>
    </plugin>
  </plugins>
</build>
```

或者直接删除 executable 参数配置，默认不配置就是 false。

镜像构建完之后就可以运行容器了，基于分层镜像新建容器、再运行该容器，如下图所示。

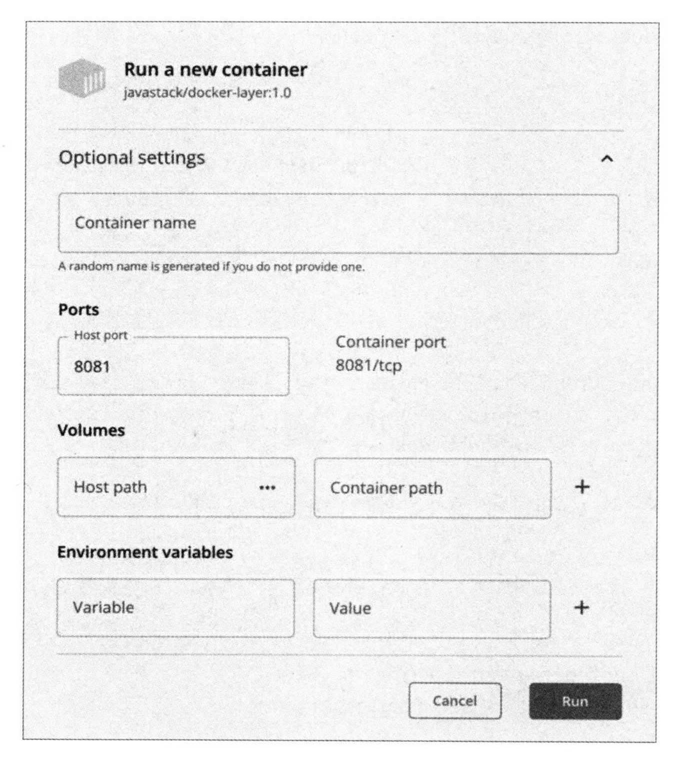

分层构建的镜像新建容器时使用 8081 端口进行区分，不能是之前整包用过的正在运行的 8080 端口，否则会报端口绑定错误。单击 Run 按钮运行容器，如下图所示。

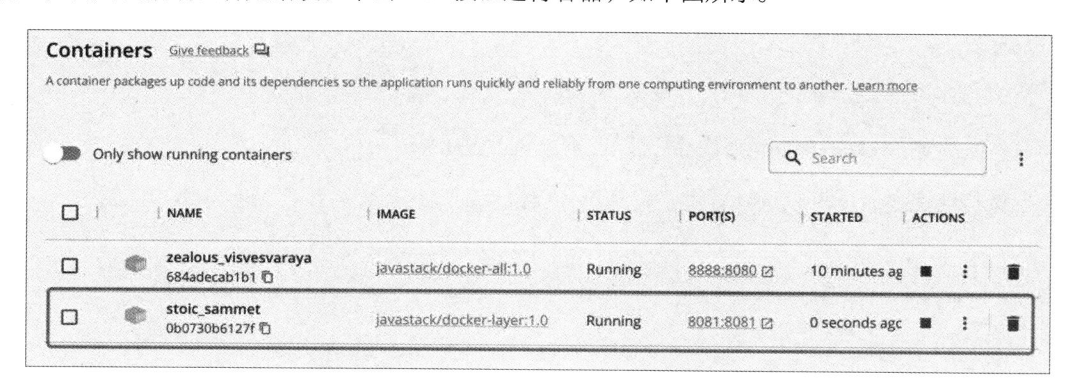

对应的新建和运行容器的命令为：

```
$ docker run -dp 8081:8081 javastack/docker-layer:1.0
7588cc29798dce2ab14ff998608dd6be543662c0b10ad12480e16bbeb14bce5d
```

容器启动完成之后，再通过以下 docker ps 命令查询所有运行中的容器：

```
$ docker ps
CONTAINER ID    IMAGE                            COMMAND                      CREATED
STATUS          PORTS                                        NAMES
7588cc29798d    javastack/docker-layer:1.0    "java org.springfram…"    4 seconds ago
Up 2 seconds    0.0.0.0:8081->8081/tcp, :::8081->8081/tcp    wonderful_bouman
d8f4f2796300    javastack/docker-all:1.0      "java -Djava.securit…"    12 minutes ago
Up 12 minutes   0.0.0.0:8080->8080/tcp, :::8080->8080/tcp    jovial_williams
```

这里可以看到整包镜像的容器和分层镜像的容器，然后分别访问 8080 和 8081 端口的 /docker/hello 接口，两个容器都可以正常响应。

前面介绍了，构建的镜像是分层的，如果对应的层没有变动就会重用缓存中的层，现在重复构建镜像验证一下，参考命令如下：

```
$ docker build -t javastack/docker-layer:1.0 .
[+] Building 1.2s (13/13) FINISHED
 => [internal] load build definition from Dockerfile                          0.0s
 => => transferring dockerfile: 37B                                           0.0s
 => [internal] load .dockerignore                                             0.0s
 => => transferring context: 2B                                               0.0s
 => [internal] load metadata for docker.io/library/eclipse-temurin:17-jre    1.1s
 => [internal] load build context                                             0.0s
 => => transferring context: 88B                                              0.0s
 => [stage-1 1/6] FROM docker.io/library/eclipse-temurin:17-jre@sha256:
87f454bb6797b918c6fcfb04d18c8bc55f87f7b209dede52a24e2385f84f9ba3             0.0s
 => CACHED [stage-1 2/6] WORKDIR application                                  0.0s
 => CACHED [builder 3/4] COPY target/*-exec.jar application.jar              0.0s
 => CACHED [builder 4/4] RUN java -Djarmode=layertools -jar application.jar extract
                                                                              0.0s
 => CACHED [stage-1 3/6] COPY --from=builder application/dependencies/ ./    0.0s
 => CACHED [stage-1 4/6] COPY --from=builder application/spring-boot-loader/ ./
                                                                              0.0s
 => CACHED [stage-1 5/6] COPY --from=builder application/snapshot-dependencies/ ./
                                                                              0.0s
 => CACHED [stage-1 6/6] COPY --from=builder application/application/ ./     0.0s
 => exporting to image                                                        0.0s
 => => exporting layers                                                       0.0s
 => => writing image sha256:c900f6909e6691d818198a27521dda3d8df396471c97c47e207eefddd0ebb3b4
                                                                              0.0s
 => => naming to docker.io/javastack/docker-layer:1.0                         0.0s
```

可以看到许多 CACHED 字样的日志，因为没有修改任何代码，所以所有的层都是从缓存中提取的，构建镜像仅用了 1.2 秒。现在任意修改一个类的代码再重新构建镜像，参考命令如下：

```
$ docker build -t javastack/docker-layer:1.0 .
[+] Building 2.8s (13/13) FINISHED
 => [internal] load build definition from Dockerfile                          0.0s
 => => transferring dockerfile: 37B                                           0.0s
 => [internal] load .dockerignore                                             0.0s
 => => transferring context: 2B                                               0.0s
 => [internal] load metadata for docker.io/library/eclipse-temurin:17-jre    1.2s
 => [internal] load build context                                             0.4s
 => => transferring context: 20.28MB                                          0.4s
 => [stage-1 1/6] FROM docker.io/library/eclipse-temurin:17-jre@sha256:
87f454bb6797b918c6fcfb04d18c8bc55f87f7b209dede52a24e2385f84f9ba3              0.0s
 => CACHED [stage-1 2/6] WORKDIR application                                  0.0s
 => [builder 3/4] COPY target/*-exec.jar application.jar                      0.1s
 => [builder 4/4] RUN java -Djarmode=layertools -jar application.jar extract 0.9s
 => CACHED [stage-1 3/6] COPY --from=builder application/dependencies/ ./    0.0s
 => CACHED [stage-1 4/6] COPY --from=builder application/spring-boot-loader/ ./
                                                                              0.0s
 => CACHED [stage-1 5/6] COPY --from=builder application/snapshot-dependencies/ ./
                                                                              0.0s
 => [stage-1 6/6] COPY --from=builder application/application/ ./            0.0s
 => exporting to image                                                        0.0s
 => => exporting layers                                                       0.0s
 => => writing image sha256:c900f6909e6691d818198a27521dda3d8df396471c97c47e207eefddd0ebb3b4
                                                                              0.0s
 => => naming to docker.io/javastack/docker-layer:1.0                         0.0s
```

此时只有 application 应用层不是从缓存中提取的，构建时间也变成了 2.8 秒，所以对于构建层、内容的多少，构建时间也是有显著区别的。

11.4.4 基于 Cloud Native Buildpacks 构建镜像

除了可以使用 Dockerfile 配置文件构建 Docker 镜像，还可以通过 Maven/Gradle 插件使用 Cloud Native Buildpacks 构建镜像。Cloud Native Buildpacks 于 2018 年由 Pivotal 和 Heroku 发起，并且在同年加入了云原生基金会，使用了现代的容器标准，比如 OCI 镜像格式。Cloud Native Buildpacks 既可以将应用构建成在任何云平台上运行的镜像，比如 Cloud Foundry，又可以构建兼容 Docker 的镜像。

添加 Spring Boot Maven 打包插件配置：

```
<build>
    <plugins>
        <plugin>
            <groupId>org.springframework.boot</groupId>
            <artifactId>spring-boot-maven-plugin</artifactId>
            <executions>
                <execution>
                    <goals>
                        <goal>build-image</goal>
                    </goals>
                </execution>
            </executions>
            <configuration>
                <!-- 分层镜像不兼容可直接执行的包 -->
                <executable>false</executable>
                <image>
                    <name>javastack/docker-cnb:${project.version}</name>
                </image>
            </configuration>
        </plugin>
    </plugins>
</build>
```

其中指定了 build-image 构建镜像目标，并且通过 image 自定义了镜像名称，默认为 docker.io/library/${project.artifactId}:${project.version} 命名规则。

回到应用根目录，直接使用 mvn 打包命令构建 Docker 镜像：

```
$ mvn clean package
```

也可以直接调用配置好的 build-image 目标构建 Docker 镜像：

```
$ mvn spring-boot:build-image
```

这两个命令默认都会先打包，然后构建镜像，首次构建时会拉取 Cloud Native Buildpacks 基础镜像，首次构建时的拉取过程会比较长。

另外，从构建日志中可以看出，构建镜像也支持分层构建：

```
[INFO]     [creator]          Creating slices from layers index
[INFO]     [creator]          dependencies (19.2 MB)
[INFO]     [creator]          spring-boot-loader (271.3 KB)
[INFO]     [creator]          snapshot-dependencies (0.0 B)
[INFO]     [creator]          application (40.5 KB)
```

但从整体上来说，使用 Cloud Native Buildpacks 方式的构建过程比较长，即使经过多次构建，后续每次构建都花了 20 多秒：

```
[INFO] Successfully built image 'docker.io/javastack/docker-cnb:1.0'
[INFO]
[INFO] ------------------------------------------------------------------------
[INFO] BUILD SUCCESS
[INFO] ------------------------------------------------------------------------
[INFO] Total time:  23.074 s
[INFO] Finished at: ...
[INFO] ------------------------------------------------------------------------
```

可以优化的一点是设置不用每次拉取基础镜像，如以下配置所示。

```
<build>
    <plugins>
        <plugin>
            <groupId>org.springframework.boot</groupId>
            <artifactId>spring-boot-maven-plugin</artifactId>
            <executions>
                <execution>
                    <goals>
                        <goal>build-image</goal>
                    </goals>
                </execution>
            </executions>
            <configuration>
                <!-- 分层镜像不兼容可直接执行的包 -->
                <executable>false</executable>
                <image>
                    <name>javastack/docker-cnb:${project.version}</name>
                    <pullPolicy>IF_NOT_PRESENT</pullPolicy>
                </image>
            </configuration>
        </plugin>
    </plugins>
</build>
```

配置项 pullPolicy 支持的值如下：

- ALWAYS：默认值，每次都拉取基础镜像。
- IF_NOT_PRESENT：不存在基础镜像时才拉取。

- NEVER：从不拉取基础镜像。

更多 Maven 插件的配置可以参考链接 16。

修改此配置后再次构建镜像，如以下日志所示。

```
[INFO] Successfully built image 'docker.io/javastack/docker-cnb:1.0'
[INFO]
[INFO] ------------------------------------------------------------------------
[INFO] BUILD SUCCESS
[INFO] ------------------------------------------------------------------------
[INFO] Total time:  13.329 s
[INFO] Finished at: ...
[INFO] ------------------------------------------------------------------------
```

优化拉取基础镜像策略后，每次构建只需要 10 多秒了。即使如此，可能还有优化的空间，但也会明显慢于基于 Dockerfile 配置文件的秒级构建，基于 Cloud Native Buildpacks 的构建方式慢了不止一点，所以，如果应用不是要部署到相关的云平台，那么使用 Docker 原生构建命令要快得多。

等构建操作完成之后，可以通过以下 docker images 命令查询所有镜像：

```
$ docker images
REPOSITORY                      ...  SIZE
javastack/docker-layer          ...  286MB
javastack/docker-all            ...  286MB
javastack/docker-cnb            ...  277MB
```

这里可以看到之前使用 Dockerfile 构建的两个镜像和使用 Cloud Native Buildpacks 构建的镜像，Cloud Native Buildpacks 构建的镜像较 Dockerfile 构建的镜像略小一些。

镜像构建完之后就可以新建容器来运行应用了，笔者未找到如何在 Maven 插件中定义镜像的端口的方法，这里先使用命令的方式基于 Cloud Native Buildpacks 镜像新建容器、再运行该容器。为了区分基于 Dockerfile 配置文件构建的镜像，这里使用 8082 端口进行映射：

```
$ docker run -dp 8082:8082 javastack/docker-cnb:1.0
e99a3425f1e44b3da049ca788dabc24d5ba25abd834a5223774d47c1124b0e84
```

容器启动完成之后，再通过以下 docker ps 命令查询所有运行中的容器：

```
$ docker ps
CONTAINER ID  IMAGE                      ... PORTS                                         ...
e99a3425f1e4  javastack/docker-cnb:1.0   ... 0.0.0.0:8082->8082/tcp, :::8082->8082/tcp ...
c0f40ca3fd6b  javastack/docker-layer:1.0 ... 0.0.0.0:8081->8081/tcp, :::8081->8081/tcp ...
```

```
d8f4f2796300  javastack/docker-all:1.0   ... 0.0.0.0:8080->8080/tcp, :::8080->8080/tcp ...
```

这里可以看到基于 Dockerfile 配置文件构建的镜像容器，还有通过 Maven 插件使用 Cloud Native Buildpacks 构建的镜像容器，然后分别访问 8080、8081、8082 端口的 /docker/hello 接口，三个容器都可以正常响应。

11.5 GraalVM 原生镜像（Spring Boot 3.0+）

11.5.1 概述

Spring Boot 3.0 应用现在可以支持生成 GraalVM 原生镜像了，这可以提供显著的内存和启动性能改进，支持 GraalVM 原生镜像也是整个 Spring 产品组合中一项重大能力的提升。

GraalVM 原生镜像直接取代了实验性的 Spring Native 项目，如下图所示。

build passing documentation

Spring Native is now superseded by Spring Boot 3 official native support, see the related reference documentation for more details.

Spring Native provides beta support for compiling Spring applications to native executables using GraalVM native-image compiler, in order to provide a native deployment option typically designed to be packaged in lightweight containers.

> Spring Boot 3.0 最低支持 GraalVM 22.3+ 和 Native Build Tools Plugin 0.9.17+。

GraalVM 是 Oracle 于 2018 年发布的一个高性能的、跨语言的通用虚拟机，号称是一个全新的全栈虚拟机，并具有高性能、跨语言交互等特性。GraalVM 主要用于提升 Java、Scala、Groovy、Kotlin 等基于 JVM 的语言的应用的执行性能，同时还支持其他像 JavaScript、Ruby、Python 和 R 等语言并为其提供运行环境。

GraalVM 在 HotSpot Java 虚拟机中添加了一个用 Java 编写的高级的 JIT 优化编译器，除了运行 Java 和基于 JVM 的语言，借助 GraalVM 的语言实现框架 Truffle，GraalVM 还能做到在 JVM 之上运行 JavaScript、Ruby、Python 等其他流行的编程语言，还支持 Java 和其他语言之间直接互相操作并在同一内存空间来回传递数据。

GraalVM 虚拟机有以下几个特性：

- 更加高效快速地运行代码。

- 能与大多数编程语言直接交互。
- 使用 Graal SDK嵌入多语言。
- 创建预编译的原生镜像。
- 提供一系列工具来监视、调试和配置所有代码。

下面通过两个示例详细介绍 GraalVM 虚拟机的优势，以下这段示例代码来自 GraalVM 官网：

```
const express = require('express');
const app = express();
app.listen(3000);
app.get('/', function(req, res) {
  var text = 'Hello World!';
  const BigInteger = Java.type(
    'java.math.BigInteger');
  text += BigInteger.valueOf(2)
    .pow(100).toString(16);
  text += Polyglot.eval(
    'R', 'runif(100)')[0];
  res.send(text);
})
```

上面这段代码同时使用了 Node.js、Java、R 三种语言，GraalVM 消除了各种编程语言之间的隔离性，多编程语言可以结合使用，并且不会有任何性能影响，零开销的互操作，这样我们就可以为应用选择最佳的编程语言组合。

下面这段命令同样来自 GraalVM 官网：

```
$ javac HelloWorld.java
$ time java HelloWorld
user 0.070s
$ native-image HelloWorld
$ time ./helloworld
user 0.005s
```

GraalVM 可以将 JVM 系语言的代码预编译成原生镜像，即通过提前处理已编译的 Java 应用来生成原生镜像文件，它是一个独立的基于特定平台的可执行文件，这样可以极大缩短 JVM 系语言的启动时间，也能减少 JVM 应用的内存占用。

11.5.2 GraalVM 应用与传统应用的区别

对于 Java 应用，GraalVM 提供了两种运行方法：

- 在 HotSpot JVM 上使用 Graal 的 JIT 编译和运行应用，应用在运行时编译，边运行边编译。
- 作为 AOT 编译的 GraalVM 原生镜像可执行文件运行，应用在运行前编译，即静态编译或者预先编译；在创建 GraalVM 原生镜像时，应用需要经过预先编译处理才能创建可执行文件，这种预先编译涉及从应用的主要入口点开始进行代码静态分析。

传统的 Java 应用必须依赖于 Java 虚拟机（JVM）运行，Spring Boot 3.0 发布后可以不需要依赖 JVM 运行应用，它提供了另一种运行和部署 Spring 应用的方式，即通过 GraalVM 将 Spring 应用编译成原生镜像的可执行文件来运行。

GraalVM 应用与传统应用的区别如下：

- 构建镜像时会进行应用的静态分析，从主入口点开始执行。
- 创建原生映像时无法访问的代码将被删除，并且不会成为可执行文件的一部分。
- GraalVM 不能直接感知代码的动态元素，比如反射、资源、序列化和动态代理等动态代码必须提前声明并配置。
- 应用的 classpath 路径在构建时是固定不变的，不能修改。
- 不支持 Bean 的延迟加载，可执行文件包含的所有内容都在启动时加载到内存中。
- 目前某些方面还存在一些限制和局限性，并不能完全兼容传统的 Java 应用。

传统的 Spring Boot 应用非常动态化，可以在运行时动态生成任意代码，Spring Boot 自动配置机制在很大程度上也取决于对运行时状态的反应，以便能正确配置。虽然可以将应用的动态代码配置并告诉 GraalVM，但这样做会抵消静态分析的大部分好处。因此，当 Spring Boot 在创建原生镜像时，应用在动态方面会受到限制，比如：

- 应用的 classpath 路径在构建时是固定不变的，不能修改。
- Bean 在运行时不能被修改，这也就意味着在 Spring 应用中，不支持 @Profile 注解及基于 Profile 配置，另外像 @ConditionalOnProperty 等条件注解也都会受到限制。

Spring AOT 在构建原生镜像时会执行预先编译并生成以下文件：

- Java 源代码。
- 字节码（用于动态代理等场景）。

- GraalVM 相关的 JSON 描述文件，包含 Resource hints（resource-config.json）、Reflection hints（reflect-config.json）、Serialization hints（serialization-config.json）、Java Proxy Hints（proxy-config.json）、JNI Hints（jni-config.json）文件。

11.5.3 创建 GraalVM 原生镜像的应用

通过前面的章节我们已经很好地了解了 GraalVM 原生镜像及 Spring AOT 引擎的工作机制，本节介绍如何创建一个基于 GraalVM 原生镜像的应用。

构建 Spring Boot 原生镜像应用主要有以下两种方式：

- 使用 Spring Boot 的 Cloud Native Buildpacks 生成一个包含原生可执行文件的轻量级容器。
- 使用 GraalVM 的原生构建工具生成一个原生可执行文件。

先创建一个 GraalVM 原生应用测试项目，建议使用继承 spring-boot-starter-parent 父项目依赖的方式创建 GraalVM 项目，它自带了两个 Profile 原生配置，如下图所示。

该 native 配置项配置了需要运行原生镜像的执行目标，可以在命令行中使用 -P 标识并激活 Profile。如果不能使用 spring-boot-starter-parent，则需要为 Spring Boot 插件的 process-aot 设置项和 Native Build Tools 插件的 add-reachability-metadata 设置项配置执行目标。

最简单方法还是去 Spring 一站式项目生成网站 start.spring.io 生成示例项目，然后添加“GraalVM Native Support”依赖项并生成项目，如下图所示。

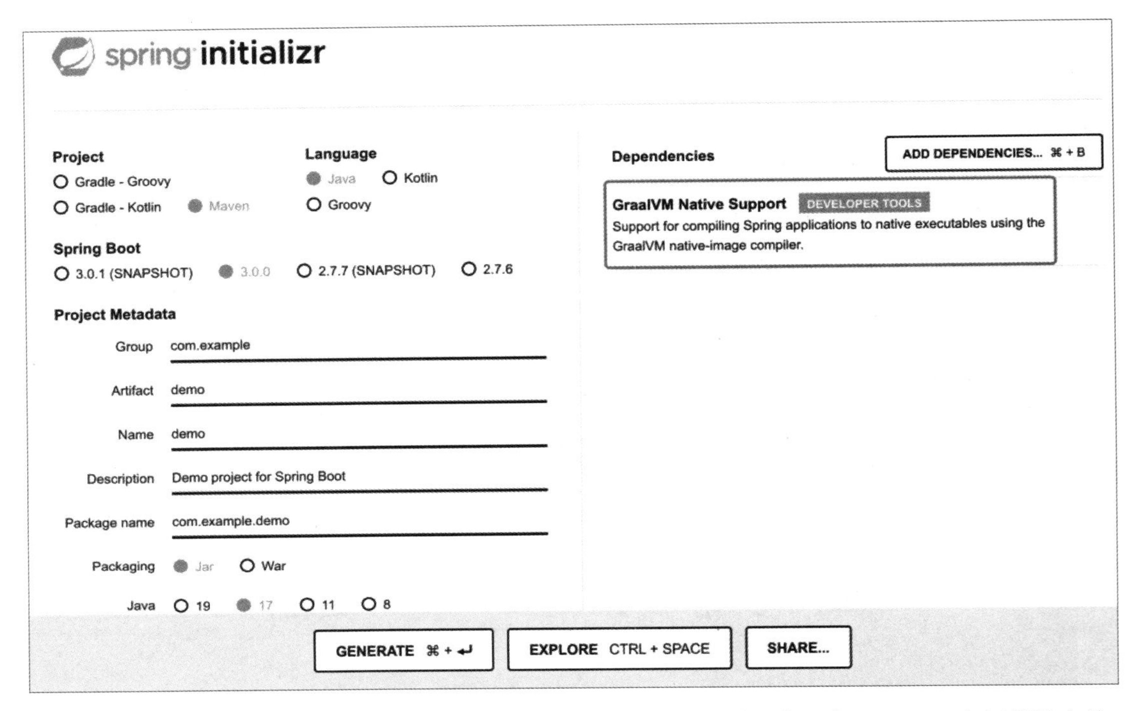

创建的项目结构和第1章介绍的生成项目结构一致，其中同样包含一个HELP.md入门帮助文件，帮助开发者了解如何创建和运行原生镜像。

与基于JVM传统的Spring Boot应用不同的是，基于GraalVM的Spring Boot应用会多依赖一个native-maven-plugin插件，如下面的配置所示。

```
<build>
    <plugins>
        <plugin>
            <groupId>org.graalvm.buildtools</groupId>
            <artifactId>native-maven-plugin</artifactId>
        </plugin>
    </plugins>
</build>
```

为了便于后续的测试，在启动类中再添加一个/hi测试接口，如下面的示例代码所示。

```
@RestController
@SpringBootApplication
public class Application {
```

```
    @RequestMapping("/hi")
    String hello() {
        return "Hello GraalVM!";
    }

    public static void main(String[] args) {
        SpringApplication.run(Application.class, args);
    }

}
```

11.5.4　构建基于 GraalVM 的原生镜像应用

1. 使用 Buildpacks 构建（部署 Docker 容器）

Spring Boot 提供了针对 Maven 和 Gradle 的原生镜像的 buildpack 的支持，只需一个命令即可将镜像放入本地运行的 Docker 守护进程，生成的镜像不会包含 JVM，而是静态编译的原生镜像。

所以，使用 Buildpacks 构建原生镜像时，本机需要安装 Docker 环境，详见 11.4.2 节。

使用 Maven 构建镜像的命令如下：

```
$ mvn -Pnative spring-boot:build-image
```

第一次构建需要下载 GraalVM 相关插件及构建所需要的基础镜像，所以会有点慢，第一次构建足足花费了 7 分钟，如下面的构建日志所示。

```
[INFO] Successfully built image 'docker.io/library/spring-boot-graalvm:1.0'
[INFO]
[INFO] ------------------------------------------------------------------------
[INFO] BUILD SUCCESS
[INFO] ------------------------------------------------------------------------
[INFO] Total time:  07:29 min
[INFO] Finished at: 2022-11-28T17:21:06+08:00
[INFO] ------------------------------------------------------------------------
```

构建完成之后，通过 docker images 命令可以查看该镜像，镜像大小为 96MB，可以使用 docker run 命令运行该镜像：

```
$ docker run --rm -p 8080:8080 docker.io/library/spring-boot-graalvm:1.0
```

启动日志如下所示。

```
  .   ____          _            __ _ _
 /\\ / ___'_ __ _ _(_)_ __  __ _ \ \ \ \
( ( )\___ | '_ | '_| | '_ \/ _` | \ \ \ \
 \\/  ___)| |_)| | | | | || (_| |  ) ) ) )
  '  |____| .__|_| |_|_| |_\__, | / / / /
 =========|_|==============|___/=/_/_/_/
 :: Spring Boot ::  (v3.0.0)
....... . . .
....... . . .
....... . . .
........ Started Application in 0.083 seconds (process running for 0.088)
```

0.083 秒便启动了一个简单的 Spring Boot 应用，启动速度真是太快了，比启动运行在传统 JVM 上的 Spring Boot 应用快了很多倍，这也见证了 GraalVM 的性能魅力了。

再访问应用的 /hi 接口，如下图所示。

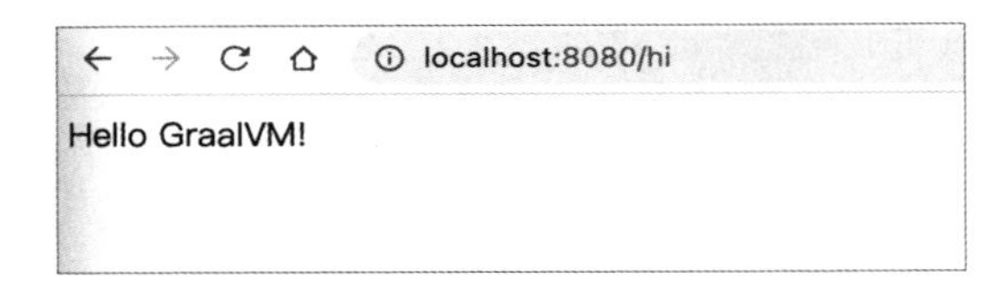

响应正常，然后停止运行该容器，再以传统的基于 JVM 的方式直接运行 main 方法，启动日志如下所示。

```
  .   ____          _            __ _ _
 /\\ / ___'_ __ _ _(_)_ __  __ _ \ \ \ \
( ( )\___ | '_ | '_| | '_ \/ _` | \ \ \ \
 \\/  ___)| |_)| | | | | || (_| |  ) ) ) )
  '  |____| .__|_| |_|_| |_\__, | / / / /
 =========|_|==============|___/=/_/_/_/
 :: Spring Boot ::  (v3.0.0)
....... . . .
....... . . .
....... . . .
........ Started Application in 2.065 seconds (process running for 2.714)
```

程序正常启动，虽然只花了 2 秒，但与基于 GraalVM 原生镜像的启动速度相比也差了很多数量级，这里只是简单的应用测试，如果是更复杂的应用，差异可能会更加明显。

2. 使用 GraalVM 工具构建（直接运行）

如果不想基于 Docker 环境构建并运行及生成原生可执行文件，则可以使用 GraalVM 的原生构建工具 Native Build Tools，它是 GraalVM 为 Maven 和 Gradle 提供的原生镜像构建插件，使用它可以执行各种 GraalVM 镜像相关的操作。

要使用 GraalVM 的原生构建工具 Native Build Tools，需要在本机安装 GraalVM 发行版，可以手动下载安装，也可以通过 SDKMAN 安装，手动下载地址为链接 17。

GraalVM 支持 Windows、macOS、Linux 等多种操作系统，读者可以根据不同的操作系统及 JDK 版本进行安装，如下图所示。

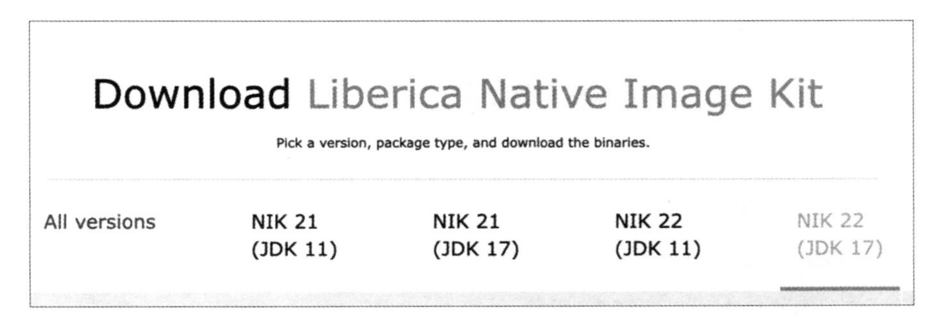

macOS 和 Linux 操作系统推荐使用 SDKMAN 安装 GraalVM，具体安装过程可见第 1 章的 Spring Boot CLI 一节，比如安装最新的基于 JDK 17 的参考命令如下：

```
$ sdk install java 22.3.r17-nik
$ sdk use java 22.3.r17-nik
```

安装完 GraalVM 之后需要确保本机的 JAVA_HOME 环境变量为 GraalVM JDK 所在的目录，可以通过 java -version 查询：

```
$ java -version
openjdk version "17.0.5" 2022-10-18 LTS
OpenJDK Runtime Environment GraalVM 22.3.0 (build 17.0.5+8-LTS)
OpenJDK 64-Bit Server VM GraalVM 22.3.0 (build 17.0.5+8-LTS, mixed mode, sharing)
```

显示 OpenJDK 及 GraalVM 相关字样说明安装成功了，然后使用 Maven 命令构建镜像：

```
$ mvn -Pnative native:compile
```

镜像构建完之后可以在当前应用的 tartget 目录中找到该原生镜像的可执行文件，如下图所示。

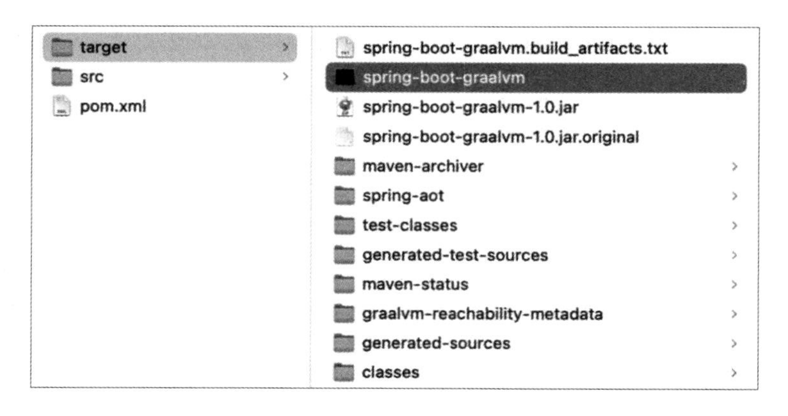

主要生成了两个文件：

- spring-boot-graalvm：应用可执行文件。
- spring-boot-graalvm.build_artifacts.txt：对应的说明文本文件。

直接运行应用可执行文件，运行命令如下：

```
$ ./spring-boot-graalvm
```

应用启动日志如下所示。

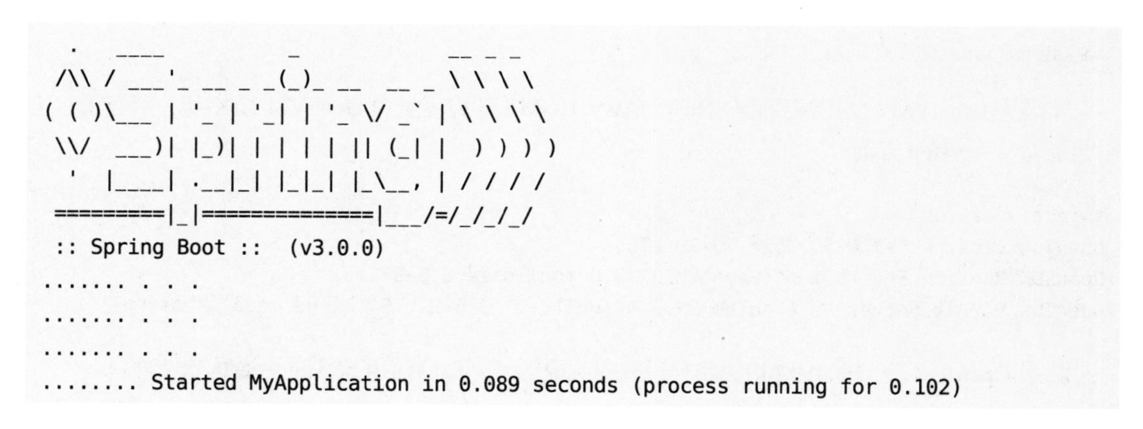

```
  .   ____          _            __ _ _
 /\\ / ___'_ __ _ _(_)_ __  __ _ \ \ \ \
( ( )\___ | '_ | '_| | '_ \/ _` | \ \ \ \
 \\/  ___)| |_)| | | | | || (_| |  ) ) ) )
  '  |____| .__|_| |_|_| |_\__, | / / / /
 =========|_|==============|___/=/_/_/_/
 :: Spring Boot ::  (v3.0.0)
....... . . .
....... . . .
....... . . .
........ Started MyApplication in 0.089 seconds (process running for 0.102)
```

和 buildpack 方式构建的应用可执行文件一样，Native Build Tools 构建的应用可执行文件一样可以达到极致的启动性能，再访问应用的 /hi 接口也是能正常响应的。

Native Build Tools 和 buildpack 虽然都使用 Maven 插件构建应用可执行文件，但是两者的构建目标和应用最终的运行环境及运行方式是不一样的，buildpack 需要基于 Docker 环境运行，而

Native Build Tools 构建的可执行包可以直接运行，即可以像 11.3.2 节介绍的基于传统的 JVM 的嵌入式容器一样直接运行。

虽然 GraalVM 能够显著提高应用的启动性能，也能节约内存的占用，但它目前还属于快速发展的技术，不是很成熟，很多类库也不支持，也存在已知的种种限制，在 Spring Boot 3.0 中也属于首发版本，要真正代替传统的 JVM 还需要很长的一段路要走。

GraalVM 原生镜像已知的使用限制参考见链接 18。

第 12 章 Spring Boot 监控与报警

Spring Boot 应用的监控是通过 Actuator 模块进行的，它也是 Spring Boot 的重要特性之一，所以 Actuator 模块一般用于生产环境的应用，使用 HTTP 端点或者 JMX 的方式提供应用的监控和管理操作，包括应用的健康检查、指标收集等，然后通过第三方监控平台抓取指标数据或者推送指标数据给监控平台。本章会按照 Actuator 模块→ Endpoints（端点）→ Metrics（指标）→监控平台的顺序介绍监控平台是如何通过这些组件完成监控的。

12.1 Spring Boot Actuator 概述

启用 Spring Boot Actuator 功能，只需要添加其启动器依赖即可，如下面的依赖配置所示。

```
<dependency>
    <groupId>org.springframework.boot</groupId>
    <artifactId>spring-boot-starter-actuator</artifactId>
</dependency>
```

具体导入的依赖如下图所示。

```
∨ spring-boot-starter-actuator : 3.0.0 [compile]
    > micrometer-core : 1.10.2 [compile]
    > micrometer-observation : 1.10.2 [compile]
    ∨ spring-boot-actuator-autoconfigure : 3.0.0 [compile]
        > jackson-databind : 2.14.1 [compile]
        > jackson-datatype-jsr310 : 2.14.1 [compile]
          spring-boot : 3.0.0 [compile]
        > spring-boot-actuator : 3.0.0 [compile]
          spring-boot-autoconfigure : 3.0.0 [compile]
    > spring-boot-starter : 3.0.0 [compile]
```

它主要包含了 micrometer 相关依赖包、actuator 自动配置模块。

12.2　Endpoints（端点）

12.2.1　概述

Endpoint 端点用来监控 Spring Boot 应用并与之交互，Spring Boot 内置了许多端点，“开箱即用”，每个端点都可以通过 HTTP 或者 JMX（Java Management Extensions）的方式暴露出去，但绝大多数情况下端点都是以 HTTP 方式暴露的，每个端点都会被映射为 /actuator/${ID} 方式，ID 即端点的 ID，比如健康端点为 /actuator/health。

既然 health 端点是以 HTTP 方式暴露的，那么就可以通过 HTTP 来访问这个端点，如下图所示。

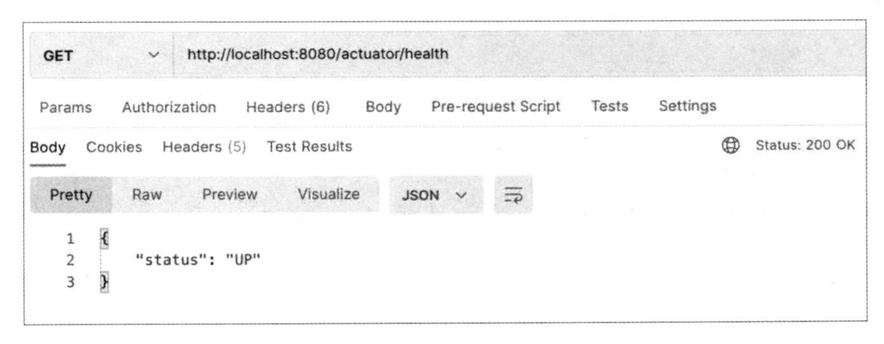

该端点返回了 UP 状态，表示应用为健康运行状态。

12.2.2　内置端点

Spring Boot 的内置端点如下表所示。

端点 ID	功能
auditevents	暴露当前应用的审计事件信息
beans	显示应用中的所有 Spring Bean
caches	暴露可用的缓存
conditions	显示配置类上评估条件及匹配成功与否的原因
configprops	显示所有的 @ConfigurationProperties 列表
env	从 Spring 环境中暴露所有 properties
flyway	显示所有 Flyway 迁移记录

续表

端点 ID	功能
health	显示健康信息
httpexchanges	显示 HTTP 交换信息（默认显示最后 100 个）
info	显示应用的基本信息
integrationgraph	显示 Spring Integration 图表
loggers	显示或者修改日志配置
liquibase	显示所有 liquibase 迁移记录
metrics	显示指标信息
mappings	显示所有 @RequestMapping 映射路径
quartz	显示 Quartz 任务调度信息
scheduledtasks	显示所有的任务调度
sessions	从 Spring Session 中检索和删除用户会话
shutdown	优雅关闭应用
startup	显示应用启动步骤数据
threaddump	线程转储，用于分析某一节点的线程堆栈状态

> **Spring Boot 3.0.0 新变化：**
> Spring Boot 3.0.0 移除了 httptrace 端点，新增了 httpexchanges 端点，只是变更了端点名称，功能还是一样的。

如果是 Web 应用，那么还内置了以下端点，如下表所示。

端点 ID	功能
heapdump	返回一个 heap dump 文件： HotSpot JVM 返回的是 HPROF-format 格式文件 OpenJ9 JVM 返回的是 PHD-format 格式文件
logfile	返回日志文件内容（需要设置 logging.file.name 或者 logging.file.path 参数）
prometheus	暴露可以被 Prometheus 服务器抓取的格式指标

如果内置的端点不满足需求，也可以自定义端点。

12.2.3 启用端点

除了 shutdown 端点，其他所有端点默认都是启用的，也可以启用或禁用某个端点，启用或禁用某个端点是通过 management.endpoint.<id>.enabled 参数配置的，比如启用 shutdown 端点，如下面的配置所示。

```
management:
  endpoint:
    shutdown:
      enabled: true
```

如果想一个个手动启用端点，则可以先禁用全部端点，再启用某个端点，如下面的配置所示。

```
management:
  endpoints:
    enabled-by-default: false
  endpoint:
    info:
      enabled: true
```

这里先禁用所有端点，然后只启用 info 端点。

需要注意的是，被禁用的端点会完全从 Spring 上下文中删除，如果只想更改端点暴露的方式，那么可以使用 include 和 exclude 配置参数替代禁用端口的方式，具体见后面的章节。

12.2.4　暴露端点

端点被启用后，并不一定能够被访问，还要看端点是否被暴露，并且暴露的方式是怎样的。因为端点可能包含敏感信息，所以需要谨慎暴露相关端点。在各大 Spring Boot 版本中暴露的端点不太一样，这里以 Spring Boot 2.7.x 为例，内置端点的默认暴露方式如下表所示。

端点 ID	JMX	Web
auditevents	Yes	No
beans	Yes	No
caches	Yes	No
conditions	Yes	No
configprops	Yes	No
env	Yes	No
flyway	Yes	No
health	Yes	Yes
heapdump	N/A	No
httptrace	Yes	No
info	Yes	No
integrationgraph	Yes	No
logfile	N/A	No

续表

端点 ID	JMX	Web
loggers	Yes	No
liquibase	Yes	No
metrics	Yes	No
mappings	Yes	No
prometheus	N/A	No
quartz	Yes	No
scheduledtasks	Yes	No
sessions	Yes	No
shutdown	Yes	No
startup	Yes	No
threaddump	Yes	No

默认以 JMX 方式暴露了所有端点，出于安全考虑，Web 方式则只暴露了 Health 端点。

> **Spring Boot 3.0.0 新变化：**
> Spring Boot 3.0.0 更改了默认暴露的端点，默认以 JMX 和 Web 方式只暴露了 Health 一个端点，更加严格、安全了。

可以使用以下参数更改端点暴露的方式，如下表所示。

配置参数	描述	默认值
management.endpoints.jmx.exposure.exclude	停止以 JMX 方式暴露端点	
management.endpoints.jmx.exposure.include	以 JMX 方式暴露端点	health
management.endpoints.web.exposure.exclude	停止以 Web 方式暴露端点	
management.endpoints.web.exposure.include	以 Web 方式暴露端点	health

多个端点以逗号分隔，也支持“*”通配符，表示所有端点，在 .yaml 配置中需要用引号包起来，如下面的配置示例所示。

```
management:
  endpoints:
    jmx:
      exposure:
        include: health,info
    web:
      exposure:
        include: "*"
        exclude: threaddump
```

这里配置了以 JMX 方式只暴露 health 和 info 端点，以 Web 方式暴露除 threaddump 端点外的所有端点，两种暴露方式是独立的，不会互相影响各自暴露的端点，如果同一种暴露方式同时配置了 include 和 exclude，则以 exclude 配置的优先。

一般通过配置参数就能暴露端点，如果想自定义实现端点的暴露策略，则可以实现 EndpointFilter 接口并注册到 Spring 容器中。

12.2.5 端点安全性

前面介绍了，端点提供的信息可能涉及敏感性，所以除了谨慎暴露所需要的端点，还需要对端点进行安全保护，如提供对应的权限访问等。

第 3 章介绍了 Spring Boot 提供的 Web 默认保护机制，只要有 Spring Security 类库依赖，Spring Boot 就会自动保护所有接口，还包括除 health 健康端点外的所有端点，具体的实现见 Actuator 的 ManagementWebSecurityAutoConfiguration 自动配置类，它注册在 spring-boot-actuator- autoconfigure 包的新的自动配置文件中，它的源码如下：

```
@AutoConfiguration(before = SecurityAutoConfiguration.class,
        after = { HealthEndpointAutoConfiguration.class, InfoEndpointAutoConfiguration.class,
                WebEndpointAutoConfiguration.class, OAuth2ClientAutoConfiguration.class,
                OAuth2ResourceServerAutoConfiguration.class,
                Saml2RelyingPartyAutoConfiguration.class })
@ConditionalOnWebApplication(type = ConditionalOnWebApplication.Type.SERVLET)
@ConditionalOnDefaultWebSecurity
public class ManagementWebSecurityAutoConfiguration {

    @Bean
    @Order(SecurityProperties.BASIC_AUTH_ORDER)
    SecurityFilterChain managementSecurityFilterChain(HttpSecurity http) throws Exception {
        http.authorizeHttpRequests((requests) -> {
            requests.requestMatchers(EndpointRequest.to(HealthEndpoint.class)).permitAll();
            requests.anyRequest().authenticated();
        });
        if (ClassUtils.isPresent("org.springframework.web.servlet.DispatcherServlet", null)) {
            http.cors();
        }
        http.formLogin(Customizer.withDefaults());
        http.httpBasic(Customizer.withDefaults());
        return http.build();
    }

}
```

然后加入 Spring Security 一站式启动器依赖：

```
<dependency>
    <groupId>org.springframework.boot</groupId>
    <artifactId>spring-boot-starter-security</artifactId>
</dependency>
```

应用重启后，再访问应用的 /actuator/health 端点，如下图所示。

如果没有登录授权，则 Spring Boot 默认会跳到登录页面，需要输入默认的用户信息才能访问端点，功能和原理与第 3 章介绍的一样，这里不再赘述。如果不想使用 Spring Boot 自动配置的默认机制，则可以注册一个 SecurityFilterChain 实现自定义的安全机制。

自定义安全机制的示例代码如下：

```
@Configuration
public class SecurityConfig {

    @Bean
    public SecurityFilterChain securityFilterChain(HttpSecurity http) throws Exception {
        return http.authorizeHttpRequests((authorize) -> {
                    authorize.requestMatchers("/").permitAll()
                            .requestMatchers(EndpointRequest.to("health")).hasRole
("ENDPOINT_ADMIN")
                            .requestMatchers(EndpointRequest.toAnyEndpoint()).permitAll();
                }).formLogin(withDefaults())
                .logout().logoutSuccessUrl("/")
                .and().build();
    }

    @Bean
    protected UserDetailsService userDetailsService() {
        InMemoryUserDetailsManager manager = new InMemoryUserDetailsManager();
        manager.createUser(User.withUsername("test").password("{noop}test")
                .roles("ENDPOINT_ADMIN", "ADMIN", "TEST").build());
        manager.createUser(User.withUsername("root").password("{noop}root")
                .roles("ADMIN").build());
        return manager;
    }

}
```

端点授权配置需要用到 EndpointRequest 类，这里配置了只有 health 端点需要 ENDPOINT_ADMIN 角色才能访问，其他接口的 URL 和端点不需要授权都能正常访问。

然后创建了两个授权用户，{noop} 表示明文密码，只有 test 用户才拥有 ENDPOINT_ADMIN 角色。重启应用，只有输入拥有 ENDPOINT_ADMIN 角色的 test 用户才能访问 health 端点，输入其他用户会返回 403 错误。

12.2.6 自定义端点映射

默认情况下通过 /actuator 根路径可以显示所有暴露出来的端点，如下图所示。

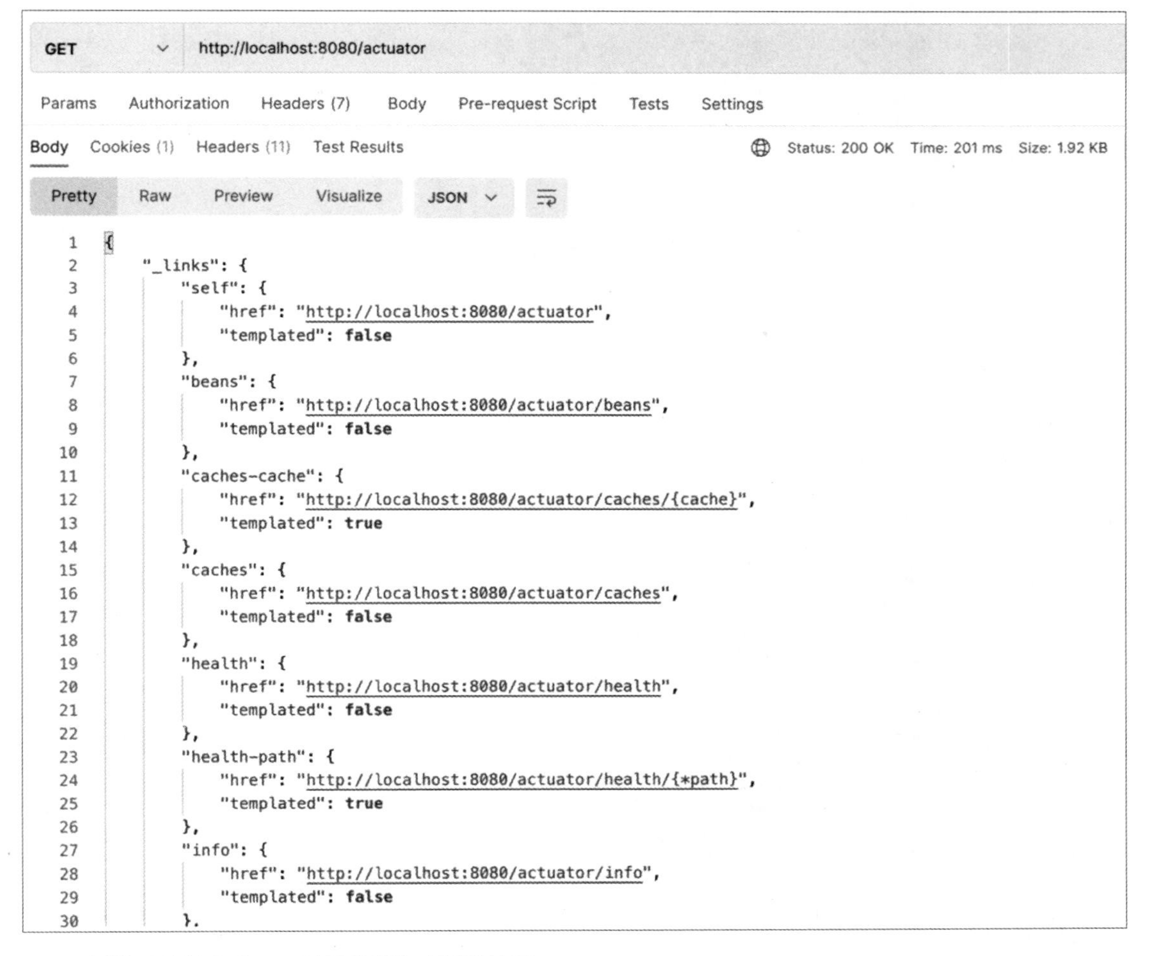

这样不是很安全，可以通过以下配置禁用：

```
management:
  endpoints:
    web:
      discovery:
        enabled: false
```

还可以更改默认的端点映射，因为默认的端点的 URL 前缀和端点 ID 所有人都知道，不是很安全，可以通过配置更改所有端点默认的前缀 URL 及端点 ID，比如以下配置示例：

```
management:
  endpoints:
    web:
      base-path: /act
      path-mapping:
        health: hth
```

这里把默认的端点前缀映射从 /actuator 改为 /act，把 health 健康端点改名为 hth，现在就能只通过 /act/hth 访问健康端点了，之前的映射就失效了。

另外，端点映射的 IP 地址和端口也可以修改，默认和 Web 应用共享一个 Server，可以把端点 Server 独立出来，配置示例如下：

```
management:
  server:
    port: 8088
    address: 127.0.0.1
```

这样表示只能通过 localhost:8088 访问端点，address 和 port 是配套使用的，address 并不是必需的。

重启应用，通过日志可以发现同时启动了两个 Server，如下图所示。

```
main] o.s.b.w.embedded.tomcat.TomcatWebServer  : Tomcat initialized with port(s): 8080 (http)
main] o.apache.catalina.core.StandardService   : Starting service [Tomcat]
main] org.apache.catalina.core.StandardEngine  : Starting Servlet engine: [Apache Tomcat/10.0.22]
main] o.a.c.c.C.[Tomcat].[localhost].[/]       : Initializing Spring embedded WebApplicationContext
main] w.s.c.ServletWebServerApplicationContext : Root WebApplicationContext: initialization completed in 1195 ms
main] o.s.s.web.DefaultSecurityFilterChain     : Will secure any request with [org.springframework.security.web.se
main] o.s.b.w.embedded.tomcat.TomcatWebServer  : Tomcat started on port(s): 8080 (http) with context path ''
main] o.s.b.w.embedded.tomcat.TomcatWebServer  : Tomcat initialized with port(s): 8088 (http)
main] o.apache.catalina.core.StandardService   : Starting service [Tomcat]
main] org.apache.catalina.core.StandardEngine  : Starting Servlet engine: [Apache Tomcat/10.0.22]
main] o.a.c.c.C.[Tomcat-1].[localhost].[/]     : Initializing Spring embedded WebApplicationContext
main] w.s.c.ServletWebServerApplicationContext : Root WebApplicationContext: initialization completed in 70 ms
main] o.s.b.a.e.web.EndpointLinksResolver      : Exposing 13 endpoint(s) beneath base path '/actuator'
main] o.s.b.w.embedded.tomcat.TomcatWebServer  : Tomcat started on port(s): 8088 (http) with context path ''
main] c.j.springboot.actuator.Application      : Started Application in 2.791 seconds (process running for 3.116)
```

这样应用的端口就和端点管理端口分开了，8080用于应用，8088用于端点管理。既然是分开的，那么管理端点的Server要开启HTTPS访问就需要通过自己的management.server.ssl.*系列参数控制。

另外，把端点Server端点改为-1还能禁止以HTTP的方式暴露所有端点，和以exclude方式排除暴露的端点是一样的效果，如以下配置所示。

```
# 两者功能是一样的
management:
  server:
    port: -1

management:
  endpoints:
    web:
      exposure:
        exclude: "*"
```

所以，通过关闭显示所有端点的页面和更改默认的端点映射与端口，也能起到保护端点安全的目的。

12.2.7 端点实现机制

内置端点一旦启用，Spring Boot就会自动配置，比如health端点的自动配置类为HealthEndpointAutoConfiguration，其源码如下：

```
@AutoConfiguration
@ConditionalOnAvailableEndpoint(endpoint = HealthEndpoint.class)
@EnableConfigurationProperties(HealthEndpointProperties.class)
@Import({ HealthEndpointConfiguration.class, ReactiveHealthEndpointConfiguration.class,
        HealthEndpointWebExtensionConfiguration.class, HealthEndpointReactiveWeb-
        ExtensionConfiguration.class })
public class HealthEndpointAutoConfiguration {

}
```

自动配置类是空的，其实是通过@Import注解导入了不同Web类型的端点配置类。

根据*EndpointAutoConfiguration这个自动配置命名规则，可以很容易地找到其他端点的自动配置类及端点类，端点自动配置类都被注册在新的org.springframework.boot.autoconfigure.AutoConfiguration.imports自动配置文件中，如下图所示。

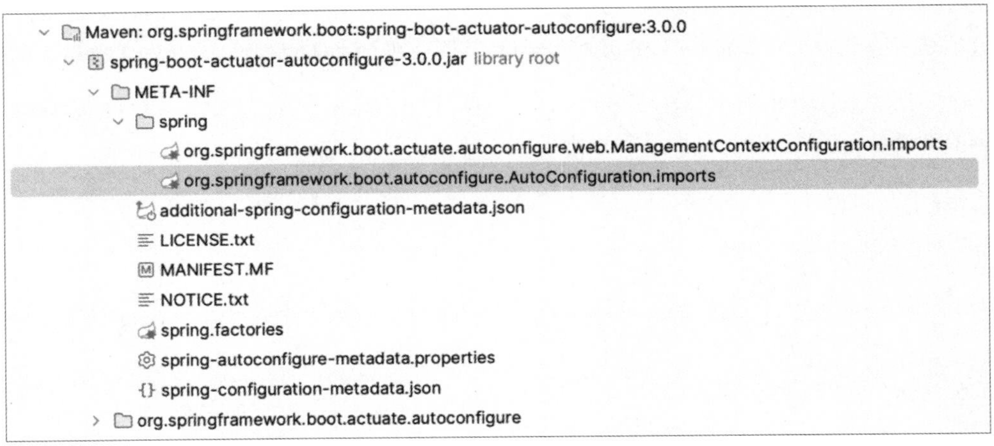

```
Maven: org.springframework.boot:spring-boot-actuator-autoconfigure:3.0.0
  spring-boot-actuator-autoconfigure-3.0.0.jar library root
    META-INF
      spring
        org.springframework.boot.actuate.autoconfigure.web.ManagementContextConfiguration.imports
        org.springframework.boot.autoconfigure.AutoConfiguration.imports
      additional-spring-configuration-metadata.json
      LICENSE.txt
      MANIFEST.MF
      NOTICE.txt
      spring.factories
      spring-autoconfigure-metadata.properties
      spring-configuration-metadata.json
    org.springframework.boot.actuate.autoconfigure
```

端点具体的实现类命名规则为 *Endpoint，比如健康端点的实现类为 HealthEndpoint，部分源码如下：

```
@Endpoint(id = "health")
public class HealthEndpoint extends HealthEndpointSupport<HealthContributor,
HealthComponent> {
    public static final EndpointId ID = EndpointId.of("health");

    ...

    @ReadOperation
    public HealthComponent health() {
        HealthComponent health = health(ApiVersion.V3, EMPTY_PATH);
        return (health != null) ? health : DEFAULT_HEALTH;
    }

    @ReadOperation
    public HealthComponent healthForPath(@Selector(match = Match.ALL_REMAINING)
String... path) {
        return health(ApiVersion.V3, path);
    }

    ...

}
```

端点实现类通过一个 @Endpoint 注解修饰，并且需要注册为 Spring 中的 Bean，在 Web 应用中支持 JMX 和 Web 两种暴露方式，也支持 @JmxEndpoint 和 @WebEndpoint 注解以不同的方式暴露，还可以通过 @EndpointJmxExtension 和 @EndpointWebExtension 注解扩展现有的端点。

通过在方法上定义不同的操作注解以实现不同 HTTP 方法与应用交互，支持的操作注解如下表所示。

操作注解	HTTP 方法
@ReadOperation	GET
@WriteOperation	POST
@DeleteOperation	DELETE

如果 Health 端点中定义了两个 @ReadOperation 注解，则说明这个端点只支持 HTTP GET 方法的请求，一个不带参数，另一个通过 @Selector 注解绑定 Path 路径上的参数。

另外，POST 请求 Content-Type 只能接受“application/vnd.spring-boot.actuator.v2+json、application/json”类型，所有方法的返回 Content-Type 则根据不同的返回类型决定，如下表所示。

方法返回	Content-Type
Void，Void	空
org.springframework.core.io.Resource	application/octet-stream
其他类型	application/vnd.spring-boot.actuator.v2+json，application/json

12.2.8　自定义端点

了解了端点实现机制，下面实现一个自定义的端点，示例如下：

```
@Component
@WebEndpoint(id = "test")
public class TestEndpoint {

    @ReadOperation
    public User getUser(@Selector Integer id) {
        return new User(id, "james", 18);
    }

    @WriteOperation
    public User updateUser(int id, @Nullable String name, @Nullable Integer age) {
        User user = getUser(id);
        user.setName(StringUtils.defaultIfBlank(name, user.getName()));
        user.setAge(ObjectUtils.defaultIfNull(age, user.getAge()));
        return user;
    }

}
```

这里创建了一个 TestEndpoint 端点类并通过 @Component 注册到 Spring 中，@WebEndpoint 注解表示只暴露为 HTTP 方式访问，端点类提供了两个方法，用于获取用户信息和更新用户信息。端点 ID 为 test，即可以通过 GET、POST 方法调用 /actuator/test 端点。

由于 @WebEndpoint 等端点注解的 enableByDefault 参数默认是 true，所以新建的端点默认是启用状态的，不需要再额外配置，虽然 @WebEndpoint 注解允许暴露为 HTTP 方式，但并未真正开启暴露。因为在之前的章节中已经把 Web 方式暴露为“*”了，所以这里也不需要再配置了。

发起 GET 请求，如下图所示。

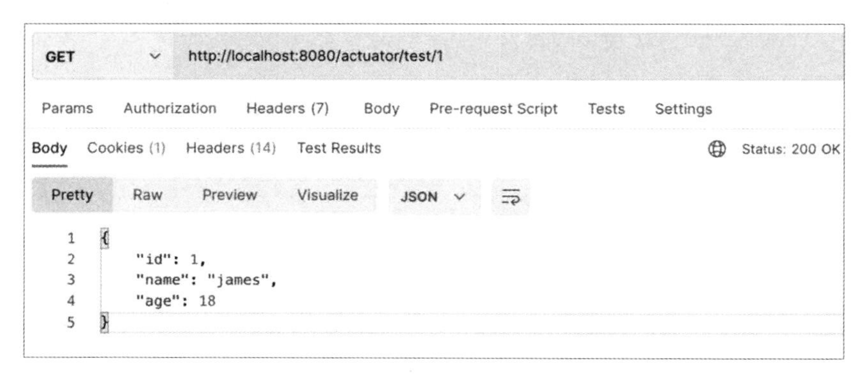

/actuator/test/1 后面的 1 是通过 @Selector 注解绑定的 URL 路径上的参数，默认只绑定单个参数，如果要绑定多个，则可以指定 @Selector(match = Match.ALL_REMAINING) 匹配所有 URL 路径上的参数。

再发起 POST 请求，如下图所示。

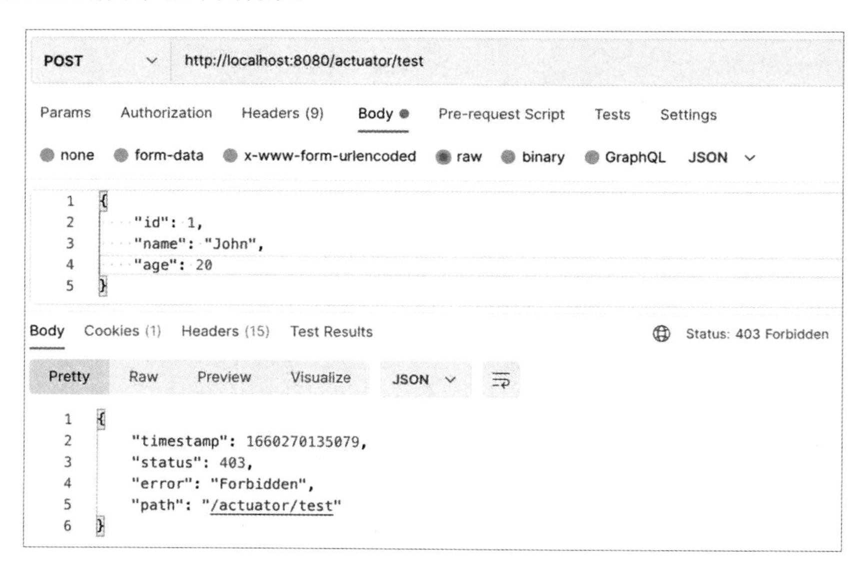

POST、DELETE 方法的请求默认会返回 403 错误，这是因为 Spring Boot 默认会开启 CSRF（跨站请求伪造）保护机制，如果不是基于浏览器的应用，则建议关闭 CSRF 保护机制，关闭示例代码如下：

```
@Bean
public SecurityFilterChain securityFilterChain(HttpSecurity http) throws Exception {
    return http.authorizeHttpRequests((authorize) -> {
                authorize.requestMatchers("/").permitAll()
                        .requestMatchers(EndpointRequest.to("health")).hasRole
("ENDPOINT_ADMIN")
                        .requestMatchers(EndpointRequest.toAnyEndpoint()).permitAll();
            })
            .csrf(csrf -> csrf.disable())
            .formLogin(withDefaults())
            .logout().logoutSuccessUrl("/")
            .and().build();
}
```

或者只对端点映射关闭 CSRF 保护机制，如以下配置示例所示。

```
@Bean
public SecurityFilterChain securityFilterChain(HttpSecurity http) throws Exception {
    return http.authorizeHttpRequests((authorize) -> {
                authorize.requestMatchers("/").permitAll()
                        .requestMatchers(EndpointRequest.to("health")).hasRole
("ENDPOINT_ADMIN")
                        .requestMatchers(EndpointRequest.toAnyEndpoint()).permitAll();
            })
//          .csrf(csrf -> csrf.disable())
            .csrf(csrf -> csrf.ignoringRequestMatchers(EndpointRequest.toAnyEndpoint()))
            .formLogin(withDefaults())
            .logout().logoutSuccessUrl("/")
            .and().build();
}
```

关闭 CSRF 保护机制后，再重启应用以 POST 方式调用自定义端点，如下图所示。

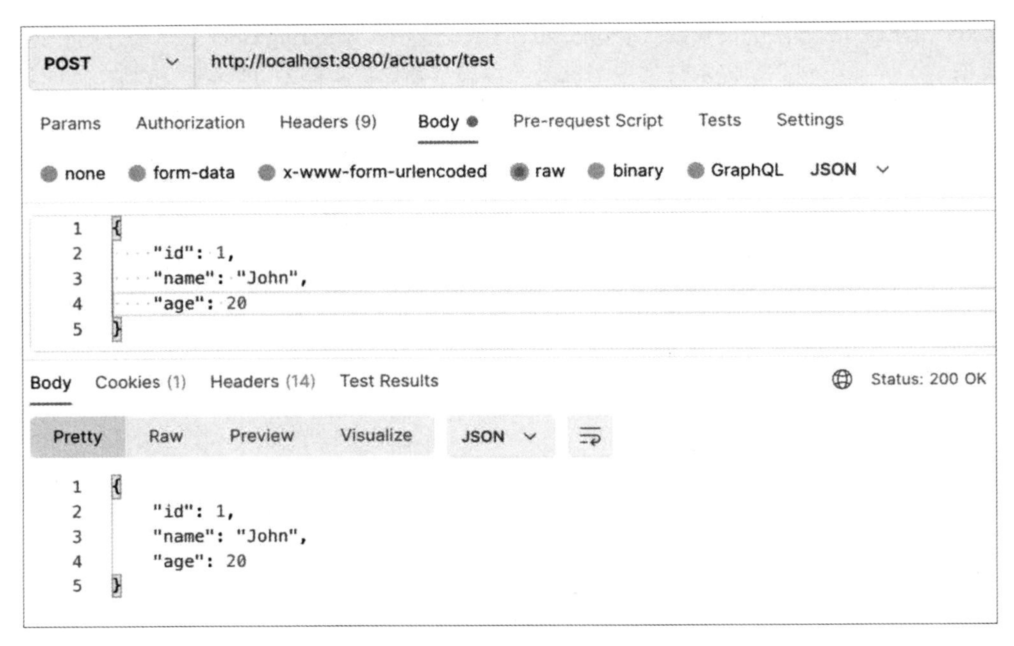

如果端点类操作方法定义了参数，则参数不能为空，否则返回400错误，通过在方法参数前添加@Nullable注解就可以允许参数为空，把name参数名改为name1后再调用，操作不会返回异常，而是返回默认值，如下图所示。

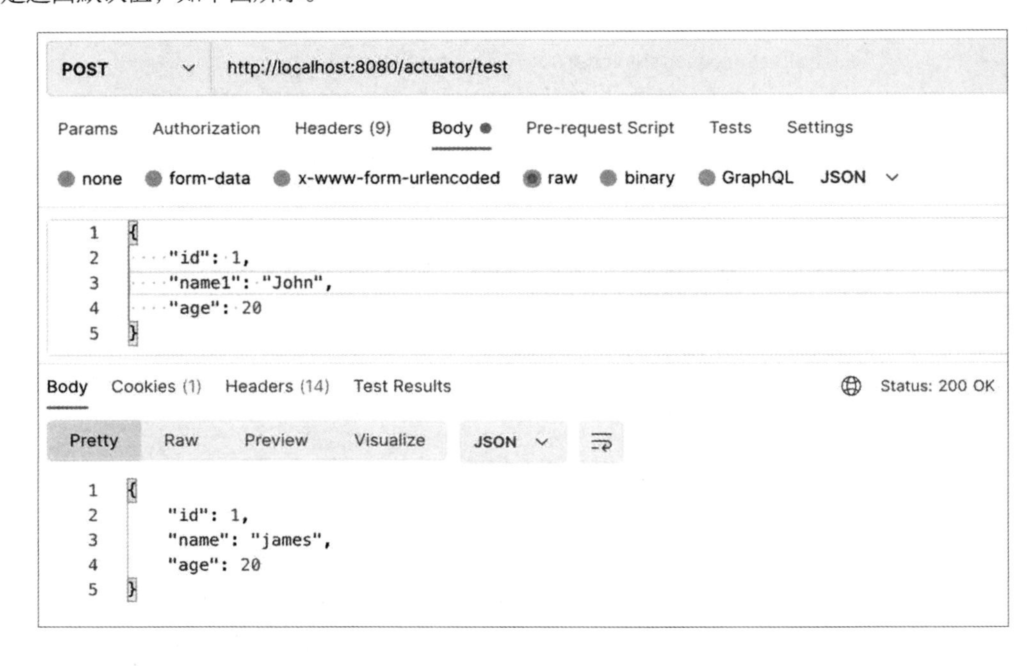

另外，端点也支持跨域调用，如以下跨域配置所示。

```
management:
  endpoints:
    web:
      cors:
        allowed-origins: "https://javastack.cn"
        allowed-methods: "GET,POST"
```

这里配置了可以允许来自 https://javastack.cn 网站所有 GET 和 POST 方法的请求，这两个配置也支持“*”通配符。

12.3 loggers（日志端点）

Spring Boot 提供了一个 loggers 日志端点，对应的端点类为 LoggersEndpoint，核心源码如下：

```
@Endpoint(id = "loggers")
@RegisterReflectionForBinding({ GroupLoggerLevelsDescriptor.class, SingleLoggerLevels-
Descriptor.class })
public class LoggersEndpoint {

    ...

    @ReadOperation
    public LoggersDescriptor loggers() {
        Collection<LoggerConfiguration> configurations = this.loggingSystem.getLogger-
Configurations();
        if (configurations == null) {
            return LoggersDescriptor.NONE;
        }
        return new LoggersDescriptor(getLevels(), getLoggers(configurations), getGroups());
    }

    @ReadOperation
    public LoggerLevelsDescriptor loggerLevels(@Selector String name) {
        Assert.notNull(name, "Name must not be null");
        LoggerGroup group = this.loggerGroups.get(name);
        if (group != null) {
            return new GroupLoggerLevelsDescriptor(group.getConfiguredLevel(),
group.getMembers());
        }
        LoggerConfiguration configuration = this.loggingSystem.getLoggerConfiguration(name);
```

```
        return (configuration != null) ? new SingleLoggerLevelsDescriptor(configuration): null;
    }

    @WriteOperation
    public void configureLogLevel(@Selector String name, @Nullable LogLevel configuredLevel) {
        Assert.notNull(name, "Name must not be empty");
        LoggerGroup group = this.loggerGroups.get(name);
        if (group != null && group.hasMembers()) {
            group.configureLogLevel(configuredLevel, this.loggingSystem::setLogLevel);
            return;
        }
        this.loggingSystem.setLogLevel(name, configuredLevel);
    }

    ...

}
```

日志端点提供了两个读操作及一个写操作，可以查看正在运行的应用的日志级别设置，也可以动态调整某个包的日志级别，该端点默认不会暴露，需要自行暴露，详见之前章节的介绍。

获取日志端点信息（GET），如下图所示。

调整某个包的日志级别（POST），如下图所示。

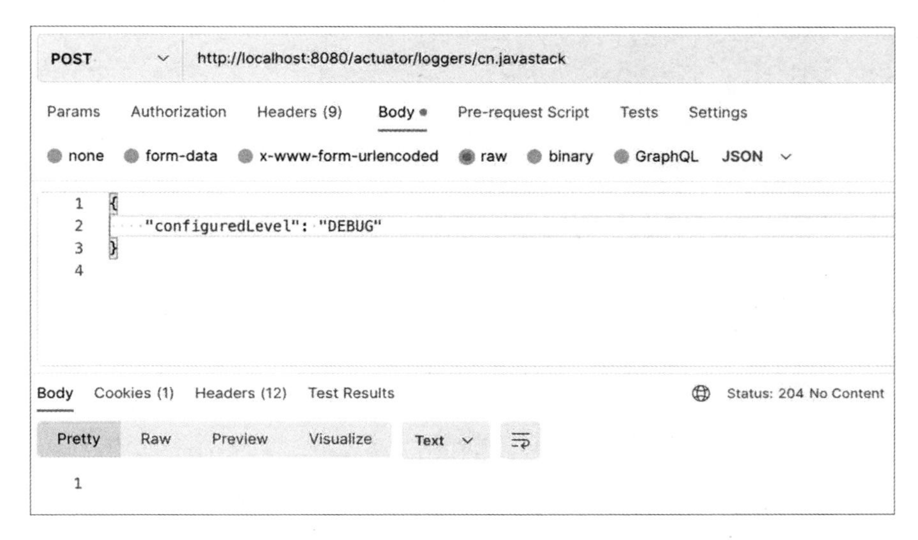

这里调整 cn.javastack 包的日志级别为 DEBUG，再重新获取日志端点信息，如下图所示。

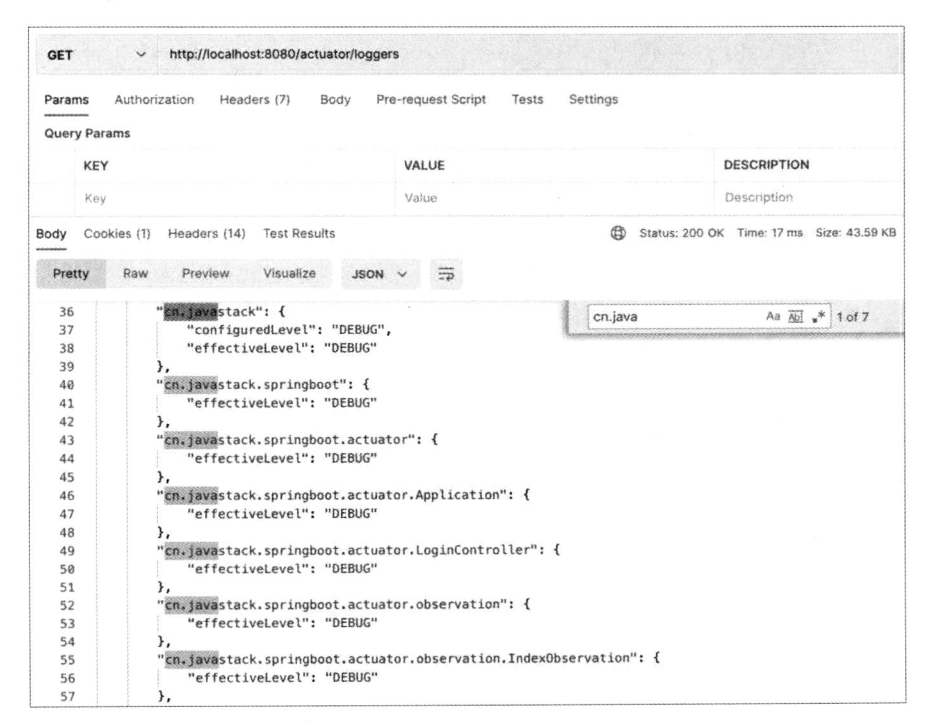

如上图所示，该包的日志级别已经设置为 DEBUG 了。

loggers日志端点支持设置的日志级别有TRACE、DEBUG、INFO、WARN、ERROR、FATAL、OFF、null（恢复默认设置）。

null表示恢复默认的包日志级别，以上cn.javastack包的日志级别已经调整为了DEBUG，发送null就会恢复默认的INFO日志级别，发送的内容如下：

```
{
    "configuredLevel": null
}
```

loggers日志端点目前只能用来查看和更新包的日志级别，以及查看所有的日志分组，但不支持查看和更新复杂的日志配置。

12.4 Observability（可观测性，Spring Boot 3.0+）

Observability（可观测性）是指从系统外部观测正在运行的系统的内部状态的能力，Spring Boot 3.0通过Micrometer和Micrometer追踪来提高应用的可观测性，支持集成Micrometer 1.10+，引入了全新的可观测API并自动配置Micrometer追踪，包括对Brave、OpenTelemetry、Zipkin和Wavefront组件的支持。

使用Micrometer可观测API完成观测后，可将观测数据主动报告给Zipkin等组件，并支持自定义追踪参数的配置，这个新特性可以更好地帮助我们了解并监控应用的健康状况。

Observability主要由以下三大组件组成：

- Logging（日志记录）。
- Metrics（度量/指标）。
- Traces（链路跟踪，Spring Boot 3.0+）。

日志就是指平常的日志输出，对于指标和跟踪，Spring Boot使用的是Micrometer Observation，如果要创建自己的Observations，则可以注入一个ObservationRegistry，并使用它添加对应的数据，如下面的示例所示。

```
@Slf4j
@RequiredArgsConstructor
@Component
public class IndexObservation {
```

```
    private final ObservationRegistry observationRegistry;

    public void observe() {
        Observation.createNotStarted("indexObservation", this.observationRegistry)
                .lowCardinalityKeyValue("area", "cn")
                .highCardinalityKeyValue("userId", "10099")
                .observe(() -> {
                    // 执行观测时的业务逻辑
                    log.info(" 开始执行业务逻辑 ...");
                });
    }

}
```

几个核心方法如下：

- createNotStarted：创建一个 Observation，但不会马上启动观测。
- lowCardinalityKeyValue：指定低基数的标签，它们将被添加到 Metrics 和 Traces 中。
- highCardinalityKeyValue：指定高基数的标签，它们将被添加到 Traces 中。
- observe：开始执行当前 Observation 观测，并传入一个 Runnable 线程用于执行业务逻辑。

observe方法接收一个 Runnable 线程参数，其源码如下：

```
default void observe(Runnable runnable) {
    start();
    try (Scope scope = openScope()) {
        runnable.run();
    }
    catch (Exception exception) {
        error(exception);
        throw exception;
    }
    finally {
        stop();
    }
}
```

它主要由以下几个过程组成：

- 启动 Observation。
- 打开一个 Scope。

- 调用线程的 Runnable.run() 方法。
- 关闭 Scope。
- 如果有错误，则向观测对象发出错误信号。
- 停止 Observation。

现在改造一下 /首页接口，让首页执行 IndexObservation 观测，如下面的示例所示。

```
@RequiredArgsConstructor
@Slf4j
@Controller
public class LoginController {

    private final IndexObservation indexObservation;

    @GetMapping("/")
    @ResponseBody
    public String index() {
        log.info("this is index page.");
        indexObservation.observe();
        return "index page.";
    }

}
```

注入 IndexObservation 并调用其 indexObservation 方法开始执行 Observation 观测，然后就会记录首页观测到的相关标签数据，可在链路跟踪中查找并展示，具体效果见后续的 Traces 链路跟踪章节。

12.5 Metrics（指标）

12.5.1 内置指标

Spring Boot 内置了各种技术的指标的实现，比如 JVM、System、Tomcat、Logger、Spring MVC 等，通过 /actuator/metrics 指标端点可以显示所有应用收集到的指标，如下图所示。

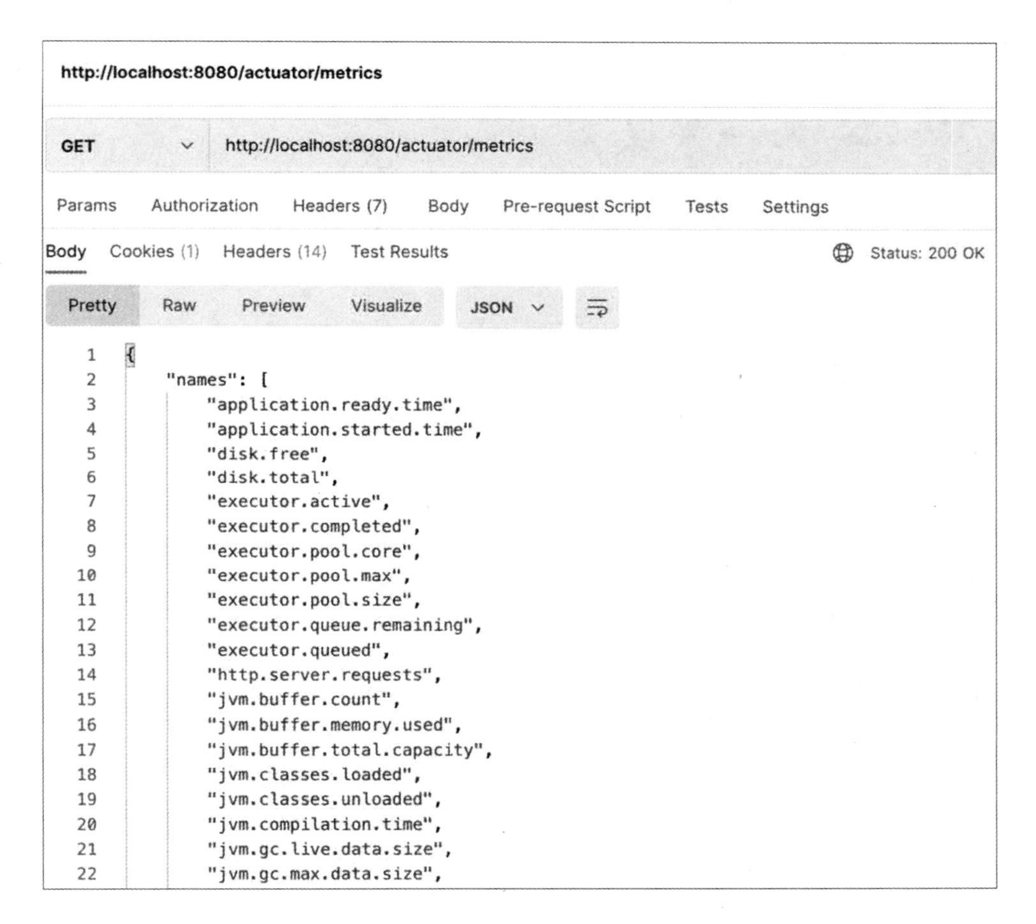

此端点默认是不暴露的，需要自行暴露。另外还可以通过 /actuator/metrics/xxx 查看详细指标信息，比如可通过 /actuator/metrics/jvm.memory.max 路径查看 JVM 最大内存，如下图所示。

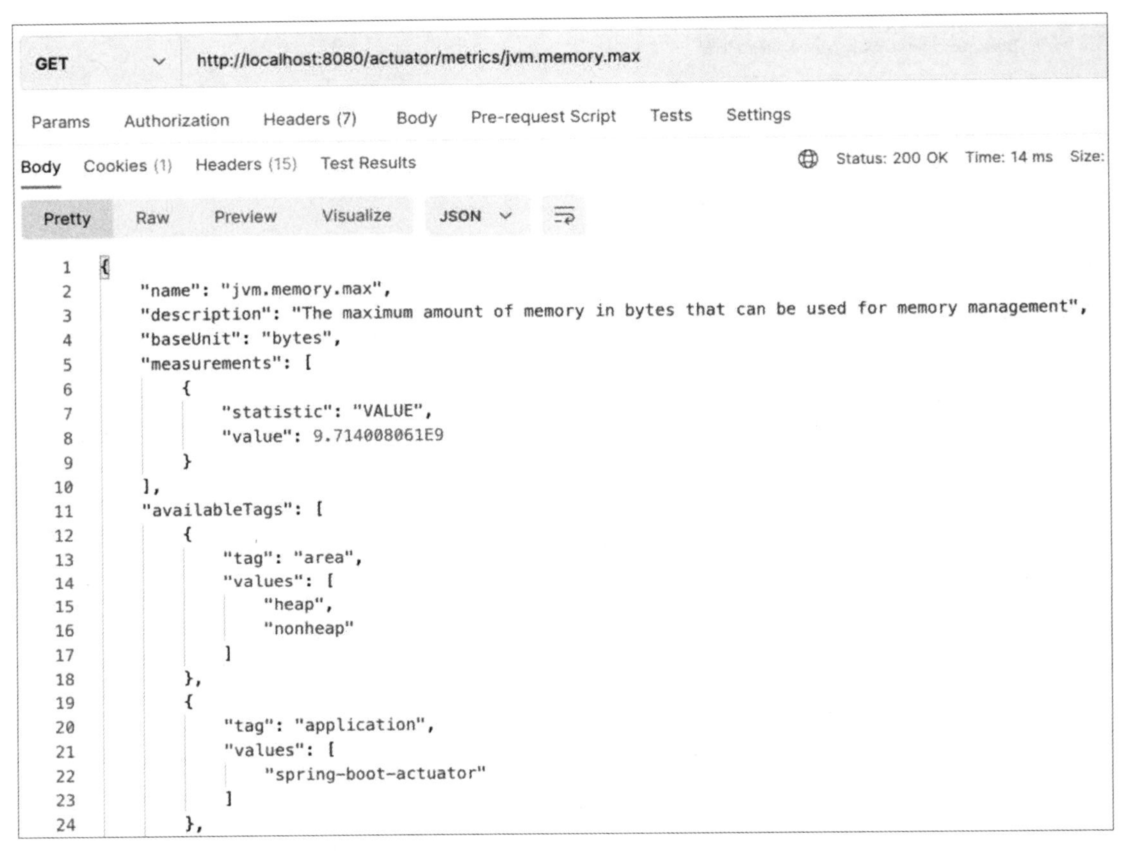

有了这些指标端点为前提，监控系统才能收集并展示相关信息。

12.5.2 自定义指标

如果 Spring Boot 提供的指标不符合要求，则可以通过注册多个 MeterBinder 实现多个自定义指标，比如下面的自定义指标示例：

```
@Configuration
public class MetricsConfig {

    @Bean
    public MeterBinder initDate(Environment env) {
        return (registry) -> Gauge.builder("init.date", this::date).register(registry);
    }
```

```
    @Bean
    public MeterBinder systemDate(Environment env) {
        return (registry) -> Gauge.builder("system.date", this::date).register(registry);
    }

    private Number date() {
        return 2022.01;
    }

}
```

这里创建了两个测试指标，通过 /actuator/metrics 或者 /actuator/prometheus 端点可以验证指标是否被收集，如下图所示。

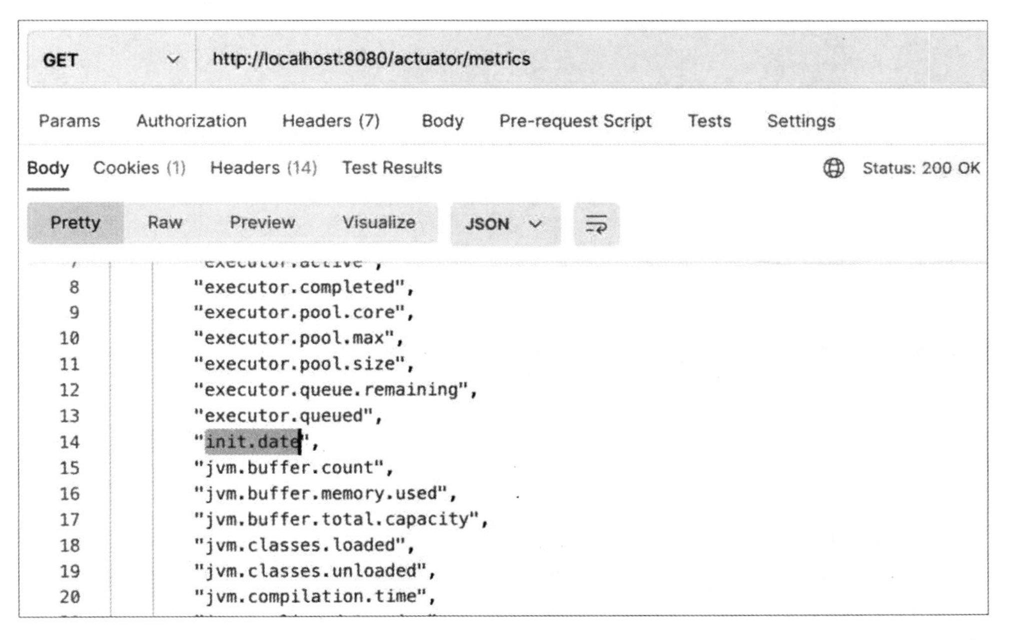

然后获取该指标的数据，如下图所示。

12.6 Traces（链路跟踪，Spring Boot 3.0+）

12.6.1 概述

Spring Boot Actuator 为 Micrometer Tracing 提供了依赖管理和自动配置，Micrometer 是当前比较流行的跟踪类库门面，Spring Boot 为以下 Tracers 链路跟踪解决方案提供了自动配置：

- 使用 OpenTelemetry 结合 Zipkin 或者 Wavefront。
- 使用 OpenZipkin Brave 结合 Zipkin 或者 Wavefront。

OpenTelemetry 是一款开源的可观测框架及数据收集中间件，可以使用它来生成、收集和导出观测数据（Metrics、Logs 和 Traces），它只是收集数据，分析和展示数据还需要结合其他中间件。

Zipkin 是由 Twitter 开源的一个分布式追踪系统，它可以收集、解决服务中的延迟问题所需的时间数据，包括收集和查找此数据并进行展示，如果日志中有 Trace ID，则直接可以展示，如下图所示。

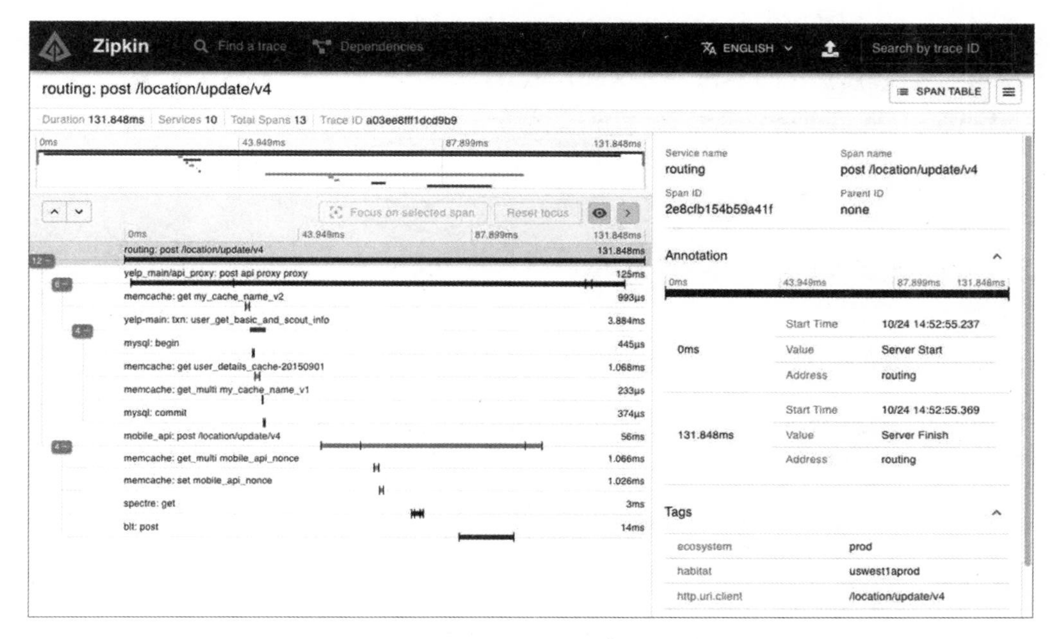

也可以根据其他属性查询数据，数据展示非常详细。

Zipkin UI 还提供了一个依赖关系图，显示有多少跟踪请求通过了每个应用，这有助于识别聚合行为，包括错误路径或对已弃用服务的调用，如下图所示。

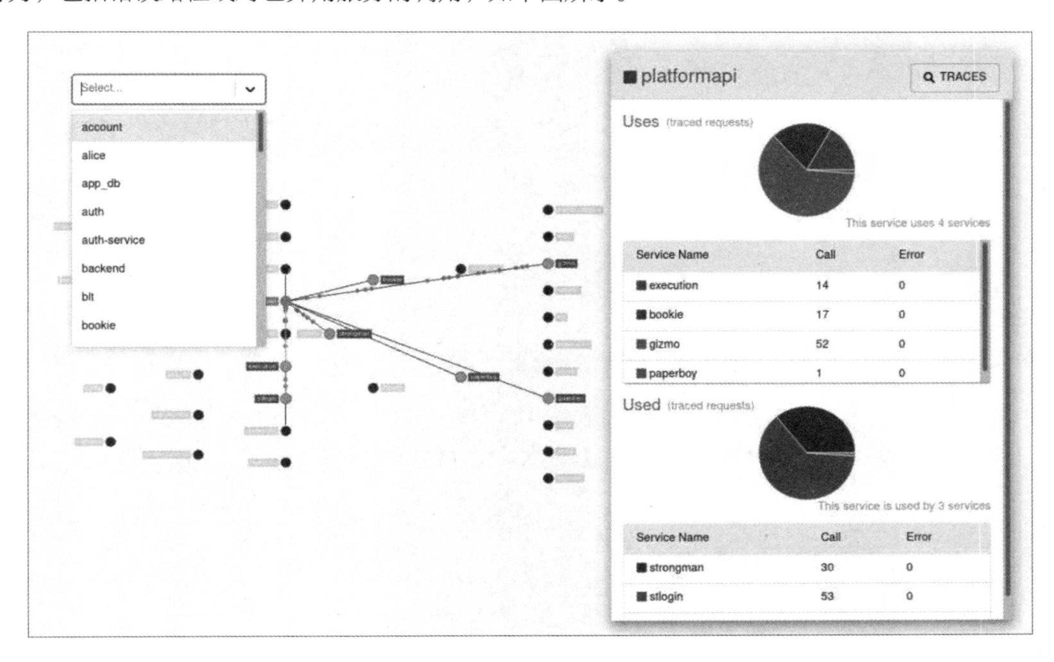

应用要使用 Zipkin 就需要向 Zipkin 发送跟踪数据，一般是通过 HTTP 或 Kafka 的方式，但也可以是 Apache ActiveMQ、gRPC 和 RabbitMQ 等，数据可以存储在内存中，也可以在 MySQL、Apache Cassandra 或 Elasticsearch 等组件中持久化存储。

12.6.2 链路跟踪环境搭建

这里使用 OpenTelemetry 结合 Zipkin 的方案来实现 Tracing 调用链跟踪，本地需要安装 Zipkin 环境。

Java 8+/Linux环境下最快安装 Zipkin 的方法的参考命令如下：

```
curl -sSL https://zipkin.io/quickstart.sh | bash -s
java -jar zipkin.jar &
```

本地可以访问链接 19 下载最新版再运行 Zipkin。

使用默认的配置，安装并运行 Zipkin 之后，访问控制台 http://localhost:9411，如下图所示。

应用需要导入相关依赖，如下面的依赖配置所示。

```
<!-- 将 Micrometer Observation API 桥接到 OpenTelemetry -->
<dependency>
    <groupId>io.micrometer</groupId>
    <artifactId>micrometer-tracing-bridge-otel</artifactId>
</dependency>

<!-- 向 Zipkin 报告调用链跟踪信息 -->
<dependency>
```

```
    <groupId>io.opentelemetry</groupId>
    <artifactId>opentelemetry-exporter-zipkin</artifactId>
</dependency>
```

添加采样参数，如下面的配置所示。

```
management:
  tracing:
    sampling:
      probability: 1.0
```

默认情况下，Spring Boot 仅对 10% 的请求进行采样（0.1），以防止 Traces 跟踪导致后端系统不堪重负，这里为了便于测试，将采样参数切换为 1.0，即对所有请求（100%）进行采样。

如果需要在日志中显示 Trace ID 和 Span ID，则可以进行以下设置：

```
logging:
  pattern:
    level: ${spring.application.name:},%X{traceId:-},%X{spanId:-}
```

参数说明如下：

- Trace ID：链路跟踪 ID，一次请求完整链路的唯一标识，由一组 Span 单元组成的一个树状结构。
- Span ID：基本工作单元 ID，每条链路都由不同的 Span ID 组成。

12.6.3 链路跟踪 / 展示

链路跟踪环境搭建好及 Zipkin 启动之后，再启动应用，并发起 HTTP 接口调用，Spring Boot 会为 HTTP 请求创建一个 Observation 观测，它桥接到 OpenTelemetry，然后通过 OpenTelemetry 向 Zipkin 报告每个链路的跟踪数据。

在 Observability可观测性一节中，已经为 / 首页实现了 IndexObservation 观测并记录了相关标签数据，现在连续调用三次 http://localhost 接口，然后回到 Zipkin 查询当前应用的调用链，如下图所示。

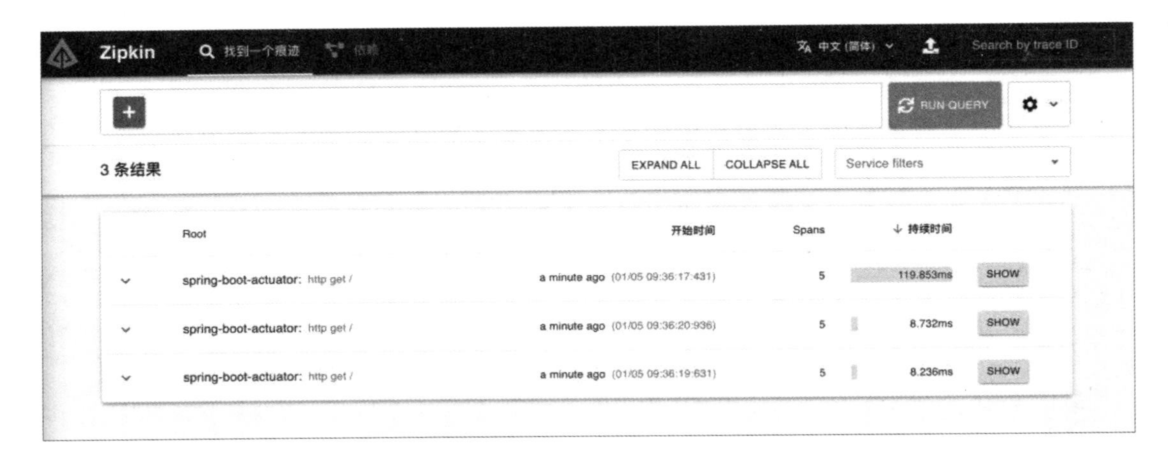

可以看到这三次的请求链路，随便单击一条进入详细链路，如下图所示。

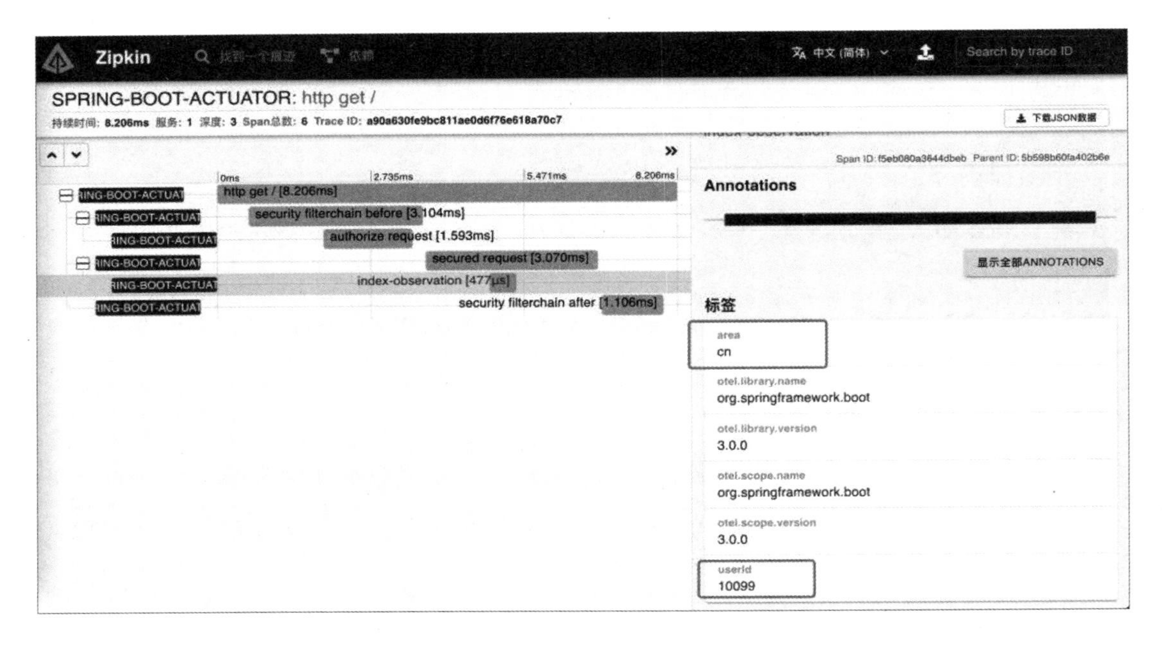

可以看到首页观测记录及携带的 area 和 userId 标签数据，后台也记录了许多调用链跟踪日志，如下图所示。

```
11f2b34ee93367c82b9f08e68ed4be,64869e2e085dd2f3 66642 --- [nio-8080-exec-2] c.j.springboot.actuator.LoginController   : this is index page.
11f2b34ee93367c82b9f08e68ed4be,1bbfe6ddf3a27658 66642 --- [nio-8080-exec-2] c.j.s.a.observation.IndexObservation     : 开始执行业务逻辑...
981ce75bafe26fdc0890f3b7db6da0,c4c45a6aa86ba925 66642 --- [nio-8080-exec-3] c.j.springboot.actuator.LoginController   : this is index page.
981ce75bafe26fdc0890f3b7db6da0,f718955db3b49bb6 66642 --- [nio-8080-exec-3] c.j.s.a.observation.IndexObservation     : 开始执行业务逻辑...
37e924e5e4942b2e2690ea5bfec674,a47d674426f09624 66642 --- [nio-8080-exec-4] c.j.springboot.actuator.LoginController   : this is index page.
37e924e5e4942b2e2690ea5bfec674,7a479684f3df9f8d 66642 --- [nio-8080-exec-4] c.j.s.a.observation.IndexObservation     : 开始执行业务逻辑...
e37d9ada7a58a0194273fcfc03f956,5d960c5a22519737 66642 --- [nio-8080-exec-5] c.j.springboot.actuator.LoginController   : this is index page.
e37d9ada7a58a0194273fcfc03f956,a38455d2b0c44318 66642 --- [nio-8080-exec-5] c.j.s.a.observation.IndexObservation     : 开始执行业务逻辑...
fbffea7a6d02930091a9b043ca81cd,dea8128e55a6645a 66642 --- [nio-8080-exec-6] c.j.springboot.actuator.LoginController   : this is index page.
fbffea7a6d02930091a9b043ca81cd,0a8b29f3145e68f4 66642 --- [nio-8080-exec-6] c.j.s.a.observation.IndexObservation     : 开始执行业务逻辑...
806e0e73d302017d5813ec879aeb69,6702204cac426ec6 66642 --- [nio-8080-exec-7] c.j.springboot.actuator.LoginController   : this is index page.
806e0e73d302017d5813ec879aeb69,3fd85238c7aa4f93 66642 --- [nio-8080-exec-7] c.j.s.a.observation.IndexObservation     : 开始执行业务逻辑...
```

可以看到每条链路输出了两条日志，也就是在 / 首页方法中输出的两条日志，其 Trace ID 相同，Span ID 不同。

Spring Boot 可观测性是针对 Spring Boot 项目的，在 Spring Cloud 微服务体系中一般推荐使用 Spring Cloud Sleuth+Zipkin 方案，但在 Spring Cloud 2022.0.0 发布后，Spring Cloud 移除了 Spring Cloud Sleuth 链路跟踪模块，如官方发布的文档所示。

- Spring Cloud Sleuth
  - This project has been removed from the release train. The core of this project has moved to Micrometer Tracing project and the instrumentations will be moved to Micrometer and all respective projects (no longer all instrumentations will be done in a single repository.

Spring Cloud Sleuth 链路跟踪模块也被 Micrometer 相关项目替代了，笔者升级到了 Spring Cloud 2022.0.0 后，发现已经不能下载 Spring Cloud Sleuth 依赖了，之前老版本的 Spring Cloud Sleuth 应用也要全部更换。

12.7 Spring Boot Admin

12.7.1 概述

Spring Boot Admin=Spring Boot+Admin，用于管理和监控 Spring Boot 应用，它并不是 Spring Boot 官方提供的，而是某个社区开源的。

开源地址为链接 20。

开源官网显示 Spring Boot Admin 的主要功能如下：

- 显示健康状况。
- 显示详细指标（如 JVM、内存、micrometer、数据源、缓存等）。

- 监控并下载日志文件。
- 显示 JVM 中的系统及环境变量。
- 显示 Spring Boot 配置属性。
- 简单的日志级别管理。
- 支持与 JMX Bean 交互。
- 显示线程堆栈。
- 下载堆信息。
- 显示 HTTP 跟踪信息。
- 显示 HTTP 端点。
- 显示计划任务。
- 显示 / 删除活动会话。
- 显示 Flyway/Liquibase迁移信息。
- 状态变更通知。

Spring Boot Admin 包括 Spring Boot Admin Server 和 Spring Boot Admin Client，Spring Boot Admin Client 即 Spring Boot 应用，以 HTTP 的方式向 Spring Boot Admin Server 注册，Spring Cloud 应用则通过 Eureka、Consul 等注册中心进行注册，Spring Boot Admin UI 只是基于 Spring Boot Actuator 端点之上的 Vue.js 应用。

12.7.2 环境搭建

Spring Boot 集成 Spring Boot Admin 需要先搭建 Spring Boot Admin Server，再搭建 Spring Boot Admin Client，Spring Boot Admin Server 用于收集 Spring Boot Admin Client 的健康数据并进行监控报警。

1. 搭建 Spring Boot Admin Server

添加 Spring Boot Admin Server 所需要的依赖，如以下依赖配置所示。

```
<dependency>
    <groupId>de.codecentric</groupId>
    <artifactId>spring-boot-admin-starter-server</artifactId>
    <version>${spring-boot-admin-starter-server.version}</version>
```

```
</dependency>
<dependency>
    <groupId>org.springframework.boot</groupId>
    <artifactId>spring-boot-starter-web</artifactId>
</dependency>
```

在启动类上添 @EnableAdminServer 注解以启用 Spring Boot Admin Server，如以下示例所示。

```
@EnableAdminServer
@SpringBootApplication
public class Application {

    public static void main(String[] args) {
        SpringApplication.run(Application.class, args);
    }


}
```

2. 搭建 Spring Boot Admin Client

添加 Spring Boot Admin Client 所需要的依赖，如以下依赖配置所示。

```
  <dependency>
      <groupId>de.codecentric</groupId>
      <artifactId>spring-boot-admin-starter-client</artifactId>
      <version>${spring-boot-admin-starter-client.version}</version>
  </dependency>
  <dependency>
      <groupId>org.springframework.boot</groupId>
      <artifactId>spring-boot-starter-actuator</artifactId>
  </dependency>
  <dependency>
      <groupId>org.springframework.boot</groupId>
      <artifactId>spring-boot-starter-security</artifactId>
  </dependency>
```

然后添加以下配置：

```
spring:
  boot:
    admin:
      client:
        url: http://localhost:8080 #1
```

```
management:
  endpoints:
    web:
      exposure:
        include: '*' #2
  info:
    env:
      enabled: true #3
```

核心参数说明如下：

- #1：指定需要注册的 Spring Boot Admin Server 的地址。
- #2：Spring Boot 3.0 默认只暴露 health 健康端点，这里全部暴露了，在生产环境中需谨慎选择。
- #3：Spring Boot 2.6+ 中的 env 信息默认是禁用的，这里需要手动开启。

然后禁用所有安全机制，让 actuator 所有端点可以被访问，如以下配置示例所示。

```
@Configuration
public class SecurityPermitAllConfig extends WebSecurityConfigurerAdapter {

    @Override
    protected void configure(HttpSecurity http) throws Exception {
        http.authorizeRequests().anyRequest().permitAll()
            .and().csrf().disable();
    }

}
```

这里为了测试，先禁用了所有安全措施，确保所有端点可以被访问，不然无法被监控到，实际项目中可自行设置安全性，具体可参考前面的端点安全性章节。

> 其他项目基础代码略，有需要的读者可以参考笔者提供的 Spring Boot 仓库的完整项目代码。

项目搭建和配置完成后，先后启动 Spring Boot Server 和 Spring Boot Client，必须先启动 Spring Boot Server，因为 Spring Boot Client 启动的时候需要向 Spring Boot Server 注册，否则会注册失败。

12.7.3 监控页面

1. 首页菜单

启动 Spring Boot Admin Server 和 Spring Boot Admin Client 后，打开 Spring Admin Server 控制台页面，地址如下：

```
http://localhost:8088/
```

打开首页，如下图所示。

服务端和客户端都运行成功了，Spring Boot Admin 3.x 比 2.x 版本的界面看起来更舒服了，单击对应的项目，可以看到更多的细节。

2. 细节菜单

细节菜单中展示了一些健康、JVM 相关信息，如下图所示。

这些信息有助于开发者掌握应用的 JVM 健康状况。

3. 性能菜单

性能菜单中可以添加并显示各种性能指标，如下图所示。

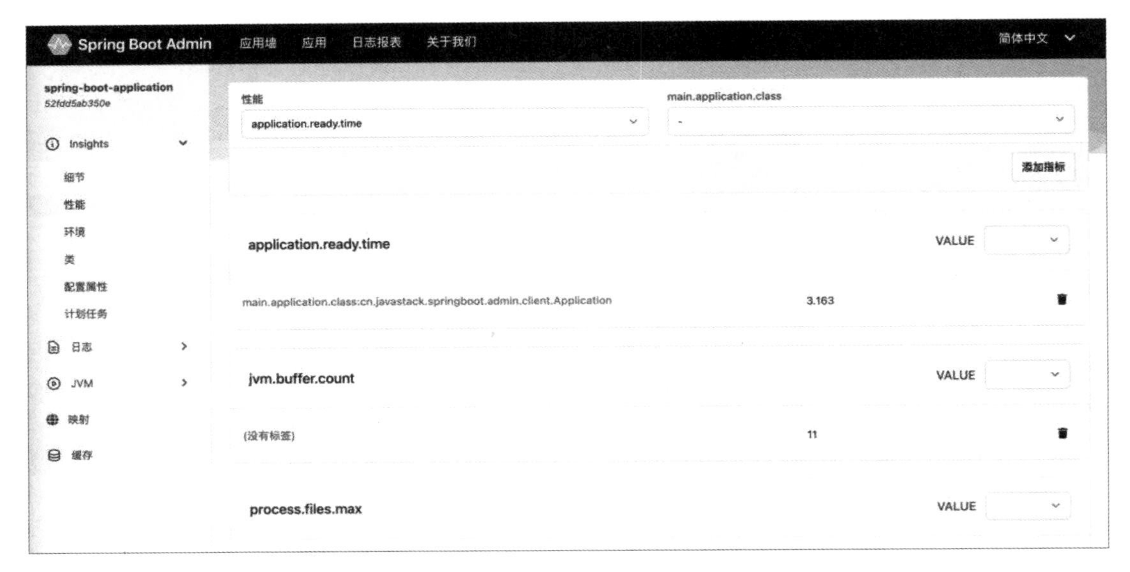

这些指标有助于我们监控更多的应用运行时的性能参数。

4. 环境菜单

环境菜单中显示所有 Servlet 初始参数、系统、环境变量参数等，如下图所示。

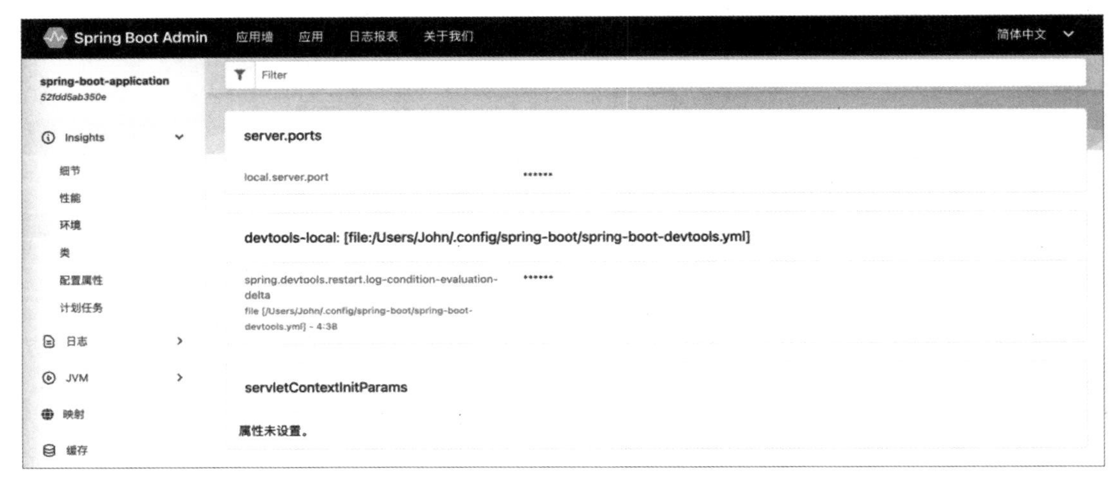

可视化搜索应用的相关参数并展示，应用的相关参数就不必用 Linux 命令查询了。

5. 类菜单

类菜单中显示所有 Spring 容器中的 Bean，以及是否单例，如下图所示。

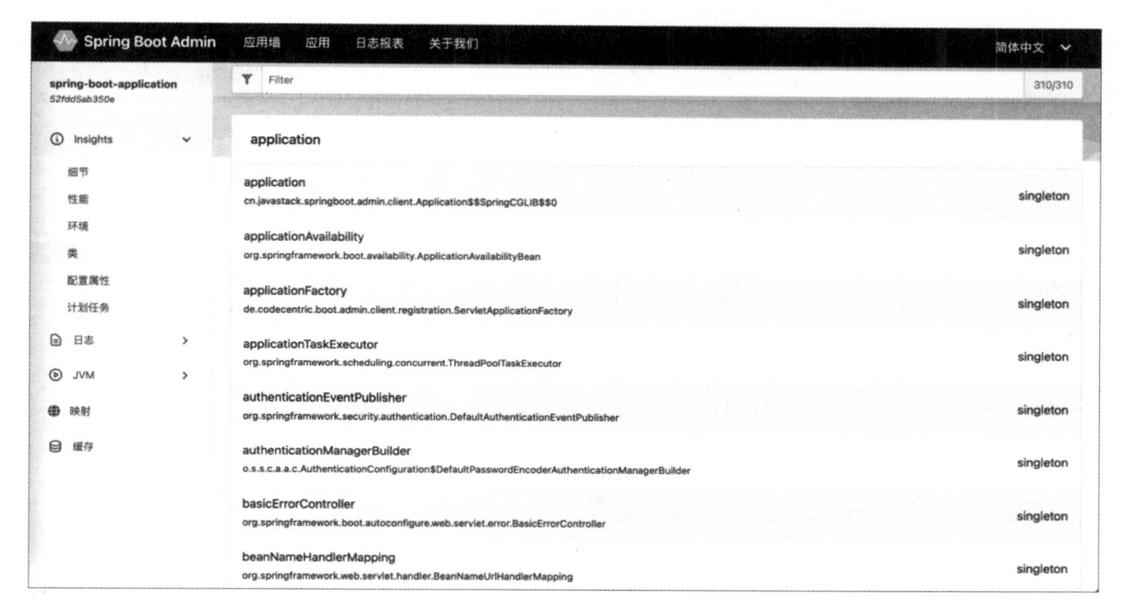

这项功能对排查疑难问题很有用。

6. 配置属性菜单

配置属性菜单中显示所有 Spring Boot 的配置参数，如下图所示。

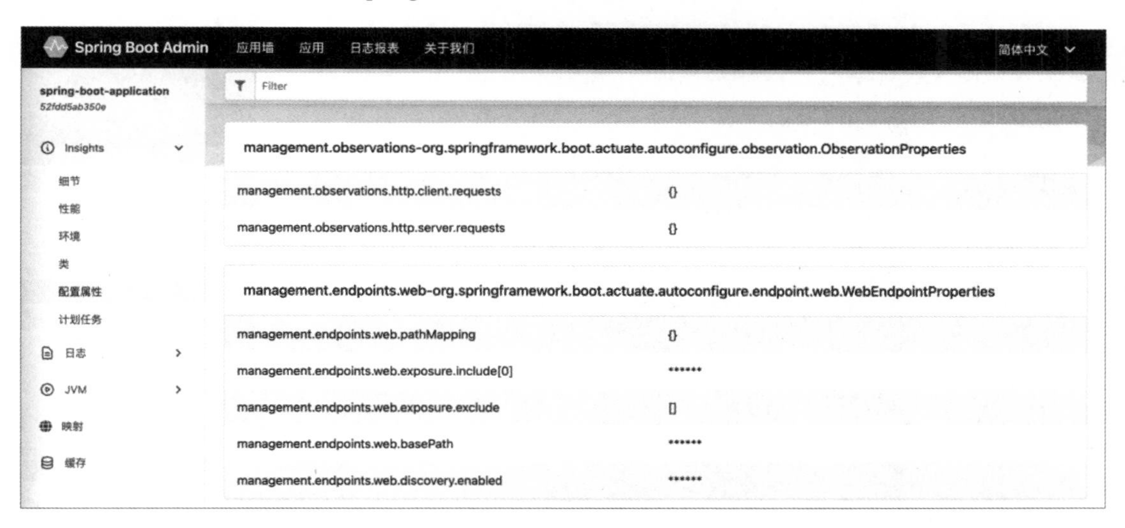

7. 计划任务菜单

计划任务菜单中显示所有的计划任务，如下图所示。

这里只是展示了任务信息及状态，并不能进行控制。

8. 日志配置菜单

日志配置菜单中显示所有类的日志级别，并能修改某个类的日志级别，如下图所示。

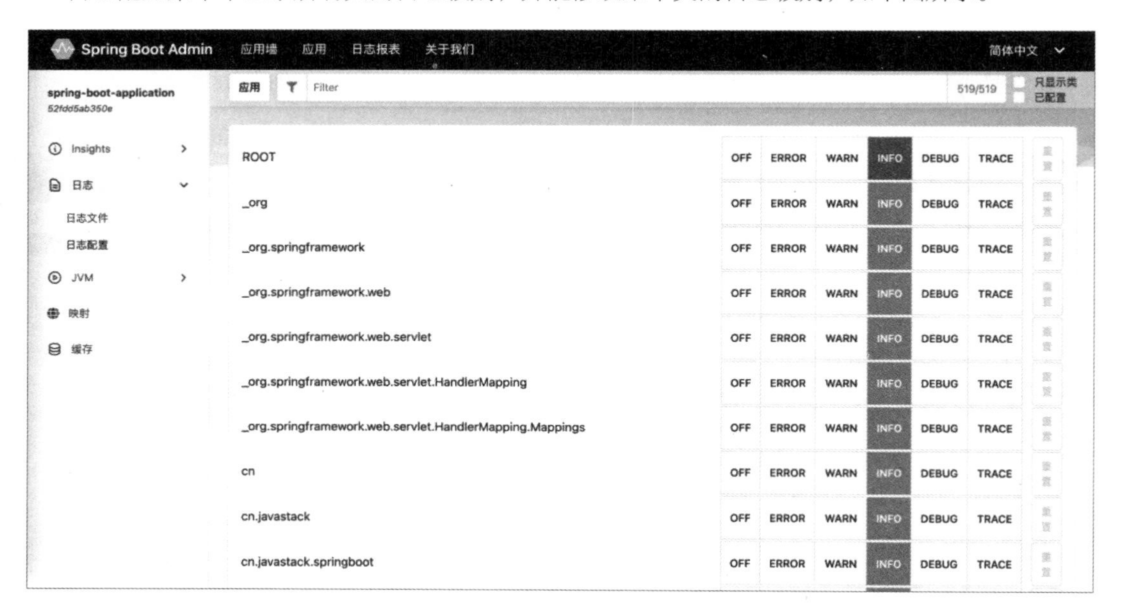

默认情况下，日志文件无法通过日志端点访问，需要在客户端应用配置中配置 logging.file. path

或 logging.file.name 参数来启用 logfile 端点，如以下配置所示。

```
logging:
  file:
    name: ./logs/sba-client.log
```

再重启客户端应用，刷新页面，如下图所示。

日志页面实时展示日志，不用再使用一堆命令查看了。

9. JVM 菜单

JVM 菜单中显示当前所有线程堆栈、堆栈，并能进行下载分析，如下面两个图所示。

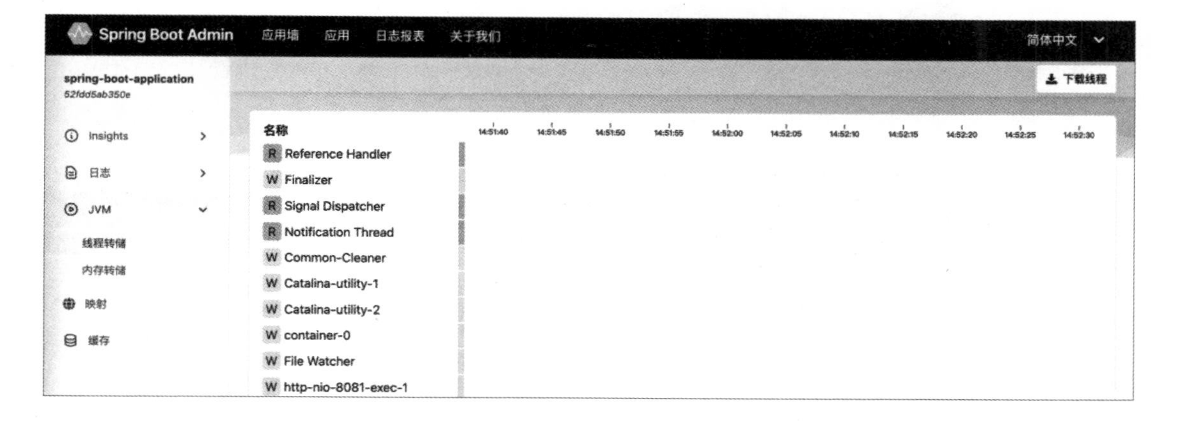

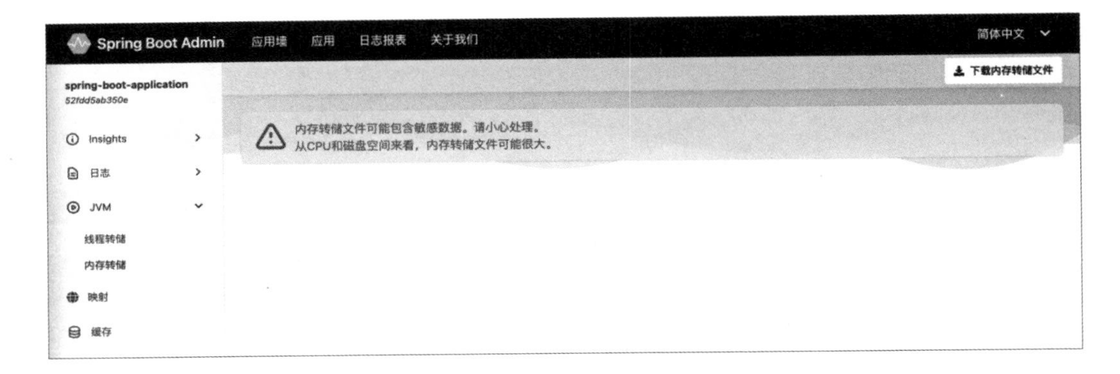

10. 映射菜单

映射菜单中显示所有的 URL 映射，如下图所示。

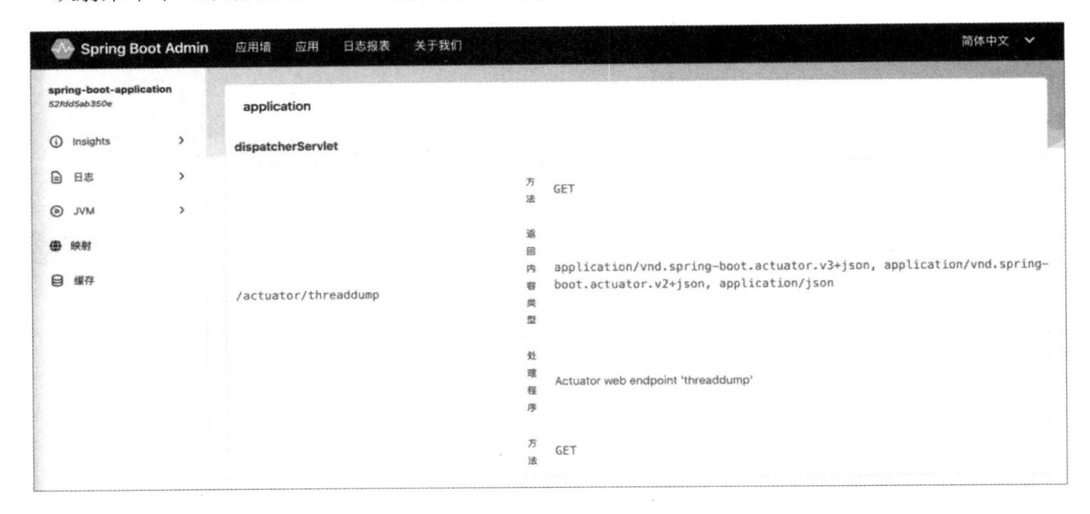

11. 缓存菜单

缓存菜单中显示所有缓存，如下图所示。

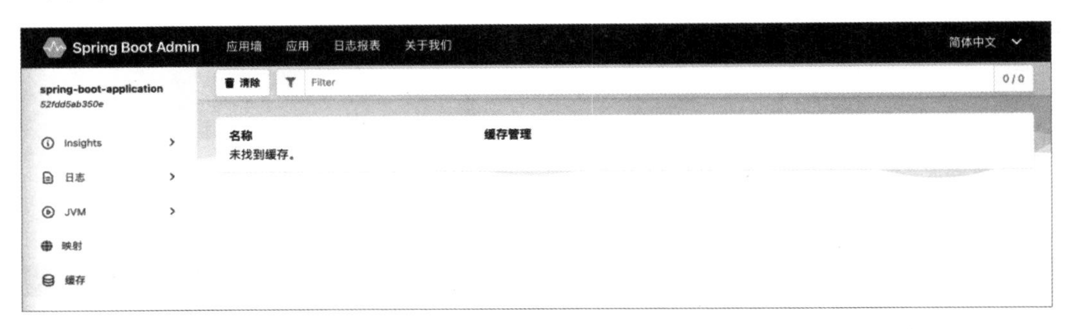

这里不仅是展示，还能清除和管理缓存。

Spring Boot Admin 用于管理和监控 Spring Boot 应用，具体就是通过客户端暴露的各种 /actuator 端点和指标实现的，并在此基础上进行了一些包装和 UI 展示。

12.7.4 监控报警

1. 浏览器通知

现在把客户端停掉，若干秒后页面会自动刷新，如下图所示。

离线实例显示 1，应用出现在离线实例中了，如果浏览器（Chrome）没有禁用通知，那么同时会发出应用上线 / 下线的通知，如下图所示。

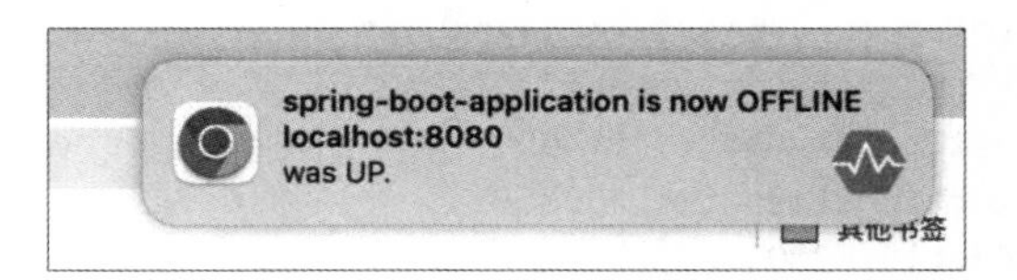

在 AdminServerAutoConfiguration 自动配置类中注册了一个状态变更触发器，如下面的源码所示。

```
@Bean(initMethod = "start", destroyMethod = "stop")
@ConditionalOnMissingBean
public StatusUpdateTrigger statusUpdateTrigger(StatusUpdater statusUpdater,
Publisher<InstanceEvent> events) {
    StatusUpdateTrigger trigger = new StatusUpdateTrigger(statusUpdater, events);
    trigger.setInterval(this.adminServerProperties.getMonitor().getStatusInterval());
    trigger.setLifetime(this.adminServerProperties.getMonitor().getStatusLifetime());
    return trigger;
}
```

默认每10秒会检测一下实例的状态，如果需要修改检测间隔时间，那么可以参考AdminServerProperties和MonitorProperties参数配置类进行配置。

2. 邮件通知

浏览器通知是比较被动的方式，由于不可能所有人都时时盯着计算机，因此这种方式不能及时通知所有需要关注告警信息的人。Spring Boot Admin支持在应用实例状态变更时报警通知，支持邮件、钉钉等多种通知方式。本节就以邮件作为示例，演示当监控到应用下线时如何及时发邮件进行报警通知。

首先在Spring Boot Admin Server项目中加入spring-boot-starter-mail邮件启动器，如下面的依赖配置所示。

```xml
<dependency>
    <groupId>org.springframework.boot</groupId>
    <artifactId>spring-boot-starter-mail</artifactId>
</dependency>
```

然后配置JavaMailSender及收件人信息，如下面的依赖配置示例所示。

```yaml
spring:
  mail:
    host: smtp.exmail.qq.com
    username: xxx@xxx.com
    password: xxx
    properties:
      "[mail.smtp.socketFactory.class]": javax.net.ssl.SSLSocketFactory
      "[mail.smtp.socketFactory.fallback]": false
      "[mail.smtp.socketFactory.port]": 465
      "[mail.smtp.connectiontimeout]": 5000
      "[mail.smtp.timeout]": 3000
      "[mail.smtp.writetimeout]": 5000
  boot:
    admin:
      notify:
        mail:
              enabled: true
          from: Spring Boot Admin <xxx@xxx.com>
          to: xxx@xxx.com # 多个以，分隔
          cc: xxx@xxx.com # 多个以，分隔
```

这里只是简单地配置了发件人、收件人、抄送人等信息，更多的配置信息可以参考AdminServ

erNotifierAutoConfiguration.mailNotifier 方法注册的 MailNotifier 实例，源码如下：

```
@Bean
@ConditionalOnMissingBean
@ConfigurationProperties("spring.boot.admin.notify.mail")
public MailNotifier mailNotifier(JavaMailSender mailSender, InstanceRepository repository,
        TemplateEngine mailNotifierTemplateEngine) {
    return new MailNotifier(mailSender, repository, mailNotifierTemplateEngine);
}
```

MailNotifier 类的源码如下：

```
public class MailNotifier extends AbstractStatusChangeNotifier {

    ...

    /**
     * recipients of the mail
     */
    private String[] to = { "root@localhost" };

    /**
     * cc-recipients of the mail
     */
    private String[] cc = {};

    /**
     * sender of the change
     */
    private String from = "Spring Boot Admin <noreply@localhost>";

    /**
     * Additional properties to be set for the template
     */
    private Map<String, Object> additionalProperties = new HashMap<>();

    /**
     * Base-URL used for hyperlinks in mail
     */
    @Nullable
    private String baseUrl;

    /**
```

```
     * Thymeleaf template for mail
     */
    private String template = "classpath:/META-INF/spring-boot-admin-server/mail/
status-changed.html";

    ...

}
```

源码中显示的所有参数都是可配置的，可以看到邮件通知默认使用的是 Spring Boot Admin 内置的 Thymeleaf 模板，如下图所示。

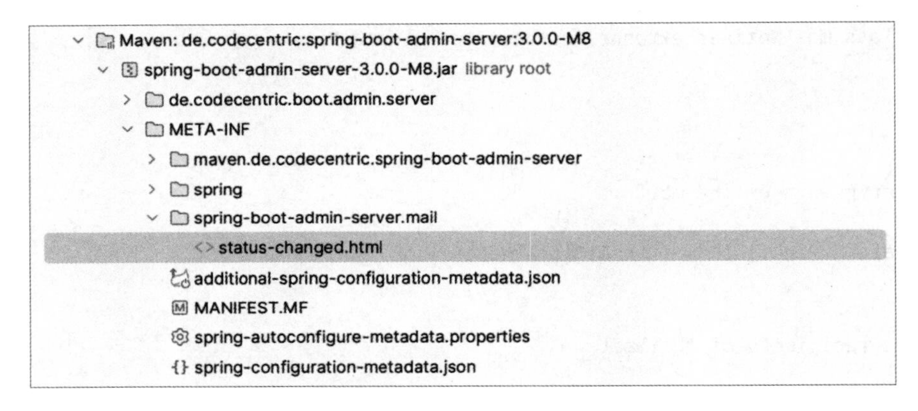

邮件通知会以 HTML 网页的形式发送，如果有需要，也可以更改此模板。配置好邮件通知参数之后，再重新启动 Spring Boot Admin Server 和 Spring Boot Admin Client，然后关闭 Spring Boot Admin Client 模拟应用掉线的场景。

从服务端控制台可以看到服务端无法连接到客户端应用的异常日志，如下图所示。

```
org.springframework.web.reactive.function.client.WebClientRequestException Create breakpoint : Connection refused: /192.168.0.108:8081; neste
    Suppressed: reactor.core.publisher.FluxOnAssembly$OnAssemblyException:
Error has been observed at the following site(s):
    *__checkpoint → Request to GET health [DefaultWebClient]
Original Stack Trace: <20 internal lines>
        at reactor.netty.http.client.HttpClientConnect$MonoHttpConnect$ClientTransportSubscriber.onError(HttpClientConnect.java:307) ~[r
        at reactor.netty.resources.DefaultPooledConnectionProvider$DisposableAcquire.onError(DefaultPooledConnectionProvider.java:158) ~
        at reactor.netty.internal.shaded.reactor.pool.AbstractPool$Borrower.fail(AbstractPool.java:477) ~[reactor-netty-core-1.0.19.jar:
        at reactor.netty.internal.shaded.reactor.pool.SimpleDequePool.lambda$drainLoop$9(SimpleDequePool.java:431) ~[reactor-netty-core-
        at reactor.netty.resources.DefaultPooledConnectionProvider$PooledConnectionAllocator$PooledConnectionInitializer.onError(Default
Caused by: io.netty.channel.AbstractChannel$AnnotatedConnectException: Connection refused: /192.168.0.108:8081
Caused by: java.net.ConnectException Create breakpoint : Connection refused
    at java.base/sun.nio.ch.Net.pollConnect(Native Method) ~[na:na]
    at java.base/sun.nio.ch.Net.pollConnectNow(Net.java:672) ~[na:na]
    at java.base/sun.nio.ch.SocketChannelImpl.finishConnect(SocketChannelImpl.java:946) ~[na:na] <10 internal lines>
```

收件人和抄送人也及时收到邮件了，如下图所示。

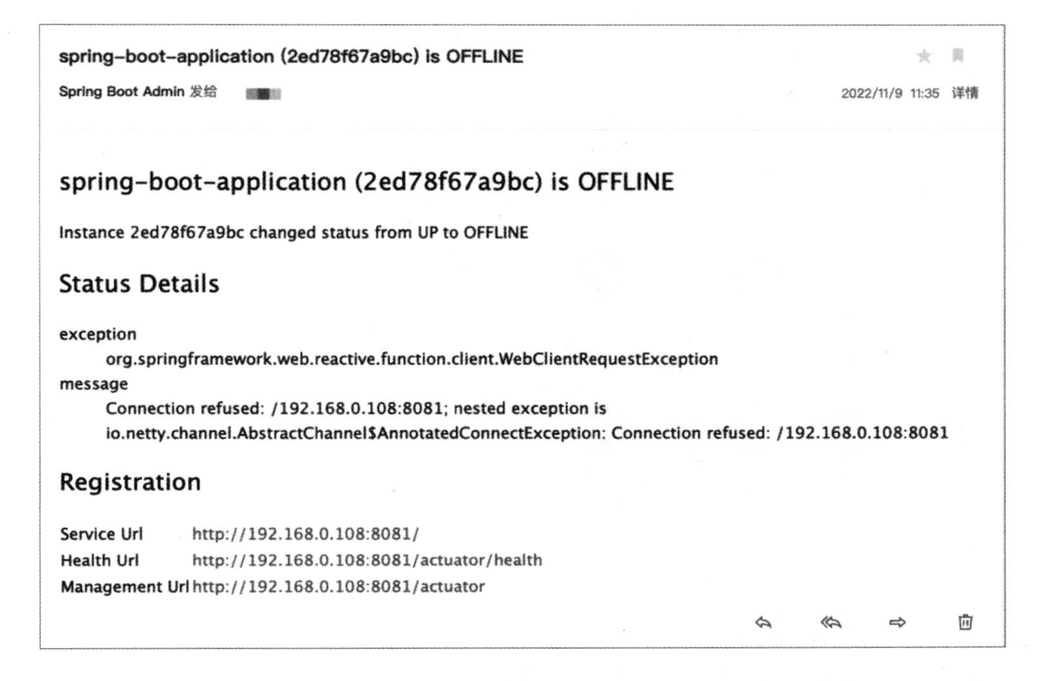

从邮件内容中可以看到应用实例下线的相关信息，为了防止敏感信息被泄露，内置的默认邮件模板不会显示实例的任何元数据信息，如果需要显示，则可以使用自定义的 Thymeleaf 模板，这里不再赘述。

> 邮件启动器的具体介绍及应用可以查看第 3 章邮件 Starter 章节。

3. 钉钉通知

首先需要在钉钉中创建和授权钉钉机器人，创建方法如下所示。

> 创建钉钉警报群，单击钉钉群设置→智能群助手→添加机器人。
> 具体过程如下面 4 个图所示。

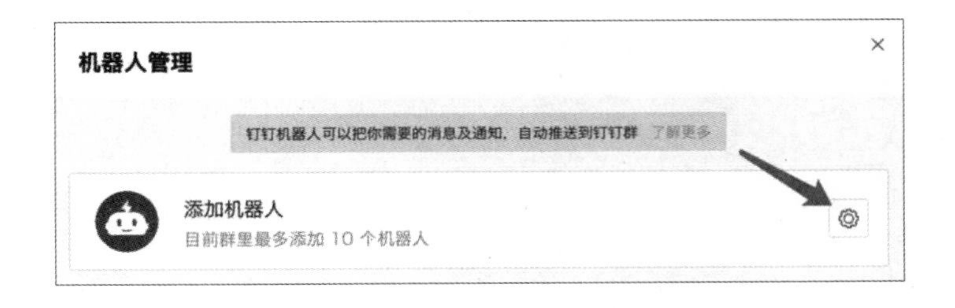

机器人管理
心知天气
自动推送天气预报和预警信息
防疫精灵
新冠疫情实况和预防咨询服务
复工宝
企业复工复产提报及相关服务
阿里云Codeup
阿里云提供的代码托管服务
GitHub
基于Git的代码托管服务
极狐GitLab
基于ROR的开源代码托管软件
JIRA
出色的项目与事务跟踪工具
Travis
出色的项目与事务跟踪工具
Trello
实时的卡片墙，管理任何事情
自定义
通过Webhook接入自定义服务

机器人管理
机器人名字:
钉钉钉
* 添加到群组:
Java技术栈
* 安全设置
说明文档
自定义关键词
警报
我已阅读并同意《自定义机器人服务及免责条款》
取消
完成

钉钉机器人创建完成后会得到一个签名用的 Secret 和 Webhook 地址，然后配置在 Spring Boot Admin Server 中，如下面的配置示例所示。

```
spring:
  mail:
    host: smtp.exmail.qq.com
    username: xxx@xxx.com
    password: xxx
    properties:
      "[mail.smtp.socketFactory.class]": javax.net.ssl.SSLSocketFactory
      "[mail.smtp.socketFactory.fallback]": false
      "[mail.smtp.socketFactory.port]": 465
      "[mail.smtp.connectiontimeout]": 5000
      "[mail.smtp.timeout]": 3000
      "[mail.smtp.writetimeout]": 5000
  boot:
    admin:
      notify:
        mail:
          enabled: true
          from: Spring Boot Admin <xxx@xxx.com>
          to: xxx@xxx.com # 多个以，分隔
          cc: xxx@xxx.com # 多个以，分隔
        dingtalk:
          enabled: true
          webhook-url: https://oapi.dingtalk.com/robot/send?access_token=xxx
          secret: xxx
          message: "警报 应用 #{instance.registration.name} - #{instance.id} 掉线了！！！
"
```

几个重要的配置参数的说明如下表所示。

参数名	描述	默认值
spring.boot.admin.notify.dingtalk.enabled	开启钉钉通知	true
spring.boot.admin.notify.dingtalk.webhook-url	钉钉机器人 Webhook 地址	
spring.boot.admin.notify.dingtalk.secret	消息签名的密钥	
spring.boot.admin.notify.dingtalk.message	要发送的通知消息，支持 SpEL 表达式	#{instance.registration.name} #{instance.id} is #{event.statusInfo.status}

具体实现逻辑见 AdminServerNotifierAutoConfiguration 和 DingTalkNotifierConfiguration 配置类，它会注册一个 DingTalkNotifier 通知实例，这里不再赘述。配置好钉钉通知参数之后，再重新启动 Spring Boot Admin Server 和 Spring Boot Admin Client，然后关闭 Spring Boot Admin Client 模拟应用掉线的场景。

不久钉钉就会收到应用掉线的警报消息，如下图所示。

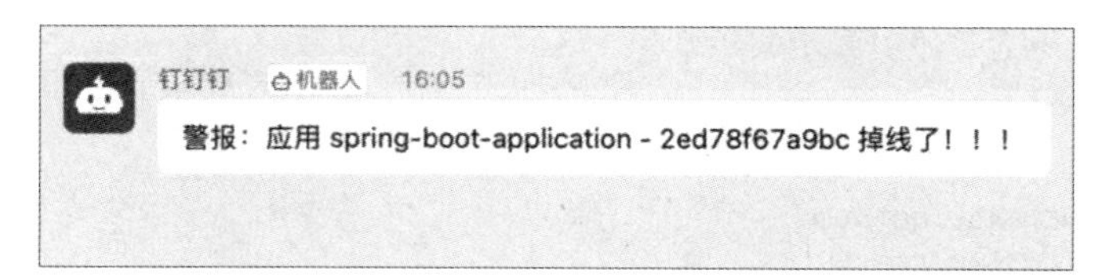

如果同时开启了多个通知方式，那么会同时收到警报信息，比如笔者同时启用了邮件和钉钉通知，应用掉线时两者会同时收到警报通知。

Spring Boot Admin 监控的功能很简单，只能监控应用的状态，而不能监控中间件的健康状况，所以不能代替专业的监控系统（比如下一节要介绍的 Prometheus+Grafana 监控系统组合），但至少 Spring Boot Admin 比较轻量级，只需要启动一个服务端应用，而不需要太多复杂的组件，所以，对于那些简单的 Spring Boot 应用的监控还是够用的。

12.8 Prometheus+Grafana

12.8.1 概述

Spring Boot Actuator 为 Micrometer 提供了依赖管理和自动配置，Micrometer 是一个支持众多监控系统的监控系统门面，具有和 Slf4j 一样的性质，Micrometer 可以支持以下监控系统的指标数据收集：AppOptics、Atlas、Datadog、Dynatrace、Elastic、Ganglia、Graphite、Humio、Influx、

JMX、KairosDB、New Relic、OpenTelemetry、Prometheus、SignalFx、Simple（in-memory）、Stackdriver、StatsD、Wavefront。

Spring Boot 自动配置了一个指标注册表 MeterRegistry，只需要在类路径中添加一个 micrometer-registry-{system} 依赖即可，自动配置类会根据类路径下不同的监控系统依赖配置指标注册表 MeterRegistry。

本节将以当前主流的 Prometheus 监控系统为基础进行介绍，Prometheus 是一个开源的应用监控系统，也是一个时序数据库，它可以从应用收集指标数据并显示结果，并且在达到指定条件时触发警报，现在已经成为 Cloud Native Computing Foundation（CNCF）云原生计算基金会的项目。

12.8.2　Prometheus 指标暴露

1. pull 方式

对于 Prometheus，Spring Boot 提供了一个 /actuator/prometheus 端点，如下图所示。

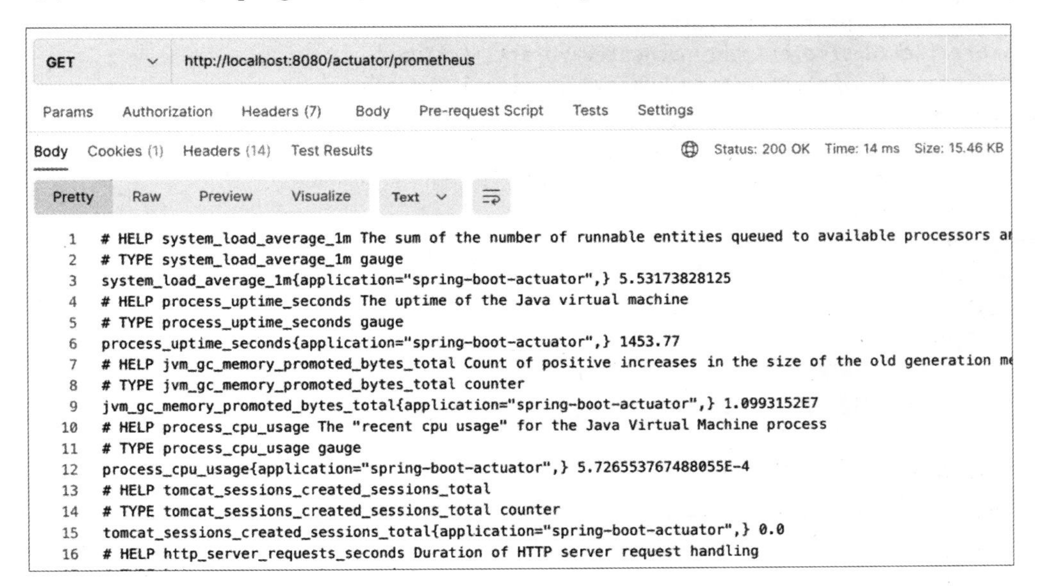

Prometheus 会通过此端点来抓取单个应用实例的指标，该端点默认是不暴露的，需要手动暴露，之前的章节中已经暴露为“*”了，所以这里不再单独暴露了。

添加基于 Prometheus 的 Micrometer 注册依赖，如下面的依赖配置所示。

```
<dependency>
    <groupId>io.micrometer</groupId>
```

```
    <artifactId>micrometer-registry-prometheus</artifactId>
    <version>${micrometer-registry-prometheus.version}</version>
</dependency>
```

引入该依赖后 Spring Boot 就会自动配置，对应的自动配置类为 PrometheusMetricsExport-AutoConfiguration，对应的参数绑定类为 PrometheusProperties，可以通过一系列 management.prometheus.metrics.export.* 参数配置 Prometheus 指标。

2. push 方式

对于 Spring Boot 提供的端点，Prometheus 是以 pull 的方式获取指标数据的，Spring Boot 也支持以 push 的方式将指标数据推送给 Prometheus，这对于时间不够长而无法被抓取的临时任务或者批处理任务特别有用。

推送指标数据需要使用 Prometheus 的 Pushgateway 模块，先引入应用的依赖：

```
<dependency>
    <groupId>io.prometheus</groupId>
    <artifactId>simpleclient_pushgateway</artifactId>
    <version>${simpleclient_pushgateway.version}</version>
</dependency>
```

默认不会启用 pushgateway 功能，需要手动配置以启用 pushgateway 功能：

```
management:
  prometheus:
    metrics:
      export:
        pushgateway:
          enabled: true
```

> **Spring Boot 3.0 新变化：**
> 以 management.metrics.export.<product> 管理的参数需要使用新的 management.<product>.metrics.export 参数形式。

在 PrometheusMetricsExportAutoConfiguration 自动配置类中会自动注册一个 PrometheusPush-GatewayManager 类型的 Bean，通过这个 Bean 可以将应用的指标推送给 Prometheus Pushgateway。

既可以通过一系列 management.prometheus.metrics.export.pushgateway.* 参数配置 Prometheus-PushGatewayManager，也可以自定义一个覆盖默认自动配置的 PrometheusPush- GatewayManager。

12.8.3 Prometheus 环境搭建

本节以 CentOS Linux 为例搭建 Prometheus 环境，首先下载并安装 Prometheus，如下面的参考命令：

```
$ wget https://github.com/prometheus/prometheus/releases/download/v2.37.0/prometheus-2.37.0.linux-amd64.tar.gz

$ tar -zxvf prometheus-2.37.0.linux-amd64.tar.gz
```

然后进入解压后的目录，打开 prometheus.yml 配置文件，修改并添加以下配置：

```
...

scrape_configs:
  # The job name is added as a label `job=<job_name>` to any timeseries scraped from this config.
  - job_name: "prometheus"
    # metrics_path defaults to '/metrics'
    # scheme defaults to 'http'.
    static_configs:
      - targets: ["localhost:9090"]
  - job_name: "javastack"
    # metrics_path defaults to '/metrics'
    metrics_path: "/actuator/prometheus"
    # scheme defaults to 'http'.
    static_configs:
      - targets: ["localhost:8080"]
```

这里配置了两个抓取任务，第一个是 Prometheus 默认提供的，可以忽略；第二个是 Spring Boot Actuator 应用，也就是我们要监控的 Spring Boot 应用，它需要使用 metrics_path 配置 Spring Boot 提供的 /actuator/prometheus 端点，这也是通过 pull 的方式获取应用指标数据的来源。

启动 Prometheus，参考命令如下：

```
$ ./prometheus --config.file=prometheus.yml &
```

Prometheus 提供了优雅关闭的方式，需要先指定参数启动，参考命令如下：

```
./prometheus --config.file=prometheus.yml --web.enable-lifecycle &
```

然后通过发送一个 POST 请求优雅关闭 Prometheus，参考命令如下：

```
curl -X POST http://localhost:9090/-/quit
```

启动 Prometheus 后默认绑定为 9090 端口，访问 Prometheus 控制台，地址如下：

```
http://localhost:9090/
```

Prometheus 首页如下图所示。

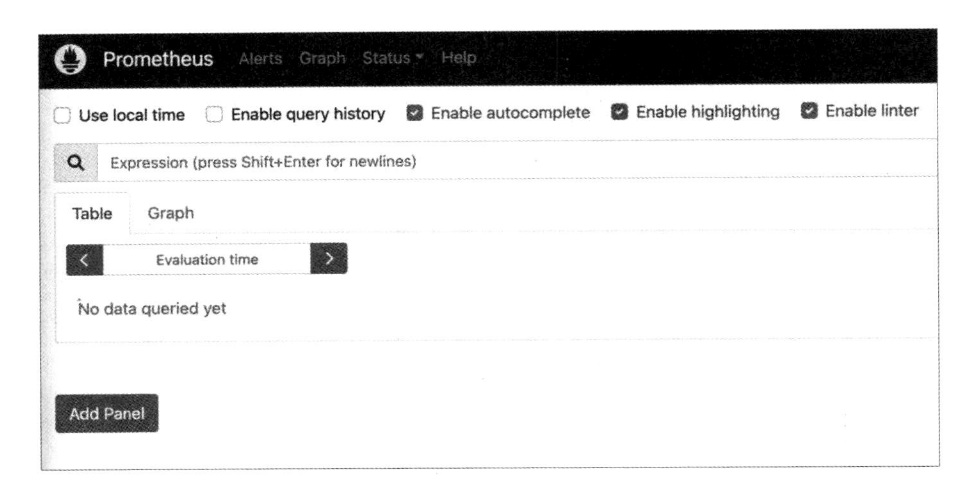

在首页，可以根据条件检索并展示一段时间内的指标数据，如下图所示。

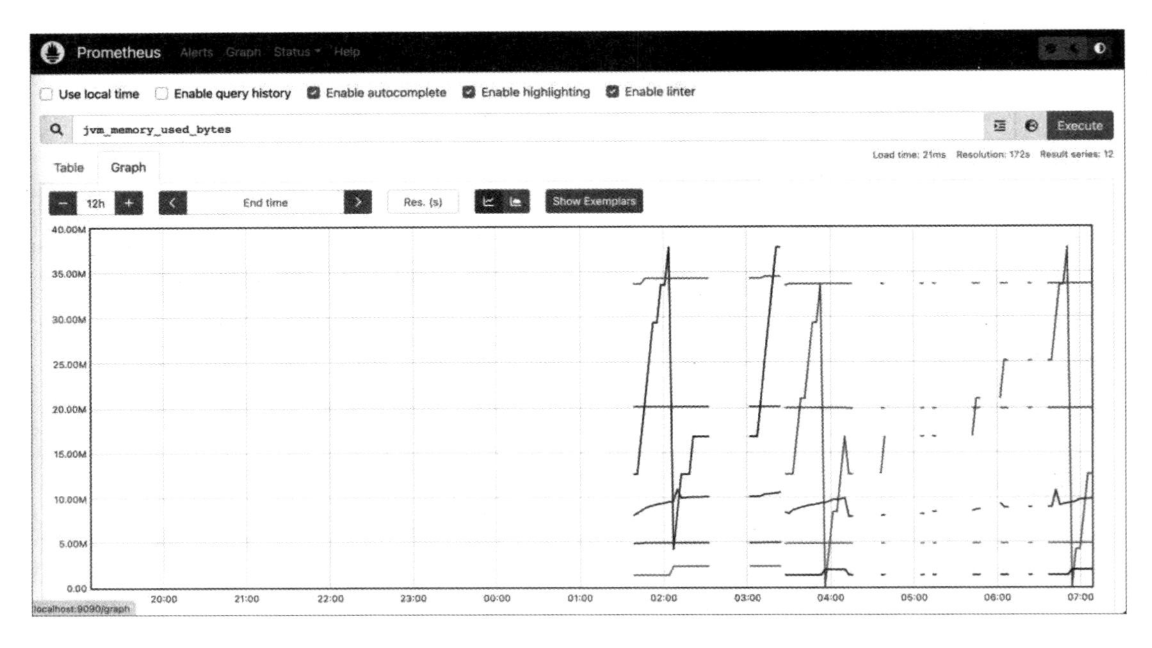

通过 Status → Targets 菜单进入监控的实例，如下图所示。

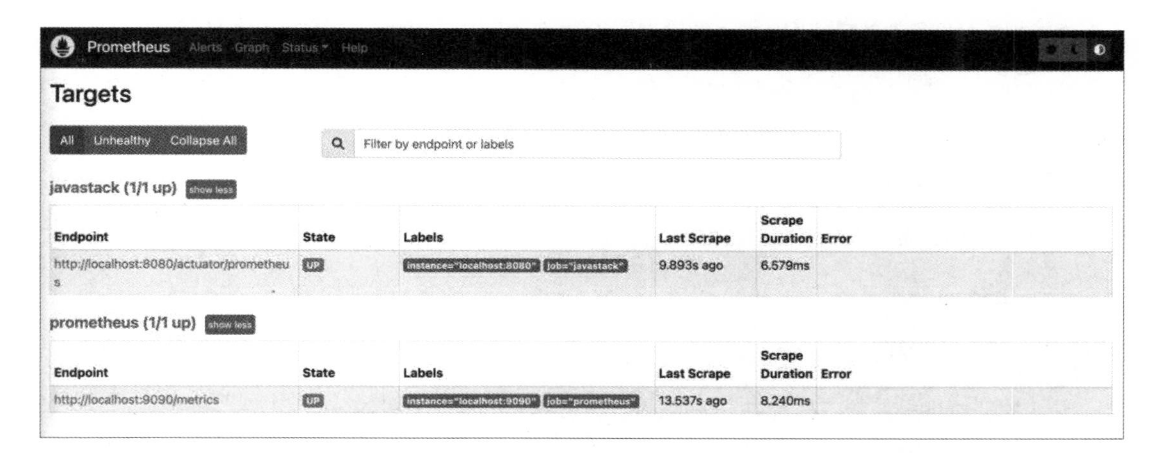

可以看到我们配置好的监控实例都是正常状态的，假如现在关闭 Spring Boot Actuator 应用，过几秒再刷新一下，可以看到 Spring Boot Actuator 应用已经是下线状态了，如下图所示。

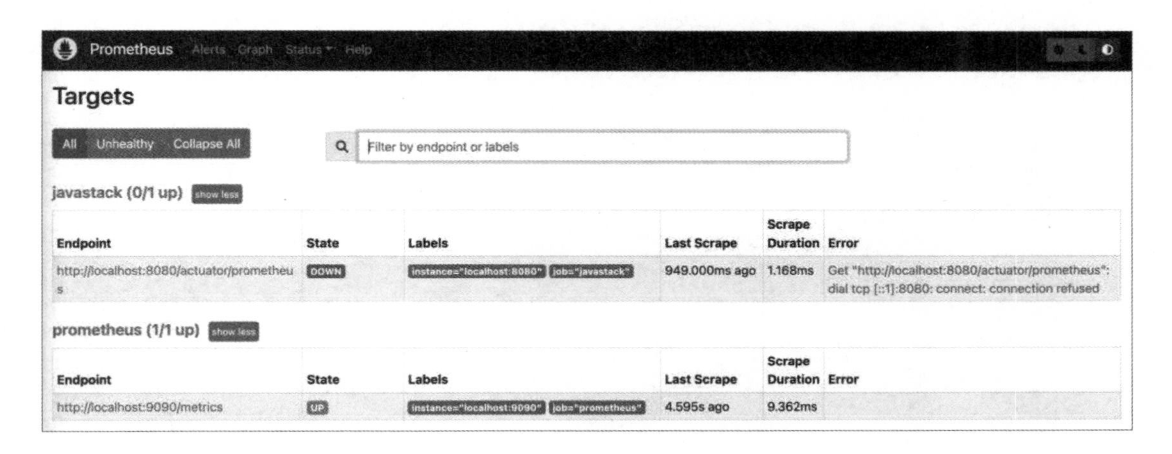

12.8.4 Grafana 数据可视化

1. Grafana 环境搭建

从上一节可以看到，Prometheus 的数据展示能力非常一般，也不是很直观，所以通常会结合 Grafana 可视化平台一起使用。Grafana 是一个开源的数据可视化平台，可以对指标数据进行可视化、查询、提醒等操作，可以帮助我们更好地了解指标数据，它支持来自不同数据源的数据，比如 Prometheus、Loki、Elasticsearch、InfluxDB、Postgres 等，并且提供了各种丰富的面板以展示不同类型的指标数据。

本节以 CentOS 为例搭建 Grafana 环境，首先下载并安装 Grafana，参考命令如下所示。

```
$ wget https://dl.grafana.com/enterprise/release/grafana-enterprise-9.0.7.linux-amd64.tar.gz$ tar -zxvf
grafana-enterprise-9.0.7.linux-amd64.tar.gz
```

进入解压后的 bin 目录启动 Grafana，参考命令如下：

```
$ ./grafana-server &
```

启动 Grafana 后默认绑定为 3000 端口，访问 Grafana 控制台，地址如下：

```
http://localhost:3000/
```

默认的登录用户名和密码都是 admin，首次登录 Grafana 后会要求修改密码，修改密码后再重新登录，登录成功后进入 Grafana 主界面，如下图所示。

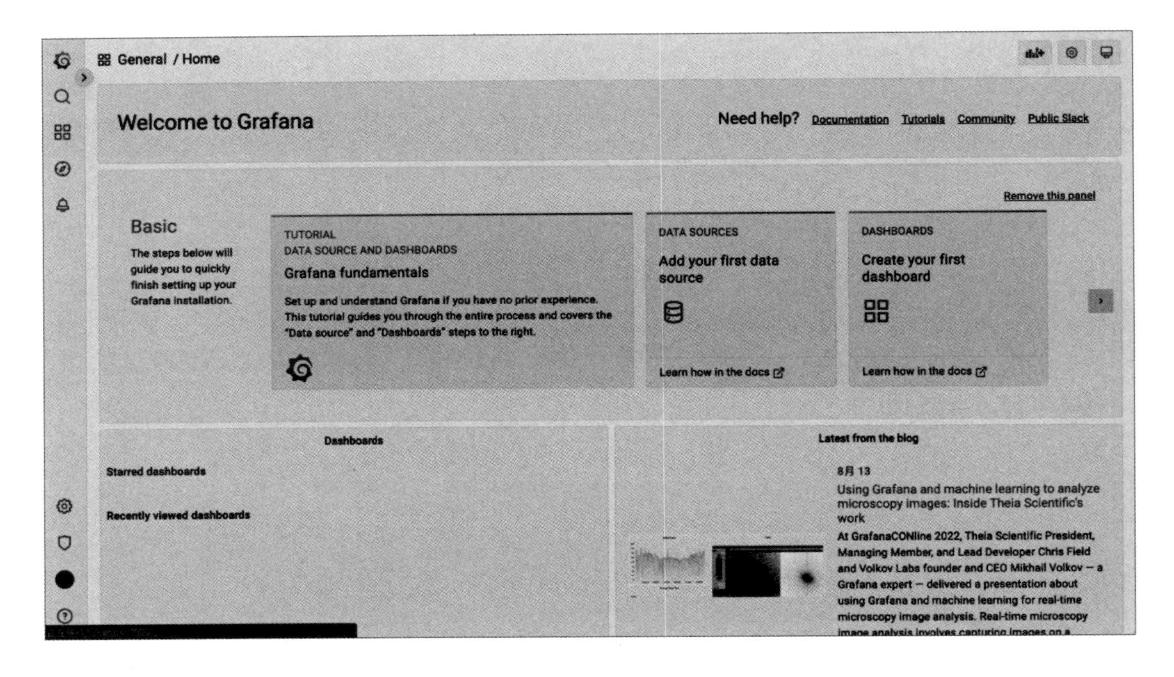

2. 添加 Prometheus 数据源

搭建好 Grafana 之后，需要 Prometheus 的指标数据来源，单击首页的 DATA SOURCES 面板项，然后选择新建一个 Prometheus 数据源，如下图所示。

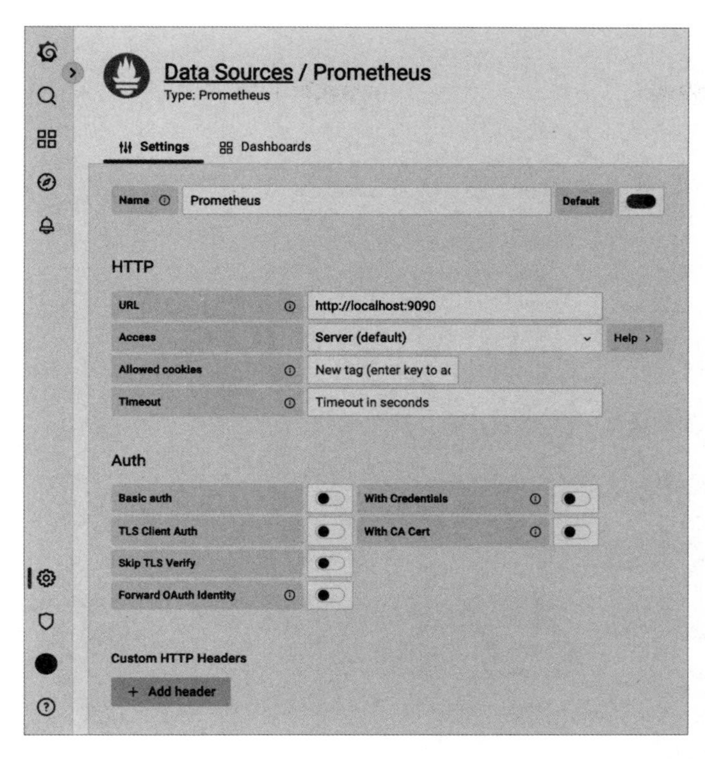

这里主要是输入 Prometheus 的 URL：http://localhost:9090。部署到生产环境中需要有 Auth 认证信息，Prometheus 数据源保存成功后，单击设置按钮→ Data sources 可以查询 / 新建数据源，如下图所示。

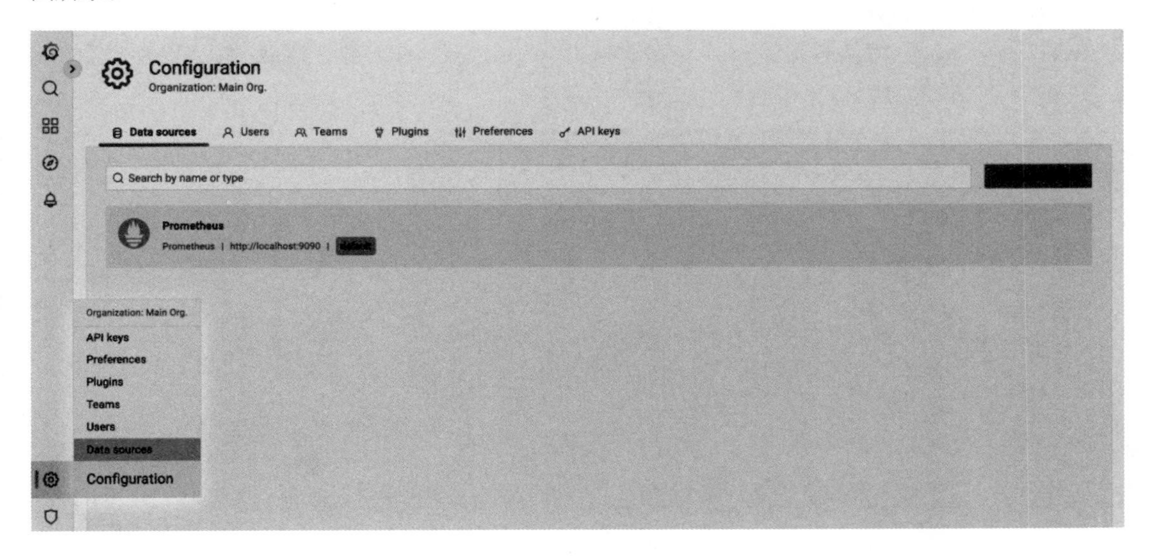

3. 添加 JVM 数据面板

Prometheus 数据源添加成功后，可以在 Grafana 官网选择一个数据面板以展示数据源中的指标数据，官方数据的面板地址为链接 21。

比如我们想展示应用的 JVM 指标数据，根据数据源和关键字条件搜索一个 JVM 数据面板，如下图所示。

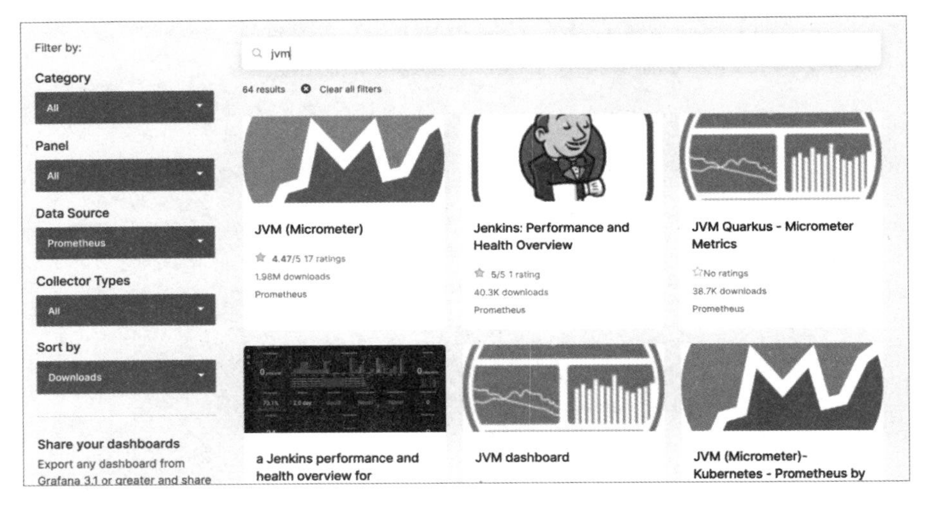

然后单击 JVM(Micrometer) 项进入该数据面板，如下图所示。

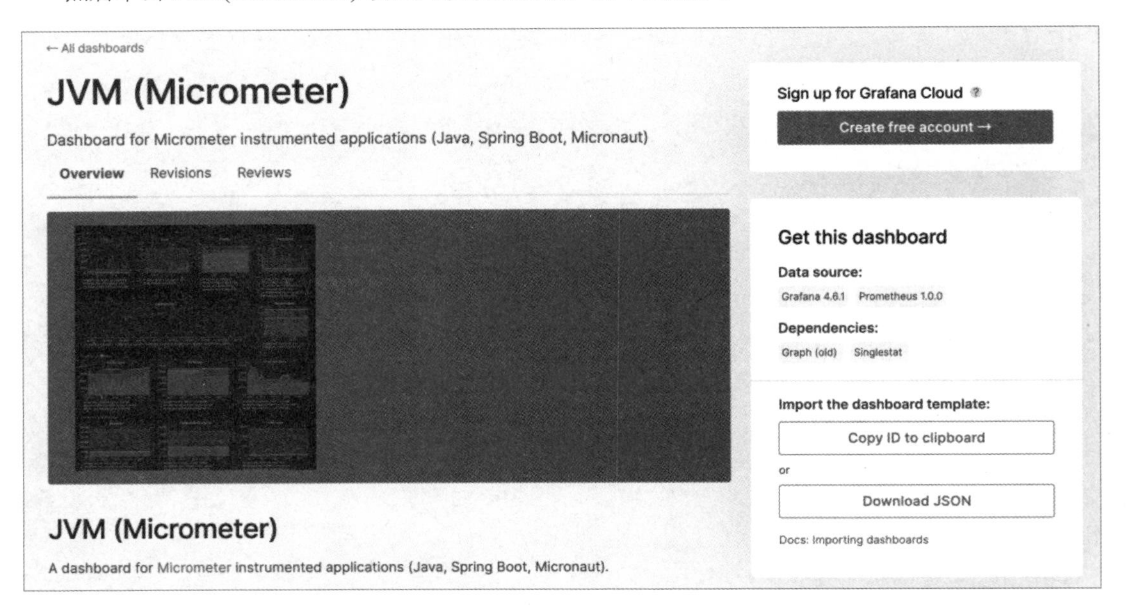

数据面板有两种方式可以集成到 Grafana 可视化平台，一是复制面板 ID 再导入 Grafana，二是下载 JSON 文件再导入 Grafana，这里使用第一种方式复制面板 ID 再导入 Grafana。

单击 Grafana 左侧面板按钮进入 Grafana 面板页面，如下图所示。

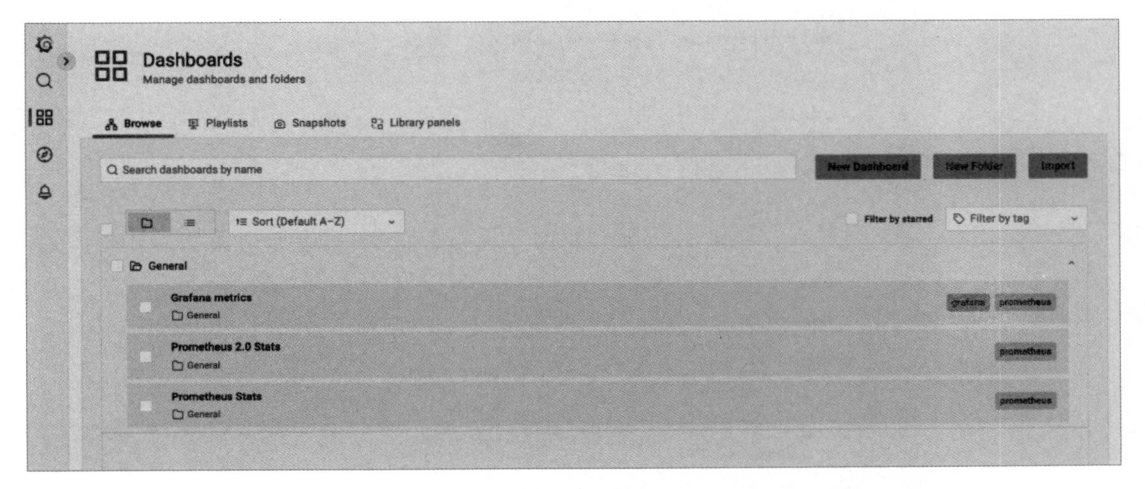

然后单击 Import 按钮，输入复制的面板 ID 后单击 Load 按钮，如下图所示。

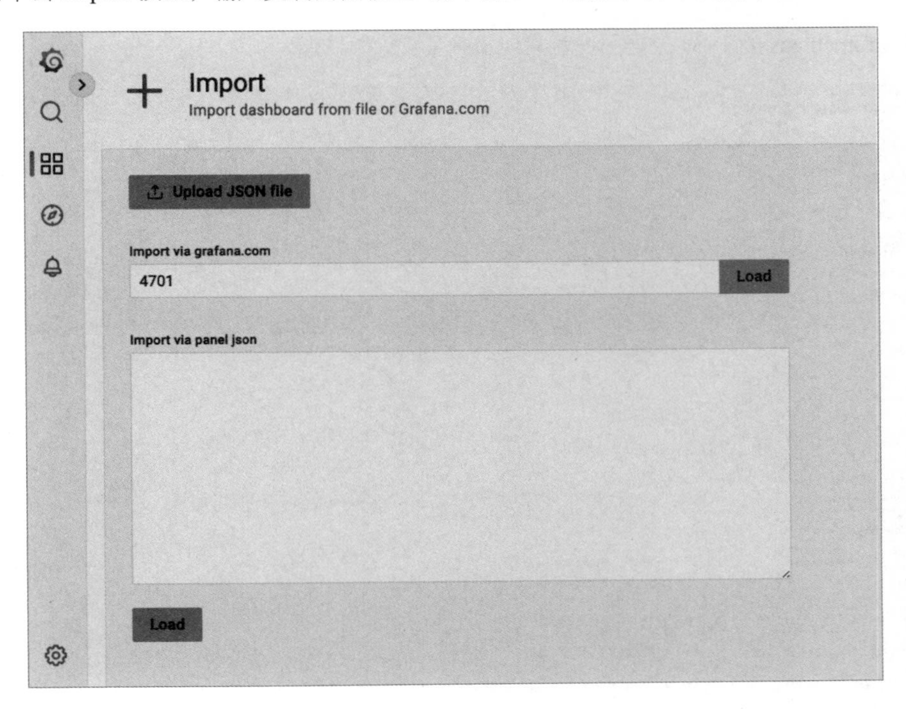

在导入界面下选择 Prometheus 数据源，然后单击 Import 按钮，如下图所示。

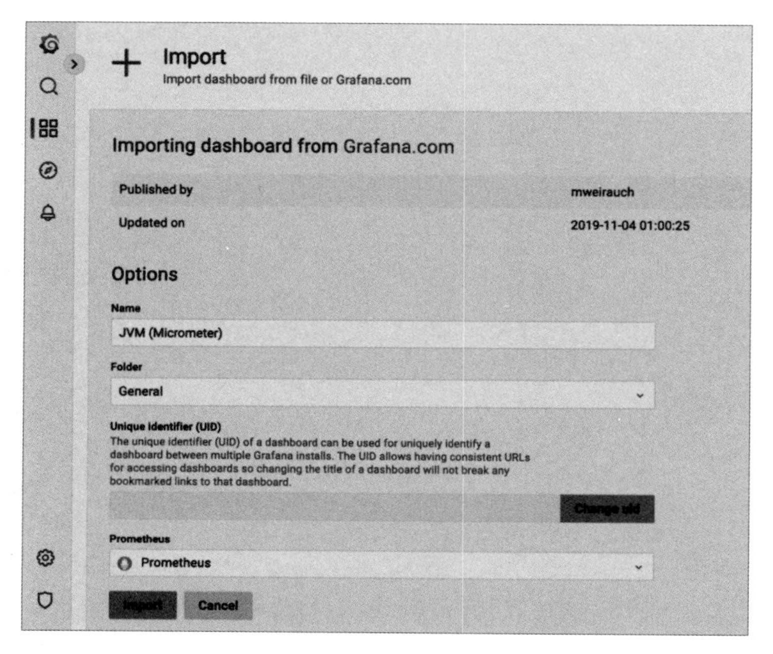

基于 Prometheus 的 JVM 数据面板就呈现出来了，如下图所示。

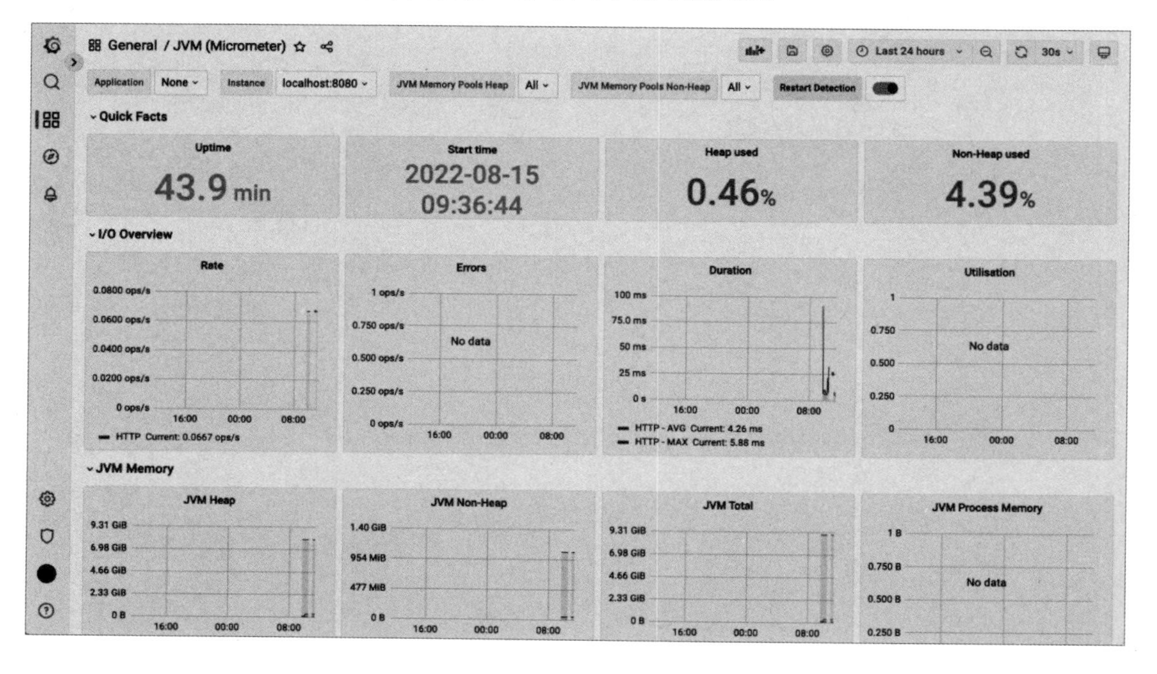

左上方的 Application显示为 None，需要在应用配置中配置一下，如下面的配置示例所示。

```
spring:
  application:
    name: spring-boot-actuator

management:
  metrics:
    tags:
      application: ${spring.application.name}
```

重启应用再刷新 Grafana 面板，如下图所示。

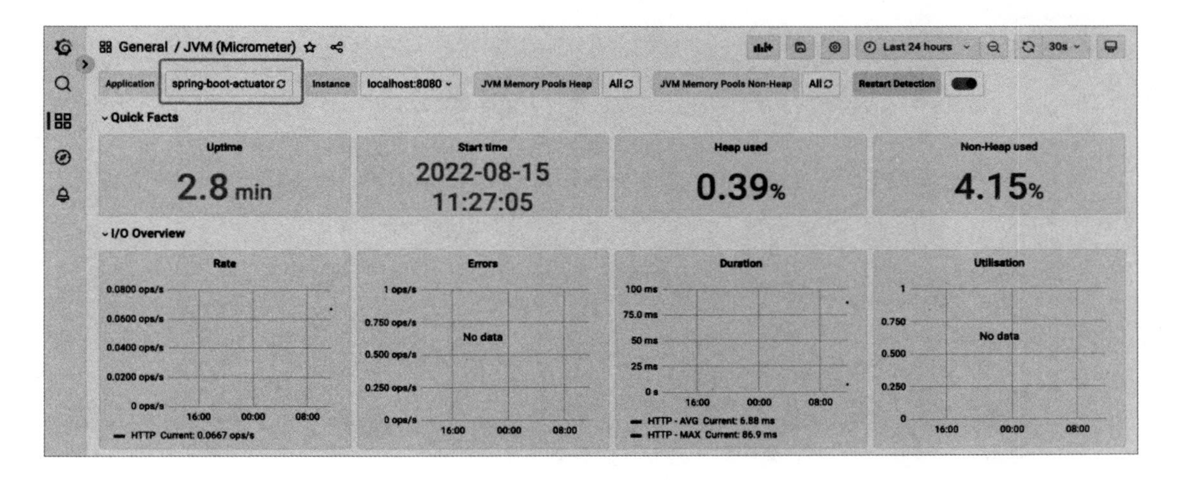

Application 名称已经正常显示了。

4. 自定义数据面板

以上使用的是官方的数据面板模板，也可以单击上方 “+” 号按钮添加一个新的自定义数据面板，如下图所示。

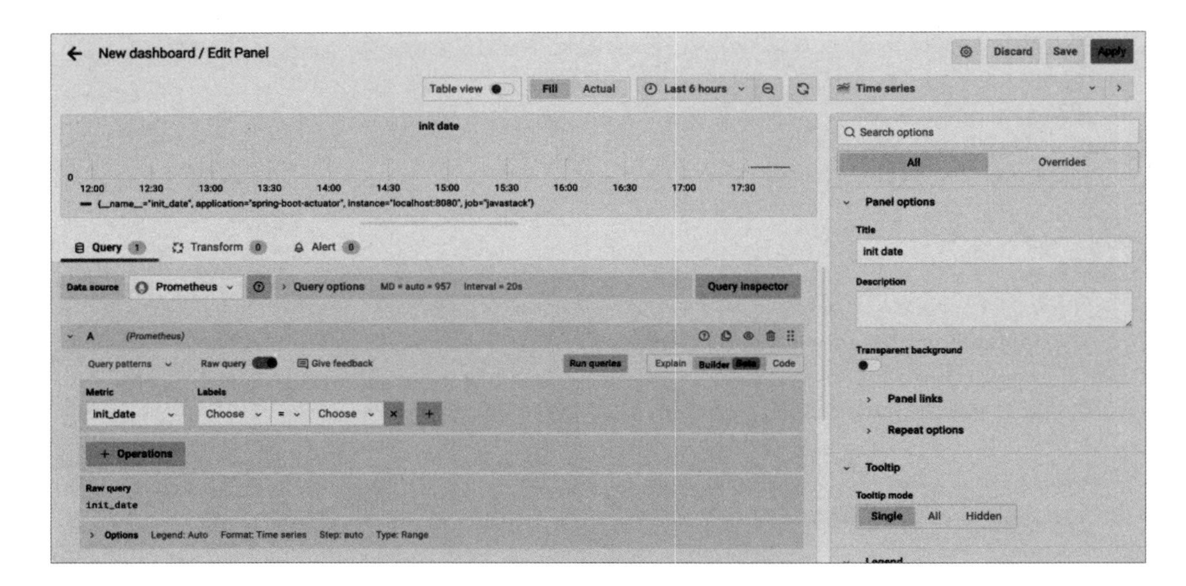

在左下方 Metric 处选择之前自定义好的 init_date 指标，然后单击 Run queries 按钮查询指标数据，单击上面 Apply 按钮可以预览面板，如下图所示。

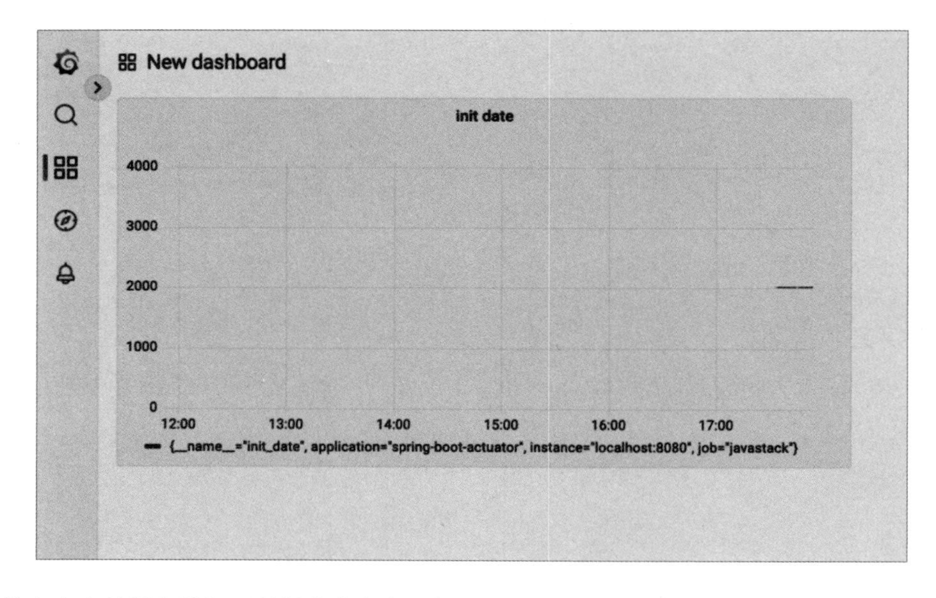

再单击上方的保存按钮可以保存自定义面板。

12.8.5 监控报警

使用 Prometheus+Grafana 监控报警一般有以下几种方案：

- 使用 Prometheus 的 alertmanager 报警管理组件，需要编写报警规则，不是很友好。
- 使用 Grafana 自带的报警功能，支持可视化配置报警规则，使用和配置简单方便。
- 使用 Grafana+ 第三方警告平台，需要注册第三方平台账号，并且可能需要付费。

读者可以根据实际需要选择合适的报警方案，下面笔者使用 Grafana 自带的报警功能来实现应用下线时警报通知。Grafana 默认为邮件警报，需要配置邮件发送信息，修改 Grafana/confi/defaults.ini 配置文件，内容如下：

```
#################################### SMTP / Emailing #####################
[smtp]
enabled = true
host = smtp.exmail.qq.com:465
user = ********
# If the password contains # or ; you have to wrap it with triple quotes.
Ex """#password;"""
password = ********
cert_file =
key_file =
skip_verify = true
from_address = ********
from_name = Grafana
ehlo_identity =
startTLS_policy =

[emails]
welcome_email_on_sign_up = false
templates_pattern = emails/*.html, emails/*.txt
content_types = text/html
```

修改 SMTP 部分的配置，然后重启 Grafana 即可。

进入 Alerting → Contact points 面板，默认只有邮件警报，这里添加一个钉钉通知，如下面两个图所示。

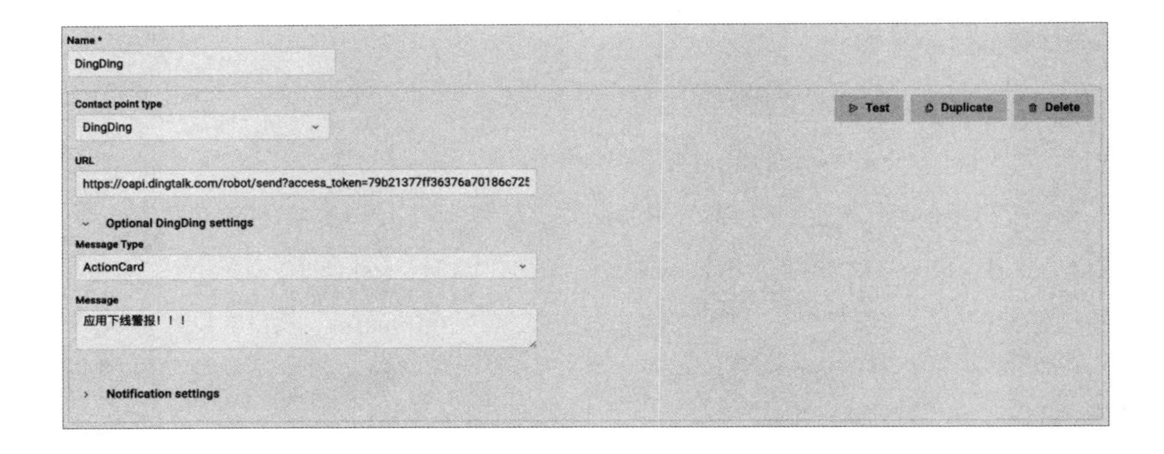

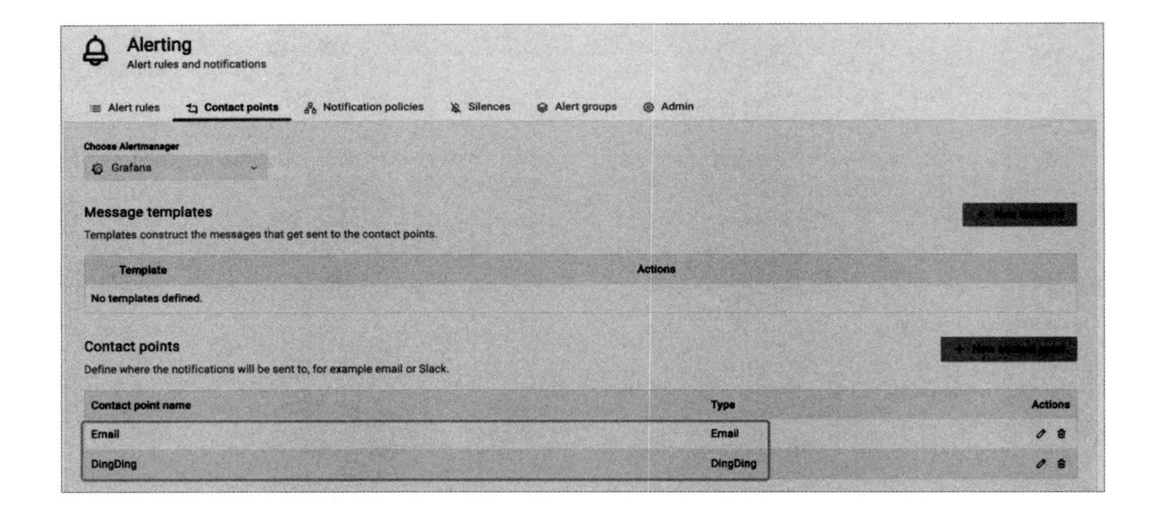

然后进入 Alerting → Notification policies 面板，这里设置所有警报规则为邮件通知方式，并且添加了一个自定义的警报匹配规则，设置标签值 type=status 的警报规则都使用钉钉通知，如下面两个图所示。

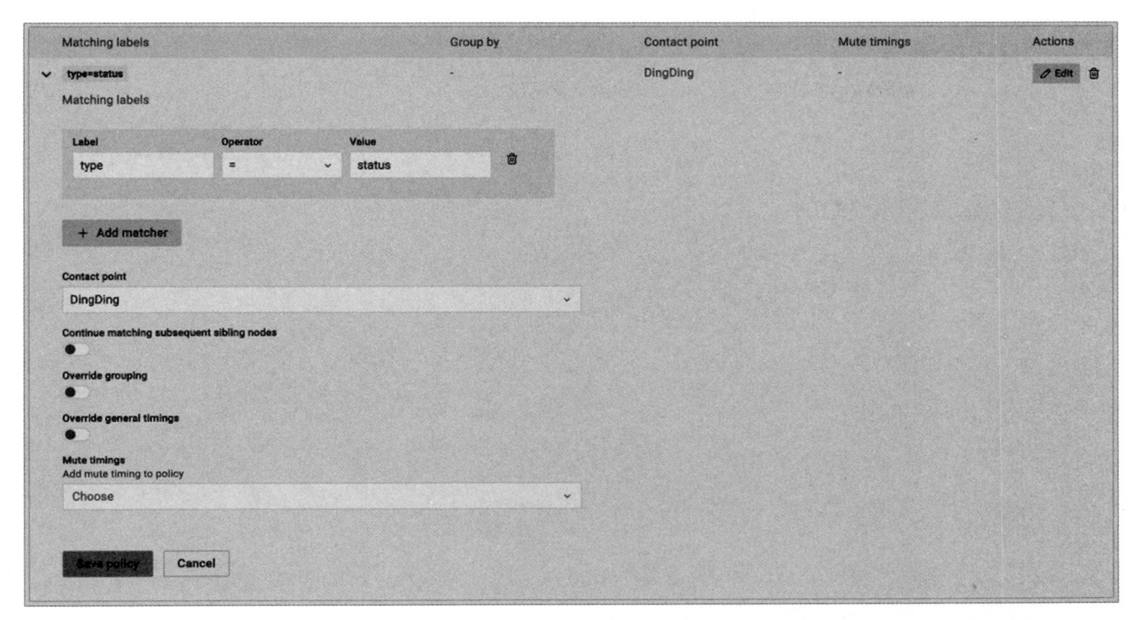

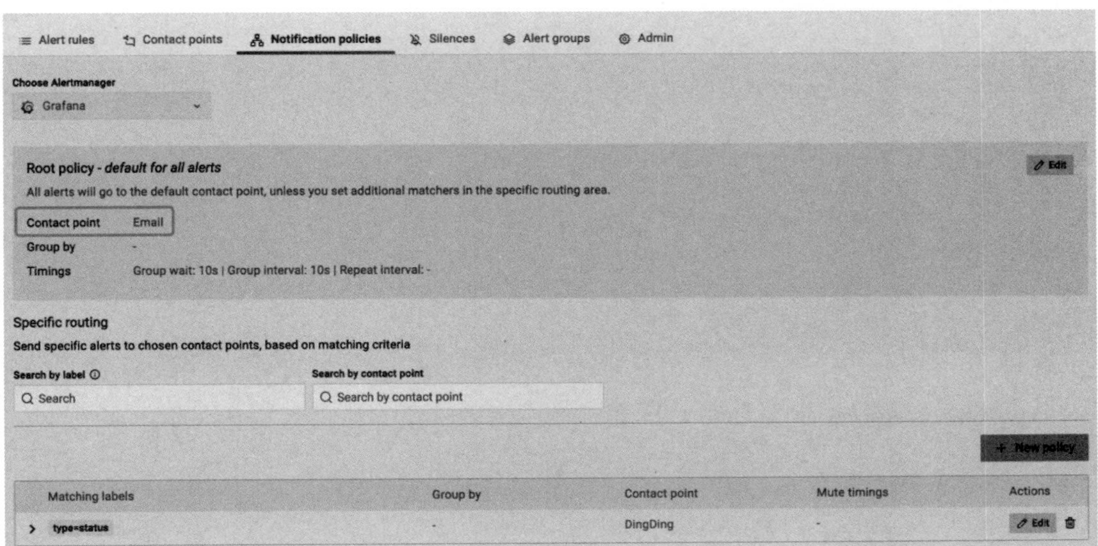

最后进入 Alerting → Alert rules 面板，添加一个应用下线警报规则，如下面的流程所示。

1）设置警报触发条件

添加一个基本指标规则 A，用于获取实例的在线状态，如下图所示。

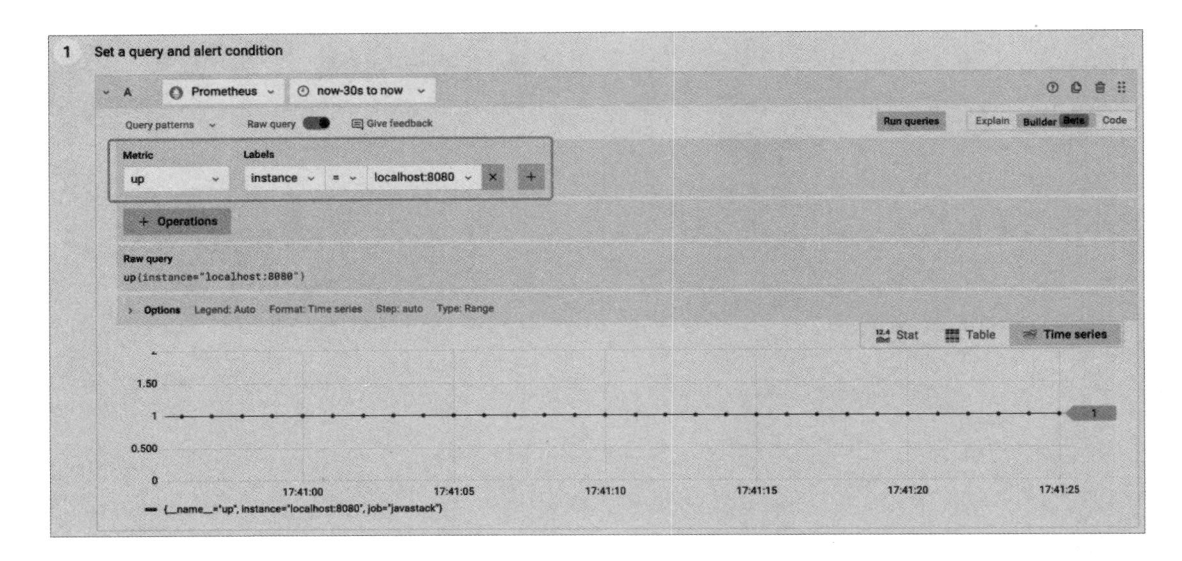

然后添加一个表达式规则 B 并设置为警报条件，如果条件 A 的最新返回值小于 1，则代表实例下线了（1：在线；0：下线），如下图所示。

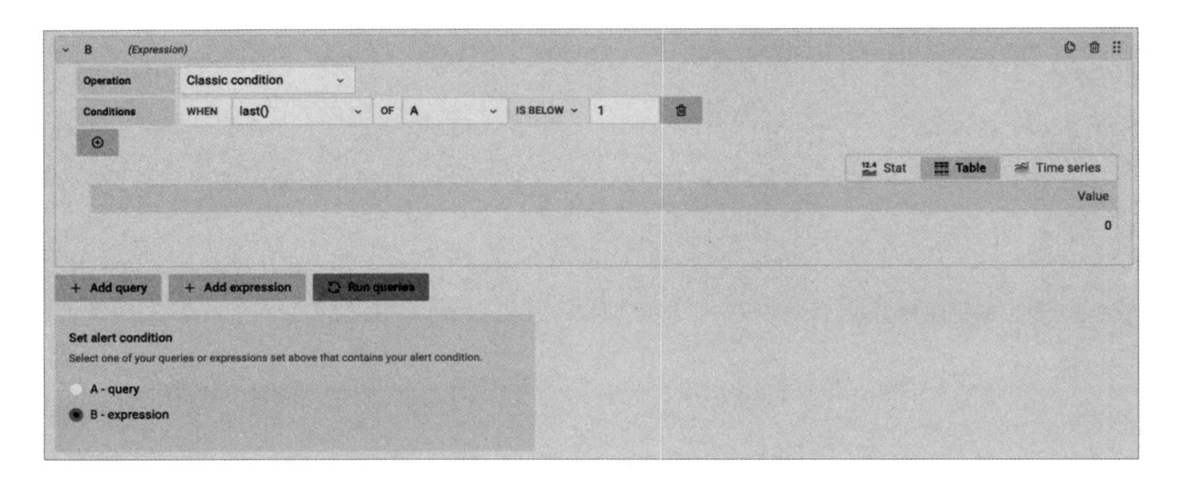

2）设置警报评估行为

设置每 10 秒检测一下警报条件，如果符合警报条件，则延迟 15 秒再进行警报通知，如下图所示。

2 Alert evaluation behavior

Evaluate

Evaluation interval applies to every rule within a group. It can overwrite the interval of an existing alert rule.

Evaluate every 10s for 15s

› Configure no data and error handling

Preview alerts

Preview based on the result of running the query, for this moment. Configuration for `no data` and `error handling` is not applied.

State	Info
Normal	

3）设置警报基本信息

设置警报基本信息，如警报名称、存储文件夹、警报分组等信息，如下图所示。

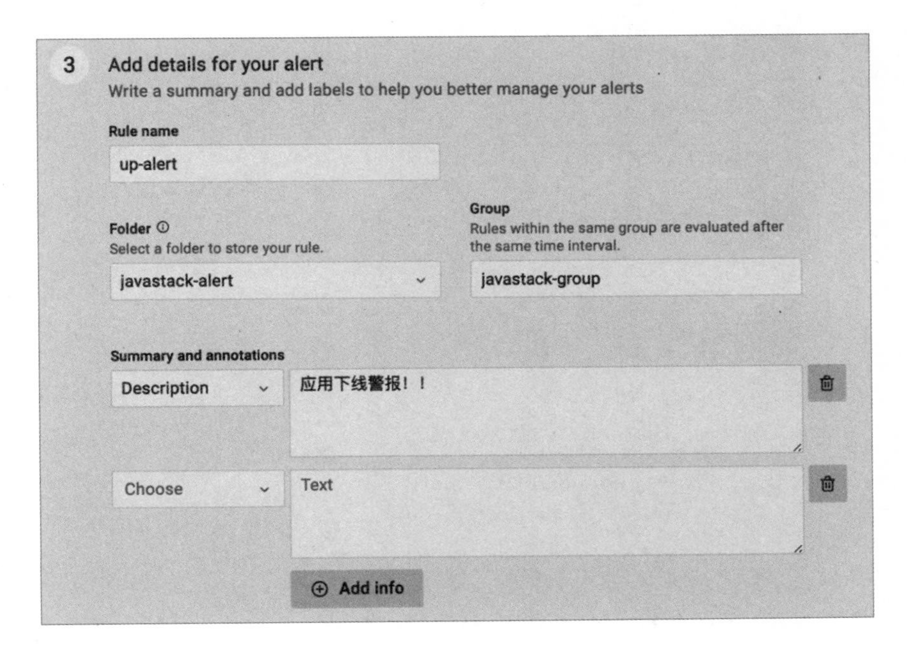

4）设置警报自定义标签

设置警报规则的自定义标签 type=status，如下图所示。

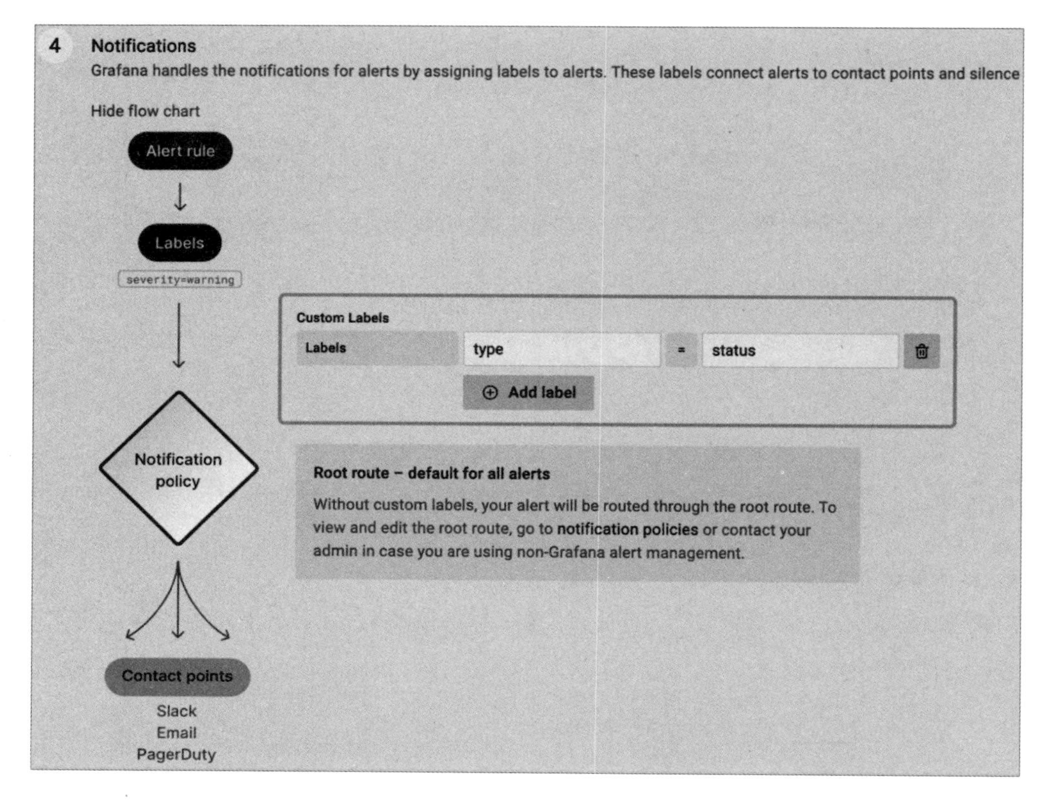

这样才能和之前设置的钉钉警报匹配规则相对应，一旦检测到应用下线就会以钉钉形式通知，而不是以默认邮件通知，钉钉警报的测试效果如下图所示。

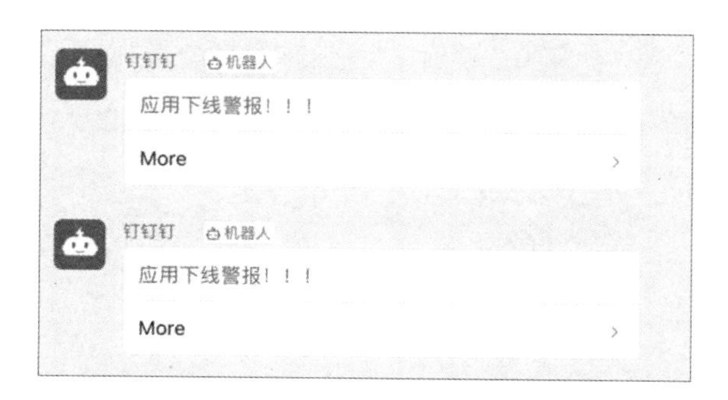

修改警报规则标签 type=111，让它不能匹配钉钉警报规则，以默认邮件形式通知，再测试应用下线，邮件警报的测试效果如下图所示。

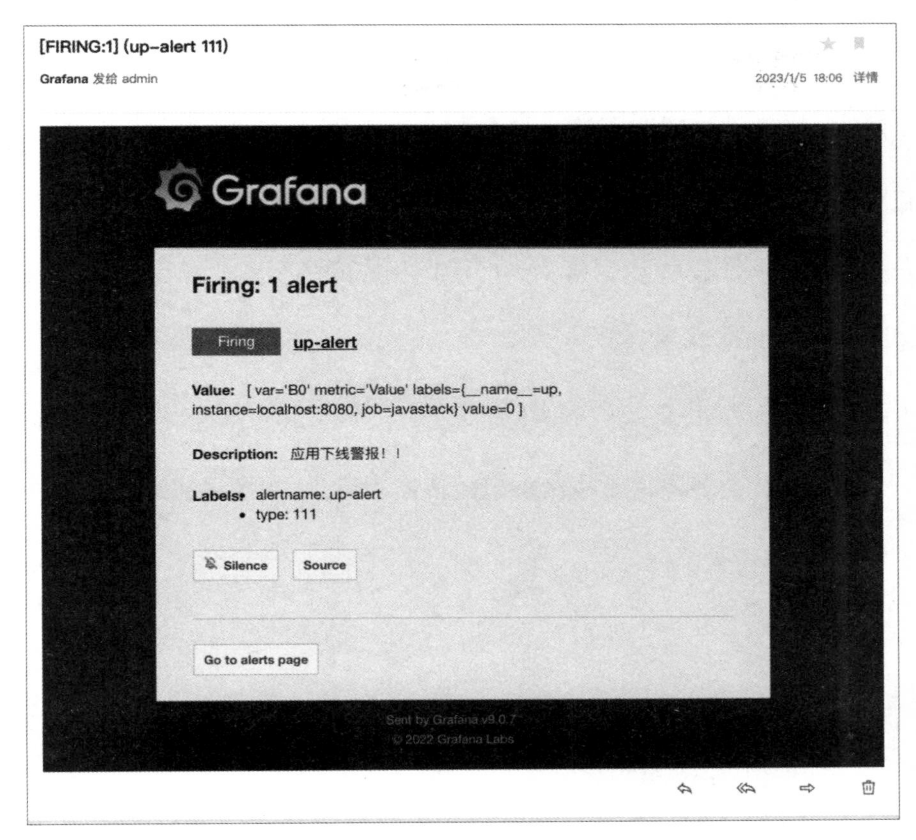

这里通过Grafana自带的警报机制演示了一个应用下线的基本警报通知，全程可视化配置操作，不需要编写复杂的警报规则，不管是运维人员，还是开发人员，都能很方便、轻松地完成警报规则的配置。

反侵权盗版声明